Advancements in Prosthodontics: Exploring Innovations in Rehabilitation Medicine

Advancements in Prosthodontics: Exploring Innovations in Rehabilitation Medicine

Guest Editor

Kelvin Ian Afrashtehfar

MDPI

Basel • Beijing • Wuhan • Barcelona • Belgrade • Novi Sad • Cluj • Manchester

Guest Editor
Kelvin Ian Afrashtehfar
Department of Reconstructive
Dentistry and Gerodontology
School of Dental Medicine
University of Bern
Berne
Switzerland

Editorial Office
MDPI AG
Grosspeteranlage 5
4052 Basel, Switzerland

This is a reprint of the Special Issue, published open access by the journal *Prosthesis* (ISSN 2673-1592), freely accessible at: https://www.mdpi.com/journal/prosthesis/special_issues/6C7K1HTKO1.

For citation purposes, cite each article independently as indicated on the article page online and as indicated below:

Lastname, A.A.; Lastname, B.B. Article Title. *Journal Name* **Year**, *Volume Number*, Page Range.

ISBN 978-3-7258-4117-2 (Hbk)
ISBN 978-3-7258-4118-9 (PDF)
https://doi.org/10.3390/books978-3-7258-4118-9

Contents

About the Editor

Kelvin Ian Afrashtehfar

Prof. Dr. Kelvin Afrashtehfar is a Canadian Board-Certified Specialist in Prosthodontics, currently serving as a research faculty member at the University of Bern (Switzerland) and adjunct clinical instructor at the Implant Dentistry–Study Consortium (IDSC) in Dubai. His clinical practice is limited to complex implant and esthetic rehabilitation, with a focus on evidence-based decision-making and interdisciplinary care. Dr. Afrashtehfar holds advanced degrees in prosthodontics and dental sciences from the University of British Columbia, McGill University, and the University of Bern. His research includes digital dentistry, dental materials, and artificial intelligence applications in prosthetic treatment planning. Recognized internationally, he has been ranked among the top 2% of scientists worldwide for several consecutive years, according to Stanford's citation-based metrics. As the Guest Editor of this Special Issue, Dr. Afrashtehfar brings together global expertise and translational insights to highlight innovative strategies in prosthodontic rehabilitation. He remains committed to bridging research and clinical practice through mentorship, collaborative projects, and scholarly contributions.

Preface

This Special Issue reprint, *Advancements in Prosthodontics: Exploring Innovations in Rehabilitation Medicine*, brings together a curated selection of research and clinical perspectives that reflect the ongoing transformation in prosthodontics through technological, digital, and biomaterial innovations. Its scope spans implant and removable prosthodontics, CAD/CAM protocols, AI-driven diagnostics, and biomechanical analyses—offering a comprehensive snapshot of emerging trends that are redefining modern oral rehabilitation.

The purpose of this collection is to present a multidisciplinary synthesis of state-of-the-art solutions addressing both functional and esthetic challenges in prosthetic dentistry. The contributing authors include clinical specialists, academic researchers, and biomedical engineers from leading institutions worldwide who collectively provide invaluable insights into cutting-edge techniques and materials.

This reprint is addressed to clinicians, researchers, educators, and postgraduate students seeking to deepen their understanding of contemporary prosthodontics and its integration with digital and regenerative medicine.

I extend my sincere gratitude to all authors for their scholarly contributions, to the reviewers for their meticulous evaluations, and to the editorial team at *Prosthesis* for their invaluable assistance in bringing this Special Issue to fruition.

Kelvin Ian Afrashtehfar
Guest Editor

prosthesis

Editorial

Artificial Intelligence in Reconstructive Implant Dentistry—Current Perspectives

Kelvin Ian Afrashtehfar [1,2,*], Moosa A. Abuzayeda [3] and Colin Alexander Murray [4]

1 Department of Reconstructive Dentistry and Gerodontology, School of Dental Medicine—ZMK Bern, University of Bern, 3010 Bern, Switzerland
2 Division of Restorative Dental Sciences, Clinical Sciences Department, College of Dentistry, Ajman University (AU), Ajman City P.O. Box 346, United Arab Emirates
3 Hamdan Bin Mohammed College of Dental Medicine (HBMCDM), Mohammed Bin Rashid University of Medicine and Health Sciences (MBRU), Dubai P.O. Box 505055, United Arab Emirates
4 Prosthodontics Unit, Department of Preventive and Restorative Dentistry, College of Dental Medicine (CDM), University of Sharjah (UoS), Sharjah P.O. Box 27272, United Arab Emirates
* Correspondence: kelvin.afrashtehfar@unibe.ch

In recent years, artificial intelligence (AI) has emerged as a transformative force in reconstructive implant dentistry. The integration of AI technologies into various aspects of dental practice, including digital data acquisition, treatment planning, and prognosis evaluation, offers unprecedented opportunities to enhance precision, efficiency, and clinical outcomes. Indeed, AI applications in implant dentistry span a broad spectrum of functionalities, from enhancing digital data acquisition and integration to providing sophisticated tools for treatment planning and prognosis evaluation. These technologies have the potential to streamline workflows, reduce human error, and improve the accuracy of clinical decisions, ultimately leading to better patient care. Therefore, this commentary synthesizes findings from recent studies by three well-regarded consultant prosthodontics and implantology professors affiliated to different Emirati academic institutions, trained in Switzerland, Canada, Germany, and the United Kingdom, to provide an overview of cutting-edge AI applications in reconstructive implant dentistry. Key areas of focus include digital data acquisition technologies, bone quality assessment, automated tissue segmentation, implant fixture identification and classification, and predictive analytics for implant planning and prognosis.

In terms of digital data acquisition and integration, AI-enhanced digital data acquisition technologies, including facial scanners (FSs), intraoral scanners (IOSs), and cone beam computed tomography (CBCT) devices, facilitate accurate data collection and integration. Revilla-León et al. [1] emphasize the role of AI in the automatic alignment, noise reduction, and segmentation of anatomical structures. These advancements streamline the creation of comprehensive virtual patient models, enhancing treatment planning accuracy. Indeed, AI-driven automated tissue segmentation significantly accelerates treatment planning. Liu et al. [2] introduced a fully automated system for segmenting oral surgery-related tissues from CBCT images, achieving a high accuracy in identifying alveolar bone, teeth, and maxillary sinus. These advancements reduce the manual segmentation effort and enhance surgical precision. For instance, Hartoonian et al. [3] reviewed AI applications in dentomaxillofacial imaging, showing the potential to improve diagnostic accuracy and treatment planning. Elgarba et al. [4] validated a cloud-based convolutional neural network for the automated segmentation of dental implants, demonstrating high performance and efficiency. A systematic review by the Afrashtehfar group demonstrated that coDiagnostiX® Digital Implant Treatment Planning Software (Dental Wings GmbH in Düsseldorf, Germany) outperforms other systems in implant treatment planning [5]. These advancements facilitate accurate and efficient clinical workflows, improving overall care quality [6–12].

Accurate bone quality assessment is crucial for successful dental implants. Lee et al. [13] demonstrated that deep learning (DL) models effectively evaluate bone quality from panoramic radiographs, correlating significantly with CBCT measurements and implant surgeons' tactile assessments. This AI application enhances objectivity and precision in bone quality evaluation, which is essential for implant stability and osseointegration. Furthermore, AI algorithms enhance implant planning by detecting edentulous areas and evaluating bone dimensions. Alqutaibi et al. [14] reported the high accuracy of AI-based diagnostic tools in implant planning, while Wu et al. [15] demonstrated the potential of AI in predicting implant prognosis. These predictive analytics tools help identify potential complications, optimizing treatment outcomes.

AI models demonstrate high accuracy in identifying and classifying dental implant fixtures from radiographs. Ibraheem [16] showed the utility of AI in implant identification, which is crucial for the continuity of care when previous records are unavailable. This capability improves clinical efficiency and reduces identification errors. Moreover, Lubbad et al. [17] compared deep learning models for classifying dental implants, and found that ConvNeXt models achieve the highest classification accuracy. Similarly, Mangano et al. [18] explored AI and augmented reality (AR) for guided implant surgery, demonstrating effective 3D planning and execution. Sakai et al. [19] developed an AI model to support implant drilling protocol decisions, showing significant accuracy in predicting appropriate protocols from CBCT images. These models increase precision in implant placement and the predictability of surgical outcomes.

The significant improvement in accuracy and efficiency AI provides is common across these studies, whether in data acquisition, segmentation, or implant identification. However, differences arise in the specific methodologies and AI models employed, such as the use of deep learning architectures like ConvNeXt [17] versus traditional machine learning algorithms [16]. Additionally, the extent of automation varies, with some studies achieving fully automated workflows [2,4], while others still require significant manual input [14,19]. Critically, while AI shows promise, challenges remain. For instance, the generalizability of AI models across diverse patient populations and varying clinical conditions needs further exploration. Studies like those by Wu et al. [15] and Alqutaibi et al. [14] conclude that there is a need for high-quality datasets and rigorous validation to increase reliability and reduce biases. Moreover, ethical considerations, including data privacy and the potential for algorithmic biases [20,21], must be addressed to fully integrate AI into clinical practice.

In conclusion, the integration of AI into reconstructive implant dentistry represents a significant advancement in the field. AI technologies offer enhanced diagnostic capabilities, streamlined workflows, and improved clinical outcomes. However, challenges such as the need for high-quality datasets, the rigorous validation of AI models, and addressing potential biases in AI algorithms remain. Future research should focus on refining these technologies, expanding their clinical applications, and ensuring their reliability and generalizability in diverse patient populations. This can lead to superior patient care and treatment success. As AI technologies continue to evolve, they hold the promise of transforming dental practice, making implant procedures more predictable and successful.

Conflicts of Interest: The authors declare that there are no conflicts of interests.

References

1. Revilla-León, M.; Gómez-Polo, M.; Sailer, I.; Kois, J.C.; Rokhshad, R. An overview of artificial intelligence based applications for assisting digital data acquisition and implant planning procedures. *J. Esthet. Restor. Dent.* **2024**, *Early View*. [CrossRef]
2. Liu, Y.; Xie, R.; Wang, L.; Liu, H.; Liu, C.; Zhao, Y.; Bai, S.; Liu, W. Fully automatic AI segmentation of oral surgery-related tissues based on cone beam computed tomography images. *Int. J. Oral Sci.* **2024**, *16*, 34. [CrossRef] [PubMed]
3. Hartoonian, S.; Hosseini, M.; Yousefi, I.; Mahdian, M.; Ahsaie, M.G. Applications of artificial intelligence in dentomaxillofacial imaging–a systematic review. *Oral Surg. Oral Med. Oral Pathol. Oral Radiol.* **2024**, *in press*. [CrossRef] [PubMed]
4. Elgarba, B.M.; Van Aelst, S.; Swaity, A.; Morgan, N.; Shujaat, S.; Jacobs, R. Deep learning-based segmentation of dental implants on cone-beam computed tomography images: A validation study. *J. Dent.* **2023**, *137*, 104639. [CrossRef] [PubMed]

5. Floriani, F.; Jurado, C.A.; Cabrera, A.J.; Duarte, W.; Porto, T.S.; Afrashtehfar, K.I. Depth distortion and angular deviation of a fully guided tooth-supported static surgical guide in a partially edentulous patient: A systematic review and meta-analysis. *J. Prosthodont.* **2024**, *Early View*. [CrossRef] [PubMed]
6. Russo, L.L.; Pierluigi, M.; Zhurakivska, K.; Digregorio, C.; Muzio, E.L.; Laino, L. Three-Dimensional Accuracy of Surgical Guides for Static Computer-Aided Implant Surgery: A Systematic Review. *Prosthesis* **2023**, *5*, 809–825. [CrossRef]
7. Afrashtehfar, K.I. Conventional free-hand, dynamic navigation and static guided implant surgery produce similar short-term patient-reported outcome measures and experiences. *Evid.-Based Dent.* **2021**, *22*, 143–145. [CrossRef] [PubMed]
8. Jurado, C.; Villalobos-Tinoco, J.; Mekled, S.; Sanchez, R.; Afrashtehfar, K. Printed Digital Wax-up Model as a Blueprint for Layered Pressed-ceramic Laminate Veneers: Technique Description and Case Report. *Oper. Dent.* **2023**, *48*, 618–626. [CrossRef] [PubMed]
9. Afrashtehfar, K.I.; Jurado, C.A.; Abu Fanas, S.H.; Del Fabbro, M. Short-term data suggests cognitive benefits in the elderly with single-implant overdentures. *Evid. -Based Dent.* **2024**, *25*, 71–72. [CrossRef]
10. Afrashtehfar, K.I.; Jurado, C.A.; Moshaverinia, A. Dynamic navigation may be used for most implant surgery scenarios due to its satisfactory accuracy. *J. Evid. Based Dent. Pract.* **2022**, *22*, 101797. [CrossRef]
11. Afrashtehfar, K.I.; Jurado, C.A.; Assaleh, N.K.; Lee, H. Autonomous robotic surgery seems promising as an accurate technology for single-tooth implant placement in the esthetic zone. *J. Evid. Based Dent. Pract.* **2023**, *23*, 101915. [CrossRef] [PubMed]
12. Russo, L.L.; Guida, L.; Mariani, P.; Ronsivalle, V.; Gallo, C.; Cicciù, M.; Laino, L. Effect of Fabrication Technology on the Accuracy of Surgical Guides for Dental-Implant Surgery. *Bioengineering* **2023**, *10*, 875. [CrossRef] [PubMed]
13. Lee, J.-H.; Yun, J.-H.; Kim, Y.-T. Deep learning to assess bone quality from panoramic radiographs: The feasibility of clinical application through comparison with an implant surgeon and cone-beam computed tomography. *J. Periodontal Implant. Sci.* **2024**, *54*, e5. [CrossRef] [PubMed]
14. Alqutaibi, A.Y.; Algabri, R.; Ibrahim, W.I.; Alhajj, M.N.; Elawady, D. Dental implant planning using artificial intelligence: A systematic review and meta-analysis. *J. Prosthet. Dent.* **2024**, in press. [CrossRef] [PubMed]
15. Wu, Z.; Yu, X.; Wang, F.; Xu, C. Application of artificial intelligence in dental implant prognosis: A scoping review. *J. Dent.* **2024**, *144*, 104924. [CrossRef]
16. Ibraheem, W.I. Accuracy of Artificial Intelligence Models in Dental Implant Fixture Identification and Classification from Radiographs: A Systematic Review. *Diagnostics* **2024**, *14*, 806. [CrossRef] [PubMed]
17. Lubbad, M.A.H.; Kurtulus, I.L.; Karaboga, D.; Kilic, K.; Basturk, A.; Akay, B.; Nalbantoglu, O.U.; Yilmaz, O.M.D.; Ayata, M.; Yilmaz, S.; et al. A Comparative Analysis of Deep Learning-Based Approaches for Classifying Dental Implants Decision Support System. *J. Imaging Inform. Med.* **2024**. [CrossRef] [PubMed]
18. Mangano, F.G.; Admakin, O.; Lerner, H.; Mangano, C. Artificial intelligence and augmented reality for guided implant surgery planning: A proof of concept. *J. Dent.* **2023**, *133*, 104485. [CrossRef] [PubMed]
19. Sakai, T.; Li, H.; Shimada, T.; Kita, S.; Iida, M.; Lee, C.; Nakano, T.; Yamaguchi, S.; Imazato, S. Development of artificial intelligence model for supporting implant drilling protocol decision making. *J. Prosthodont. Res.* **2023**, *67*, 360–365. [CrossRef]
20. Kim, C.S.; Samaniego, C.S.; Sousa Melo, S.L.; Brachvogel, W.A.; Baskaran, K.; Rulli, D. Artificial intelligence (A.I.) in dental curricula: Ethics and responsible integration. *J. Dent Educ.* **2023**, *87*, 1570–1573. [CrossRef]
21. Ueda, D.; Kakinuma, T.; Fujita, S.; Kamagata, K.; Fushimi, Y.; Ito, R.; Matsui, Y.; Nozaki, T.; Nakaura, T.; Fujima, N.; et al. Fairness of artificial intelligence in healthcare: Review and recommendations. *Jpn. J. Radiol.* **2024**, *42*, 3–15. [CrossRef] [PubMed]

prosthesis

MDPI

Article

Comparison between Bone-Level and Tissue-Level Implants in Immediate-Loading Full-Arch Rehabilitations: A Retrospective Multi-Center 1-Year Follow-Up Study

Francesco Pera [1,†], Massimo Carossa [1,*,†], Francesco Bagnasco [2], Armando Crupi [1], Giulia Ambrogio [1], Gaetano Isola [3], Maria Menini [2,‡] and Paolo Pesce [2,‡]

1 Department of Surgical Sciences, C.I.R. Dental School, University of Turin, 10126 Turin, Italy
2 Department of Surgical Sciences (DISC), University of Genoa, 16132 Genoa, Italy
3 Unit of Periodontology, Department of General Surgery and Surgical-Medical Specialties, University of Catania, 95124 Catania, Italy
* Correspondence: massimo.carossa@unito.it
† These authors contributed equally to this work.
‡ These authors also contributed equally to this work.

Abstract: The objective of the present retrospective multi-center study was to analyze the outcomes of bone-level (BL) implants and tissue-level (TL) implants in immediate-loading full-arch rehabilitations. Patients who were previously rehabilitated with full-arch immediate-loading rehabilitations with either BL or TL implants were considered. Data regarding implant survival rate, marginal bone loss (MBL), peri-implant probing depth (PPD), plaque index (PI), and bleeding on probing (BOP) were recorded, and the 1-year follow-up data were statistically analyzed between the two groups. In total, 38 patients were evaluated for a total implant number of 156 (n = 80 TL implants and n = 76 BL implants). An implant survival rate of 97.37% was recoded for the BL group while an implant survival rate of 100% was noted for the TL group. A total MBL of 1.324 ± 0.64 mm was recorded for BL implants, while a total MBL of 1.194 ± 0.30 mm was recorded for TL implants. A statistically significant difference was highlighted regarding MBL at the mesial aspect ($p = 0.01552$) of the implants, with BL implants presenting with higher MBL. Within the range of acceptable healthy values, a statistically significant difference was also highlighted regarding BOP ($p < 0.00001$), with TL implants presenting higher values. No statistically significant difference ($p > 0.05$) was recorded for any of the other variables analyzed. Within the limitations of the present retrospective study, both TL and BL implants seem to provide good clinical outcomes after a 12-month observational period when employed in immediate-loading full-arch rehabilitation.

Keywords: dental implants; immediate loading; full-arch; bone-level; tissue-level; abutments

1. Introduction

Nowadays, immediate-loading full-arch implant rehabilitation represents the elective treatment plan for the fixed rehabilitation of patients suffering from edentulism or with residual terminal dentition [1], offering them a transformative solution with profound implications for both their oral health and quality of life [2–4]. Unlike traditional delayed implant techniques that involve prolonged waiting periods, immediate-loading full-arch implant rehabilitation allows for the insertion of dental implants and rehabilitation with a fixed full-arch prostheses within 24–48 h after surgery [5,6]. This groundbreaking approach not only provides patients with an immediate restoration of their smile and oral function but also significantly reduces treatment time. However, despite high long-term survival rates [1,5], complications continue to be undesirable events [7–9], and therefore, research on the topic remains prominent.

Traditionally, implants were initially proposed in the morphology of Branemark implants as bone-level (BL) implants presenting an external connection [10]. This connection has been widely used and studied [11,12]. It is reported to present different advantages, such as an optimal passive-fit with the prosthesis [13] and better management facility in case of multiple implants [14]. However, over the years, some criticism has been raised, linked to the fact that this connection type may exhibit slight micro-movements between components, potentially affecting long-term stability [15,16] and increasing the risk of complications such as screw loosening and bacterial micro-leakage [17].

To avoid these possible complications, internal connections, also commonly adopted as BL implants, were later introduced, aiming to improve the implant–prosthetic mechanical stability by minimizing micro-movements between the implant components [18]. This stability was reported to be particularly crucial in full-arch rehabilitations where multiple implants need to work together to distribute the load effectively [14]. Furthermore, the internal connection led to the development of the platform switching concept [19] in which decreasing the diameter of the abutment in relation to the connection diameter provides increased space for the peri-implant soft tissue. As a consequence, the sealing around the implant's neck is improved, with the goal to better preserve the marginal bone level [19].

To date, different articles have investigated and compared the usage of BL implants with external and internal connections in immediate-loading full-arch rehabilitation [20,21]. Menini et al. [20] and Pera et al. [21] followed for 1 year and 3 years of follow-up, respectively, 20 full-arch rehabilitations supported by internal or external connections. According to their findings, no variations in the peri-implant soft and hard tissue were highlighted between the two connection designs, and therefore, both the designs can be considered clinically reliable for this type of rehabilitation.

Furthermore, another implant design called tissue-level (TL) implant with a convergent collar was introduced in contrast to the above-mentioned traditional BL implants [22,23]. Unlike their BL counterparts, where the most coronal part of the implant is positioned at the bone level, TL implants are characterized by their collar, which emerges at or just above the level of the mucosal tissues. Therefore, this implant design is composed altogether by the implant body that is placed into the bone and by the collar that serves as a trans-mucosal component. Among its advantages, this implant design is reported to avoid the presence of possible micro-gaps in the trans-mucosal area [24] and to increase soft tissue sealing, minimizing irritation and inflammation of the surrounding gums while promoting healthy soft tissue integration and long-term stability [23]. The increased soft tissue sealing is obtained by moving the prosthetic platform at the coronal level of the soft tissue and, therefore, the possible damages of the tissues during the prosthetic procedures are avoided [24]

Currently, few articles are available on the employment of TL implants in immediate-loading full-arch rehabilitations [24,25]. According to the available results, this implant design appears to be a viable option, even for the rehabilitation of fully edentulous patients.

However, to the authors knowledges, while different articles compared TL implants and BL implants in single- [23,26] and multi-unit [27] rehabilitations, no previous articles are available comparing these two implant designs in immediate-loading full-arch rehabilitations.

Therefore, the first objective of the present article was to retrospectively compare the outcomes of BL implants and TL implants in immediate-loading full-arch rehabilitations. The second objective was to analyze possible factors influencing marginal bone loss (MBL) including implant diameters and lengths, type of abutment, jaw distribution, and implant inclination. The first null hypothesis was that no clinical outcome differences are present between the two implant designs. The second null hypothesis was that no differences in MBL exist between the different subgroups analyzed in the study.

2. Materials and Methods

Patients who were previously rehabilitated with full-arch immediate-loading rehabilitation with either BL or TL implants at the University of Turin and University of Genoa were evaluated for the present study at the 1-year follow-up. The present research was performed following the Declaration of Helsinki. All the participants signed an informed consent form. The present study was approved by the local ethical committee of the University of Genoa (protocol n. 527) and of the University of Turin (protocol n. 0130929). The present study was reported following the Strengthening the Reporting of Observational Studies in Epidemiology (STROBE) guidelines.

2.1. Patient Selection

All the patients initially presented with residual dentition with unfavorable prognosis, either in the mandibular or in the maxilla, and were seeking immediate fixed rehabilitations. Bone availability was evaluated based on ortopantomography and Tc cone beam. After the clinical and radiological evaluation, patients who were found eligible were then rehabilitated with an immediate-loading implant-supported full-arch rehabilitation.

Patients who met the following inclusion/exclusion criteria were enrolled in the present study.

Inclusion criteria: Age $\geq$ 18 years; previously rehabilitated with immediate-loading full-arch rehabilitation with BL or TL implants; systemically healthy. Exclusion criteria: smokers; requirement of bone regeneration procedures; presence of diabetes; intake of drugs that could possibly interfere with bone remodeling and healing; previous radiotherapy of head and neck area; inability to attend the control visit.

2.2. Study Design

Firstly, implants were divided into two primary groups based on the division between BL implants (Group 1) and TL implants (Group 2).

Secondly, macro-topography of the implants—including implant length and diameter—jaw distribution (mandible vs. maxilla), implant inclination (tilted vs. axial), and abutment type with different inclinations were considered as subgroups.

2.3. Surgery Procedures

The workflow adopted (Columbus Bridge Protocol, CBP), including the surgical and prosthetic aspects, is reported in detail in previously published articles [5].

All the surgeries were performed by two experienced surgeons (one per center) specialized in implant surgery. Patients underwent professional oral hygiene on the day prior to surgery, including scaling and root planing to decrease the bacterial load of the mouth. Pre-operative antibiotic coverage with Amoxicillina 875 mg + Clavulanic acid 125 mg every 12 h for 6 days was prescribed [28,29], beginning one day before the surgery appointment. Chlorhexidina digluconate solution was provided to the patient to rinse for one minute prior to start the surgery.

A dose of 4% articaine with 1:10.000 adrenaline (Alfacaina SP; Dentsply Italy, Rome, Italy) was used to locally perform anesthesia. Patients who presented with residual terminal dentition underwent teeth extractions, and residual sockets were carefully debrided. A full thickness mucoperiostal flap was elevated. Four to six implants, based on the bone availability, were then inserted. Implant sites were prepared with dedicated drills following the manufacturer's instructions. BL implants (Syra or Shelta implants, Sweden & Martina, Due Carrare, Padova, Italy) or TL implants (Prama, Sweden & Martina, Due Carrare, Padova, Italy) were used (Figure 1).

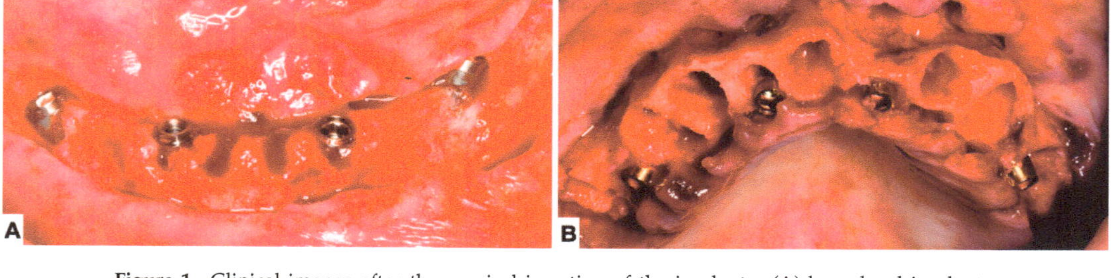

Figure 1. Clinical images after the surgical insertion of the implants: (**A**) bone-level implants; (**B**) tissue-level implants.

The two frontal implants were inserted straight, and the two posterior implants were tilted when necessary to avert the anatomical boundaries (alveolar nerve and sinus) following the CBP [5]. The length and diameter of the inserted implants were decided according to the bone availability evaluated on X-rays (ortopantomography and Tc cone beam) acquired prior to the surgery. BL implants were all connected to either straight or angulated abutments (PAD, Sweden & Martina, Due Carrare, Padova, Italy), while TL implants were connected to angulated abutments (PAD 330-303, Sweden & Martina, Due Carrare, Padova, Italy) in the posterior tilted implants and left with no abutment in the frontal straight implants (Figure 2).

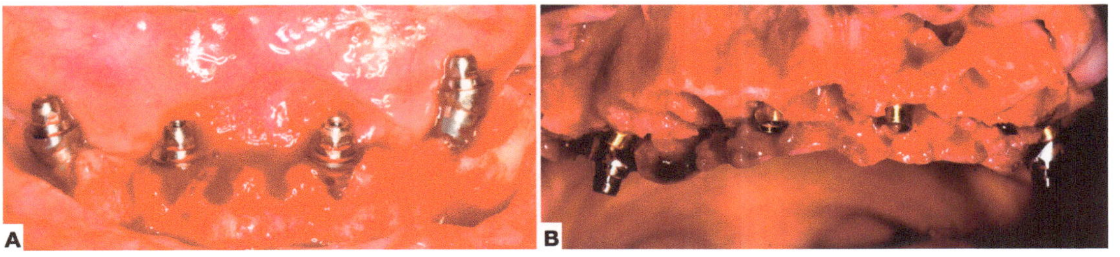

Figure 2. Clinical images after insertion of the abutments. (**A**) Bone-level implants; the two posterior tilted implants are linked to angulated abutments, while the two straight frontal ones are linked to straight abutments. (**B**) Tissue-level implants; the two posterior tilted implants are linked to angulated abutments, while the two straight frontal ones are left without abutments.

Sutures were made using silk multifilament (PERMA-HAND SILK 4-0, Ethicon, Somerville, NJ, USA). Impressions were made using open tray and impression plaster (BF-Plaster Dental, Turin, Italy). Post-operative instructions including soft diet and hygienic instructions were provided to the patients. Provisional screw-retained full-arch prosthesis made of resin with a metal framework was delivered to the patients within 24–48 h after the surgery. Peri-apical X-rays were acquired. Patients returned for suture removal one week after the surgery (Figure 3).

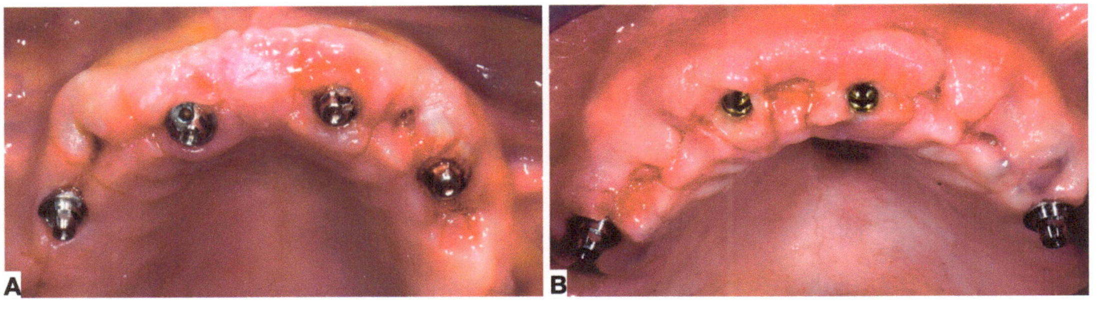

Figure 3. Clinical images at the sutures removal appointment one week after the surgery: (**A**) bone-level implants; (**B**) tissue-level implants.

Six months after the surgery, a new analogic impression (open-tray) was taken, and final composite with metal framework screw-retained prostheses was then fabricated and delivered. Patients were then evaluated for the present study 12 months after surgery and follow-up periapical X-rays were acquired.

2.4. Outcomes

The following clinical outcomes were considered:

- Implant survival rate;
- MBL assessed 12-months after surgery (T12). Digital intraoral periapical radiographs acquired using the parallel approach were used to assess MBL following the methods described in previous published articles [30,31]. The bone level was calculated as the distance between the head of the implant and the most coronal bone at both the mesial and distal aspect of the implants. Both the X-rays taken immediately following surgery (T0) and the ones taken at T12 were used. The MBL resulted as difference between T12 and T0;
- Plaque index (PI), peri-implant probing depth (PPD), and bleeding on probing (BOP) were evaluated as peri-implant soft tissue parameters at the 12-month follow-up. A periodontal UNC 15 probe (Hu-Friedy, Chicago, IL, USA) was used to measure PI, PD, and BI at 4 locations for each implant. PI and BOP were expressed as number of surfaces per implant presented with plaque or bleeding.

All the measurements were performed by two calibrated and trained clinicians per center.

2.5. Statistical Analysis

Data regarding MBL, PPD, BOP, and PI were analyzed to investigate any differences between the two groups (BL and TL implants). T-test for independent means was used to compare variables that were normally distributed. For all the other variables that did not meet the requirement of normal distribution, the Mann–Whitney *U* nonparametric test was adopted. All the subgroups were then analyzed to investigate any differences in MBL among them, both within and between the primary groups. Results were considered statistically significant with $p < 0.05$. All analyses were performed using SAS Software version 9.4 (SAS Institute Inc., Cary, NC, USA).

3. Results

A total of 38 (n = 38) patients (mean age at the control visit 62.9 years, 24 males 63.16%) were recalled for a total implant number of 156 (n = 156). Of these, 80 implants (n = 80) were TL implants, while 76 were BL implants (n = 76). Ten patients were rehabilitated at the University of Genova, and twenty-eight patients were rehabilitated at the University of Turin. Two posterior BL implants failed within the first six months. The failure was ascribed to insufficient osseointegration. Therefore, a total of 154 implants (n = 154) were considered at the 12-month follow-up (n = 80 TL implants and n = 74 BL implants). An implant survival rate of 97.37% was recoded for the BL group, while an implant survival rate of 100% was recorded for the TL one.

A total MBL of 1.324 ± 0.64 mm (mesial 1.412 ± 0.75 mm and distal 1.264 ± 0.81) was recorded for BL implants, while a total MBL of 1.194 ± 0.30 mm (mesial 1.165 ± 0.38 mm and distal 1.222 ± 0.37 mm) was noted for TL implants. Table 1 reports mean ± standard deviation and statistical results in regard to MBL, PPD, BOP, and PI between BL and TL implants.

A statistically significant difference was recorded in regard to MBL at the mesial aspect of the implants ($p = 0.01552$) with BL implants presenting with a statistically higher MBL compared to TL implants. A statistically significant difference was also highlighted regarding BOP ($p < 0.00001$) with TL implants presenting with higher BOP values. No statistical significance different ($p > 0.05$) was recorded for any of the other variables analyzed.

Table 2 shows the distribution and analysis of the total MBL between the subgroups (abutment type, implant inclination, jaw distribution, lengths and diameters) for both BL and TL groups.

Table 1. The table shows the analysis of the findings for each variable among the two groups (BL and TL implants). *T*-test for independent means was adopted for Distal MBL, Total MBL, and PPD since they were normally distributed. For all the other variables which did not meet the requirement of normally distribution, the Mann–Whitney *U* nonparametric test was adopted. Significant statistical differences are highlighted with *.

Bone-Level/ Tissue-Level	Variable	Mean (mm)	Standard Deviation	*p*-Value
BL TL	Mesial MBL	1.412 1.165	0.75 0.38	* 0.01552
BL TL	Distal MBL	1.264 1.222	0.81 0.37	0.8839
BL TL	Total MBL	1.324 1.194	0.64 0.30	0.10302
BL TL	BOP	0.905 1.7	1.05 1.15	* <0.00001
BL TL	PI	1.892 1.938	1.51 1.27	0.61708
BL TL	PPD	2.155 2.066	0.46 0.44	0.22004

Table 2. MBL comparison between subgroups for both BL and TL groups and tested through Mann–Whitney *U* nonparametric test.

Parameter	Variable	N	Mean (mm)	Std Dev	Median	*p*-Value
		MBL Bone-Level Implants				
Abutment	0° 17° 30°	25 16 33	1.13 1.39 1.44	0.69 0.52 0.65	1 1.25 1.5	0.1386
Implant inclinations	Tilted Upright	34 40	1.39 1.27	0.59 0.69	1.5 1.25	0.26
Jaw distribution	Mandible Maxilla	20 54	1.11 1.40	0.75 0.59	1 1.5	0.083
Lengths (mm)	11.5 13 15	1 4 69	2 1.5 1.3	- 1.08 0.62	2 1.75 1.25	0.33
Diameters (mm)	3.8 4.25	15 59	1.12 1.38	0.76 0.61	1 1.25	0.25
		MBL tissue-level implants				
Abutment	None 17° 30°	40 24 16	1.16 1.21 1.27	0.31 0.25 0.32	1.2 1.25 1.21	0.482
Implant inclinations	Tilted Upright	40 40	1.23 1.16	0.28 0.31	1.25 1.20	0.2485
Jaw distribution	Mandible Maxilla	36 44	1.14 1.24	0.26 0.32	1.175 1.25	0.2945
Lengths (mm)	10 11.5 13 15	8 18 27 27	1.04 1.30 1.12 1.16	0.23 1.28 1.25 1.15	0.925 1.28 1.25 1.15	0.1114
Diameters (mm)	3.8 4.25	48 32	1.18 1.21	0.28 0.32	1.21 1.25	0.89

Table 3 shows the analysis of the MBL between the two groups (BL and TL implants) by each subgroups' parameters.

Table 3. MBL is compared in the two groups (BL and TL) by each subgroup parameter and tested through Mann–Whitney *U* nonparametric test.

Parameter	Variable	MBL BL Group Mean (SD)	MBL TL Group Mean (SD)	*p*-Value
Abutment	17°	1.39 (0.52)	1.21 (0.25)	0.4272
	30°	1.44 (0.65)	1.27 (0.32)	0.1139
Implant inclinations	Tilted	1.39 (0.59)	1.23 (0.28)	0.069
	Upright	1.27 (0.69)	1.16 (0.31)	0.4556
Jaw distribution	Mandible	1.11 (0.75)	1.14 (0.26)	0.6674
	Maxilla	1.40 (0.59)	1.24 (0.32)	0.085
Lengths (mm)	11.5	2	1.30 (1.28)	0.1397
	13	1.5 (1.08)	1.12 (1.25)	0.1835
	15	1.3 (0.62)	1.16 (1.15)	0.242
Diameters (mm)	3.8	1.12 (0.76)	1.18 (0.28)	0.8777
	4.25	1.38 (0.61)	1.21 (0.32)	0.1438

No statistically significant difference was highlighted ($p > 0.05$) for any of the subgroups analyzed.

4. Discussion

The first objective of the present article was to retrospectively compare the outcomes of BL implants and TL implants in immediate-loading full-arch rehabilitations after 12-months of functional follow-up. For this purpose, patients who were previously treated with immediate-loading full-arch rehabilitations using either BL or TL implants were evaluated, and data about implant survival rate, MBL (mesial, distal and total), PPD, BOP, and PI were collected and analyzed. Based on the results, some statistically significant differences were highlighted between the groups. Therefore, the first null hypotheses were rejected.

An implant survival rate of 97.37% was recorded for the BL group while an implant survival rate of 100% was recorded for the TL group. A slightly less non-significant total MBL was recorded in favor of TL implants (1.194 ± 0.30 mm) against BL ones (1.324 ± 0.64 mm), while a statistically significant difference was highlighted when considering MBL at the mesial aspect of the implants (TL 1.165 ± 0.38 mm, BL 1.412 ± 0.75 mm, $p = 0.01552$). The aforementioned results regarding the implant survival rate and MBL for both groups are in agreement with those reported in the literature regarding full-arch implant-supported rehabilitation after the 12-month follow-up [32–35]. The lower MBL detected for TL implants may be attributed to the different position of the implant–abutment interface and to the possibility of using TL implants without an abutment. This topic has surfaced recently, with different articles highlighting how the implant–abutment interfaces as well as the mechanical procedure of screwing and unscrewing at the trans-mucosal level may be related to increased risks of bacterial contamination and, therefore, bone loss [17,36]. Indeed, the present results are in agreement with other studies that compared BL and TL implants in different types of rehabilitations and found a lower MBL in favor of TL implants [23,26]. One of the possible main advantages of TL implants may be the possibility of using them without abutment, as documented in previously published articles [24,25]. When an abutment is used, two possible micro-gaps are present: one between the abutment and the implant and one between the abutment and the prosthesis. In the present study, the two frontal implants were functionalized without using an abutment and, therefore, this may represent a possible reason for the lower MBL detected. However, further studies are required to more deeply investigate the topic.

In an interesting study by Afrashtehfar et al. [37], the authors compared the reliability of bone height measurements between BL and TL implants. According to their results, no statistically significant difference was highlighted between the two implant designs and, therefore, the measurement of bone loss between them can be considered reliable.

Furthermore, a statistically significant difference was also highlighted with BOP ($p < 0.00001$), which was calculated as the number of surfaces per implant with bleeding after probing, with TL implants showing higher values than BL ones. This result is in contrast with those reported in the literature, where TL implants are reported to possibly improve soft tissue health [24]. However, it must be noted that the increased BOP values recorded in the present study were not correlated with any increased PPD nor MBL nor any sign of peri-implantitis and were within clinical and radiological health guidelines in accordance with the last Workshop of Periodontology [38]. TL implants with a convergent collar, contrary to those with a divergent one [39], are described in articles with follow-up ranging between 18 months and 60 months [23,26,40] to improve the space and thickness of the soft tissue and thus promote peri-implant health [23,26]. Therefore, assuming that even in the present study the values were within clinical health guidelines for both groups, a longer follow-up period and further studies are required to confirm the result. Indeed, the main limitation of the present retrospective study is represented by the short-term follow-up period. Further studies with medium- and long-term follow-ups are required to further understand the differences between BL and TL implants in immediate-loading full-arch rehabilitation.

The second objective of the present study was to analyze possible factors influencing MBL, including implant diameters and lengths, type of abutments, jaw distribution, and implant inclination. Based on the results, no statistically significant difference ($p > 0.05$) was recorded for any of the analyzed subgroups, both within and between BL and TL implants. Therefore, the second null hypotheses was accepted. However, it must be noted that an additional limitation of the research is linked to the fact that most of the subgroups analyzed were imbalanced. This limitation is inherent to the retrospective design of the study, where randomization among the subgroups was not possible. However, the present results may indicate that as long as the CBP is followed within the implant range of the study, such as minimum implant length of 10 mm and minimum diameter of 3.8; all the other variables do not seem to influence the MBL. This result, together with the high implant survival rate recorded, are in agreement with the articles that analyzed the outcomes of the CBP in the medium- and long-term observational periods [5]. Indeed, in accordance with the literature, the CBP is reported to provide an implant survival rate higher than 92.25% even after a 10-year observational period post load [1].

In conclusion, research on different implant designs as well as new materials and protocols is consistently advancing [41–44]. To the authors' knowledge, the present article represents the first study reporting data on the comparison between BL and TL implants with a convergent collar in immediate-loading full-arch rehabilitations. The data observed in this study seem to indicate that both of the implant designs may be a good option for this type of rehabilitation. However, further research is required to confirm the results.

5. Conclusions

Within the limitations of the present retrospective study, both TL and BL implants seem to provide good clinical outcomes after a 12-month observational period when employed in immediate-loading full-arch rehabilitation. Further clinical trials with longer observational times are required to confirm the results and further understand the possibility of different clinical outcomes between the two implant designs in this type of implant rehabilitation.

Author Contributions: Conceptualization, F.P., M.C., M.M. and P.P.; methodology F.P., M.C., M.M. and P.P.; software, F.B., A.C., G.A. and G.I.; formal analysis, F.P., M.C., F.B., M.M. and P.P.; data curation, F.B., A.C., G.A. and G.I.; writing—original draft preparation, M.C.; writing—review and editing, M.C. and P.P.; supervision, F.P., M.C., M.M. and P.P.; project administration, F.P., M.C., M.M. and P.P.; All authors have read and agreed to the published version of the manuscript.

Funding: This research received no external funding.

Institutional Review Board Statement: The present study was approved by the local ethical committee of the University of Genoa (protocol n. 527) and of the University of Turin (protocol n. 0130929).

Informed Consent Statement: Informed consent was obtained from all subjects involved in the study.

Data Availability Statement: The datasets generated during and/or analyzed during the current study are available from the corresponding author on reasonable request.

Acknowledgments: The authors wish to acknowledge Sweden & Martina (Due Carrare, Padova, Italy for providing part of the surgical materials.

Conflicts of Interest: The authors declare no conflict of interest.

References

1. Barootchi, S.; Askar, H.; Ravidà, A.; Gargallo-Albiol, J.; Travan, S.; Wang, H.L. Long-term Clinical Outcomes and Cost-Effectiveness of Full-Arch Implant-Supported Zirconia-Based and Metal-Acrylic Fixed Dental Prostheses: A Retrospective Analysis. *Int. J. Oral Maxillofac. Implant.* **2020**, *35*, 395–405. [CrossRef] [PubMed]
2. Schropp, L.; Isidor, F.; Kostopoulos, L.; Wenzel, A. Patient experience of, and satisfaction with, delayed-immediate vs. delayed single-tooth implant placement. *Clin. Oral Implant. Res.* **2004**, *15*, 498–503. [CrossRef] [PubMed]
3. Catapano, S.; Ortensi, L.; Mobilio, N.; Grande, F. The New Elderly Patient: A Necessary Upgrade. *Prosthesis* **2021**, *3*, 99–104. [CrossRef]
4. Ortensi, L.; Ortensi, M.; Minghelli, A.; Grande, F. Implant-Supported Prosthetic Therapy of an Edentulous Patient: Clinical and Technical Aspects. *Prosthesis* **2020**, *2*, 140–152. [CrossRef]
5. Tealdo, T.; Menini, M.; Bevilacqua, M.; Pera, F.; Pesce, P.; Signori, A.; Pera, P. Immediate versus delayed loading of dental implants in edentulous patients' maxillae: A 6-year prospective study. *Int. J. Prosthodont.* **2014**, *27*, 207–214. [CrossRef] [PubMed]
6. De Bruyn, H.; Raes, S.; Ostman, P.O.; Cosyn, J. Immediate loading in partially and completely edentulous jaws: A review of the literature with clinical guidelines. *Periodontology 2000* **2014**, *66*, 153–187. [CrossRef] [PubMed]
7. Wittneben, J.G.; Buser, D.; Salvi, G.E.; Bürgin, W.; Hicklin, S.; Brägger, U. Complication and failure rates with implant-supported fixed dental prostheses and single crowns: A 10-year retrospective study. *Clin. Implant Dent. Relat. Res.* **2014**, *16*, 356–364. [CrossRef]
8. Sailer, I.; Karasan, D.; Todorovic, A.; Ligoutsikou, M.; Pjetursson, B.E. Prosthetic failures in dental implant therapy. *Periodontology 2000* **2022**, *88*, 130–144. [CrossRef]
9. Catapano, S.; Ferrari, M.; Mobilio, N.; Montanari, M.; Corsalini, M.; Grande, F. Comparative Analysis of the Stability of Prosthetic Screws under Cyclic Loading in Implant Prosthodontics: An In Vitro Study. *Appl. Sci.* **2021**, *11*, 622. [CrossRef]
10. Brånemark, P.; Adell, R.; Breine, U.; Hansson, B.O.; Lindstrom, J.; Ohlsson, A. Intra-osseous anchorage of dental prostheses. I. Experimental studies. *Scand. J. Plast. Reconstr. Sug.* **1969**, *3*, 81–100. [CrossRef]
11. Balshi, T.J.; Wolfinger, G.J.; Slauch, R.W.; Balshi, S.F. A retrospective analysis of 800 Brånemark System implants following the All-on-Four protocol. *J. Prosthodont.* **2014**, *23*, 83–88. [CrossRef] [PubMed]
12. Palacios-Garzón, N.; Mauri-Obradors, E.; Roselló-LLabrés, X.; Estrugo-Devesa, A.; Jané-Salas, E.; López-López, J. Comparison of Marginal Bone Loss Between Implants with Internal and External Connections: A Systematic Review. *Int. J. Oral Maxillofac. Implant.* **2018**, *33*, 580–589. [CrossRef] [PubMed]
13. Manzella, C.; Burello, V.; Bignardi, C.; Carossa, S.; Schierano, G. A method to improve passive fit of frameworks on implant-supported prostheses: An in vivo study. *Int. J. Prosthodont.* **2013**, *26*, 577–579. [CrossRef] [PubMed]
14. Gil, F.J.; Herrero-Climent, M.; Lazaro, P.; Rios, J.V. Implant-abutment connections: Influence of the design on the microgap and their fatigue and fracture behavior of dental implants. *J. Mater. Sci. Mater. Med.* **2014**, *25*, 1825–1830. [CrossRef] [PubMed]

15. Almeida, E.O.; Freitas, A.C., Jr.; Bonfante, E.A.; Marotta, L.; Silva, N.R.; Coelho, P.G. Mechanical testing of implant-supported anterior crowns with different implant/abutment connections. *Int. J. Oral Maxillofac. Implant.* **2013**, *28*, 103–108. [CrossRef] [PubMed]
16. Maeda, Y.; Satoh, T.; Sogo, M. In vitro differences of stress concentrations for internal and external hex implant-abutment connections: A short communication. *J. Oral Rehabil.* **2006**, *33*, 75–78. [CrossRef] [PubMed]
17. Canullo, L.; Penarrocha-Oltra, D.; Soldini, C.; Mazzocco, F.; Penarrocha, M.; Covani, U. Microbiological assessment of the implant-abutment interface in different connections: Cross-sectional study after 5 years of functional loading. *Clin. Oral Implant. Res.* **2015**, *26*, 426–434. [CrossRef] [PubMed]
18. Vetromilla, B.M.; Brondani, L.P.; Pereira-Cenci, T.; Bergoli, C.D. Influence of different implant-abutment connection designs on the mechanical and biological behavior of single-tooth implants in the maxillary esthetic zone: A systematic review. *J. Prosthet. Dent.* **2019**, *121*, 398–403. [CrossRef]
19. Lazzara, R.J.; Porter, S.S. Platform switching: A new concept in implant dentistry for controlling postrestorative crestal bone levels. *Int. J. Periodontics Restor. Dent.* **2006**, *26*, 9–17.
20. Menini, M.; Pesce, P.; Bagnasco, F.; Carossa, M.; Mussano, F.; Pera, F. Evaluation of internal and external hexagon connections in immediately loaded full-arch rehabilitations: A within-person randomised split-mouth controlled trial. *Int. J. Oral Implantol.* **2019**, *12*, 169–179.
21. Pera, F.; Menini, M.; Bagnasco, F.; Mussano, F.; Ambrogio, G.; Pesce, P. Evaluation of internal and external hexagon connections in immediately loaded full-arch rehabilitations: A within-person randomized split-mouth controlled trial with a 3-year follow-up. *Clin. Implant Dent. Relat. Res.* **2021**, *23*, 562–567. [CrossRef]
22. Taheri, M.; Akbari, S.; Shamshiri, A.R.; Shayesteh, Y.S. Marginal bone loss around bone-level and tissue-level implants: A systematic review and meta-analysis. *Ann. Anat.* **2020**, *231*, 151525. [CrossRef] [PubMed]
23. Canullo, L.; Menini, M.; Covani, U.; Pesce, P. Clinical outcomes of using a prosthetic protocol to rehabilitate tissue-level implants with a convergent collar in the esthetic zone: A 3-year prospective study. *J. Prosthet. Dent.* **2020**, *123*, 246–251. [CrossRef] [PubMed]
24. Carossa, M.; Alovisi, M.; Crupi, A.; Ambrogio, G.; Pera, F. Full-Arch Rehabilitation Using Trans-Mucosal Tissue-Level Implants with and without Implant-Abutment Units: A Case Report. *Dent. J.* **2022**, *10*, 116. [CrossRef] [PubMed]
25. Pera, F.; Pesce, P.; Menini, M.; Fanelli, F.; Kim, B.C.; Zhurakivska, K.; Mayer, Y.; Isola, G.; Cianciotta, G.; Crupi, A.; et al. Immediate loading full-arch rehabilitation using transmucosal tissue-level implants with different variables associated: A one-year observational study. *Minerva Dent. Oral Sci.* **2023**, *72*, 230–238. [CrossRef] [PubMed]
26. Canullo, L.; Menini, M.; Bagnasco, F.; Di Tullio, N.; Pesce, P. Tissue-level versus bone-level single implants in the anterior area rehabilitated with feather-edge crowns on conical implant abutments: An up to 5-year retrospective study. *J. Prosthet. Dent.* **2022**, *128*, 936–941. [CrossRef] [PubMed]
27. Menini, M.; Dellepiane, E.; Deiana, T.; Fulcheri, E.; Pera, P.; Pesce, P. Comparison of Bone-Level and Tissue-Level Implants: A Pilot Study with a Histologic Analysis and a 4-Year Follow-up. *Int. J. Periodontics Restor. Dent.* **2022**, *42*, 535–543. [CrossRef]
28. El-Kholey, K.E. Efficacy of two antibiotic regimens in the reduction of early dental implant failure: A pilot study. *Int. J. Oral Maxillofac. Surg.* **2014**, *43*, 487–490. [CrossRef]
29. Tan, W.C.; Ong, M.; Han, J.; Mattheos, N.; Pjetursson, B.E.; Tsai, A.Y.; Sanz, I.; Wong, M.C.; Lang, N.P.; ITI Antibiotic Study Group. Effect of systemic antibiotics on clinical and patient-reported outcomes of implant therapy—A multicenter randomized controlled clinical trial. *Clin. Oral Implant. Res.* **2014**, *25*, 185–193. [CrossRef]
30. Pera, F.; Menini, M.; Alovisi, M.; Crupi, A.; Ambrogio, G.; Asero, S.; Marchetti, C.; Canepa, C.; Merlini, L.; Pesce, P.; et al. Can Abutment with Novel Superlattice CrN/NbN Coatings Influence Peri-Implant Tissue Health and Implant Survival Rate Compared to Machined Abutment? 6-Month Results from a Multi-Center Split-Mouth Randomized Control Trial. *Materials* **2023**, *16*, 246. [CrossRef]
31. Menini, M.; Pesce, P.; Delucchi, F.; Ambrogio, G.; Canepa, C.; Carossa, M.; Pera, F. One-stage versus two-stage technique using two splinted extra-short implants: A multicentric split-mouth study with a one-year follow-up. *Clin. Implant Dent. Relat. Res.* **2022**, *24*, 602–610. [CrossRef] [PubMed]
32. Testori, T.; Del Fabbro, M.; Capelli, M.; Zuffetti, F.; Francetti, L.; Weinstein, R.L. Immediate occlusal loading and tilted implants for the rehabilitation of the atrophic edentulous maxilla: 1-year interim results of a multicenter prospective study. *Clin. Oral Implant. Res.* **2008**, *19*, 227–232. [CrossRef] [PubMed]
33. Capelli, M.; Zuffetti, F.; Del Fabbro, M.; Testori, T. Immediate rehabilitation of the completely edentulous jaw with fixed prostheses supported by either upright or tilted implants: A multicenter clinical study. *Int. J. Oral Maxillofac. Implant.* **2007**, *22*, 639–644.
34. Albrektsson, T.; Zarb, G.A. Current interpretations of the osseointegrated response: Clinical significance. *Int. J. Prosthodont.* **1993**, *6*, 95–105.
35. Albrektsson, T.; Zarb, G.; Worthington, P.; Eriksson, A.R. The long-term efficacy of currently used dental implants: A review and proposed criteria of success. *Int. J. Oral Maxillofac. Implant.* **1986**, *1*, 11–25.
36. Tallarico, M.; Canullo, L.; Caneva, M.; Ozcam, M. Microbial colonization of the implant-abutment interface and its possible influence on the development of peri-implantitis: A systematic review with meta-analysis. *J. Prosthodont. Res.* **2017**, *61*, 233–241. [CrossRef]

37. Afrashtehfar, K.I.; Brägger, U.; Hicklin, S.P. Reliability of Interproximal Bone Height Measurements in Bone- and Tissue-Level Implants: A Methodological Study for Improved Calibration Purposes. *Int. J. Oral Maxillofac. Implant.* **2020**, *35*, 289–296. [CrossRef]
38. Caton, J.G.; Armitage, G.; Berglundh, T.; Chapple, I.L.C.; Jepsen, S.; Kornman, K.S.; Mealey, B.L.; Papapanou, P.N.; Sanz, M.; Tonetti, M.S. A new classification scheme for periodontal and peri-implant diseases and conditions—Introduction and key changes from the 1999 classification. *J. Clin. Periodontol.* **2018**, *45*, S1–S8. [CrossRef]
39. Agustín-Panadero, R.; Martínez-Martínez, N.; Fernandez-Estevan, L.; Faus-López, J.; Solá-Ruíz, M.F. Influence of Transmucosal Area Morphology on Peri-Implant Bone Loss in Tissue-Level Implants. *Int. J. Oral Maxillofac. Implant.* **2019**, *34*, 947–952. [CrossRef]
40. Canullo, L.; Tallarico, M.; Pradies, G.; Marinotti, F.; Loi, I.; Cocchetto, R. Soft and hard tissue response to an implant with a convergent collar in the esthetic area: Preliminary report at 18 months. *Int. J. Esthet. Dent.* **2017**, *12*, 306–323.
41. Lepidi, L.; Grande, F.; Baldassarre, G.; Suriano, C.; Li, J.; Catapano, S. Preliminary clinical study of the accuracy of a digital axiographic recording system for the assessment of sagittal condylar inclination. *J. Dent.* **2023**, *135*, 104583. [CrossRef] [PubMed]
42. Grande, F.; Pozzan, M.C.; Marconato, R.; Mollica, F.; Catapano, S. Evaluation of Load Distribution in a Mandibular Model with Four Implants Depending on the Number of Prosthetic Screws Used for OT-Bridge System: A Finite Element Analysis (FEA). *Materials* **2022**, *15*, 7963. [CrossRef] [PubMed]
43. Grande, F.; Celeghin, G.; Gallinaro, F.; Mobilio, N.; Catapano, S. Comparison of the accuracy between full-arch digital scans and scannable impression materials: An in vitro study. *Minerva Dent. Oral Sci.* **2023**, *72*, 168–175. [CrossRef] [PubMed]
44. Montanari, M.; Grande, F.; Lepidi, L.; Piana, G.; Catapano, S. Rehabilitation with implant-supported overdentures in preteens patients with ectodermal dysplasia: A cohort study. *Clin. Implant Dent. Relat. Res.* 2023; *early view.* [CrossRef]

prosthesis

MDPI

Article

Mechanical Behavior of Dental Restorations: A Finite Element Pilot Study of Implant-Supported vs. Multiunit-Supported Restorations

Eduardo Anitua [1], Patricia Truchuelo Díez [2,*], Jorge Pesquera Velasco [2], Naiara Larrazabal [1], Mikel Armentia [1] and Jesús Seco-Calvo [2]

[1] R&D Department, Biotechnology Institute I mas D S.L., 01510 Miñano, Spain; eduardo@fundacioneduardoanitua.org (E.A.); naiara.larrazabal@bti-implant.es (N.L.); mikel.armentia@bti-implant.es (M.A.)
[2] Institute of Biomedicine (IBIOMED), Physiotherapy Department, Universidad de León, Campus de Vegazana, 24071 León, Spain; jpesquera@leonformacion.es (J.P.V.); jesus.seco@unileon.es (J.S.-C.)
* Correspondence: ptruchuelo@gmail.com

Abstract: Implant-supported-screw-retained prostheses are highly popular. Some of the most frequent complications are connected with the mechanical properties of the fixing elements. These include abutment screw loosening or even screw fracture. Using an intermediate abutment can offer several advantages. However, few studies detail how this affects the mechanical behavior of dental restorations. This study focuses on understanding the mechanical behavior of implant-supported restorations with a transepithelial component compared to direct implant-supported restoration. It was carried out using the finite element method (FEM) and was experimentally validated. The results showed that in the case of transepithelial-supported restoration, the prosthetic screw mounted over the transepithelial component suffered higher stress than the one screwed directly into the implant. After applying a cyclic fatigue load, it was experimentally proven that, in the transepithelial-supported restorations, the fuse changed from being the screw that went into the implant to being the upper one. In conclusion, we can state that the use of an intermediate abutment in dental restoration not only provides better protection for the rest of the dental restoration but also allows for easier repair in the event of a fracture. This can potentially lead to more efficient procedures and improved patient outcomes.

Keywords: finite element analysis; dental-implant abutment design; mechanical complications; prosthesis failure; cyclic fatigue; preload; mechanical stress; implant-supported single crown; intermediate abutment

1. Introduction

Dental implants are a widely used option for oral rehabilitation when one or more natural teeth are missing. Although implants have a high survival rate, marginal bone loss frequently occurs [1,2]. This compromises the long-term prognosis of implants since early marginal bone loss appears to increase the risk of peri-implantitis [1]. The prevalence of peri-implantitis has been reported in several studies, varying from approximately 10% to 12.8%, and it is a significant problem for dental teams today and in the foreseeable future [2]. Moreover, periimplantitis leads to excessive bone loss [2], soft tissue recession, implant exposure, aesthetic problems [1], and even implant loss [2].

Implant-supported prostheses can be either screw- or cement-retained. They have traditionally been composed of an implant, an abutment, and a screw that joins both pieces and provides structural integrity to the restoration. This type of dental restoration is known as implant-supported, directly attached, or direct-to-implant restoration. The screw-retained prostheses are popular because they are easily retrieved for maintenance [3]. Prefabricated titanium abutments are the most common type used because they have a simple fabrication technique and are less expensive compared with other types of abutments [4].

In the traditional implant prosthesis, the prosthetic screw was intentionally designed as the weakest link within the system. Specifically, in this type of restoration, a punctual excessive occlusal force (overload) or the succession of moderate loads over time (fatigue) can lead to the mechanical failure of the screw and, as a consequence, to the failure of the dental restoration [5–7]. However, in the case of any mechanical stress challenging the prosthesis, the fact that the screw would absorb the stress without endangering the bone–implant interface may be seen as an advantage [8].

The topic under study is of great relevance because some of the most common complications in implant prosthodontics have been mechanical complications. Abutment screw loosening has been reported as the most common prosthetic complication and has been understood to be that which precedes the more challenging abutment screw or even implant fracture [9,10]. The incidence of abutment screw fracture has been examined by a number of research studies, including Katsavochristou and Koumoulis [9]. This research concluded that the incidence of screw loosening falls within the range of 7% to 11%. Interestingly, the occurrence of abutment screw fracture remains much lower, at precisely 0.6% [9]. Due to the lack of standardized study designs and the diversity of implant prosthetic components, the data should be viewed with caution but should still be utilized for the individual evaluation of each implant system.

Several researchers have investigated the most common mechanical complications in single dental implants, as well as how complication rates are influenced by various clinical factors. For Lee et al. [11], the incidence of mechanical complications was 18.1%. The rates of occurrence of abutment screw loosening [ASL], abutment screw fracture [ASF], ceramic fracture [CF], repeated ASL, and repeated CF were 12.7%, 1.4%, 4.1%, 1.8%, and 0.9%, respectively. Excessive or parafunctional mastication dynamics (e.g., high occlusal force, bruxism, and clenching) and anatomic characteristics (e.g., alveolar bone resorption, presence of the inferior alveolar nerve or maxillary sinus floor, and bone quality) can cause occlusal overloading and/or non-axial loading, increasing the risk of mechanical complications in the posterior region.

Mechanical complications continue to be reported in the literature, and their clinical management can often be very challenging for the clinician as there is no consensus on ideal management [9]. Rescuing the fragment of a fractured abutment screw without damaging the remaining implant components has often been found to be impossible. If this rescue is not achieved, it may be necessary to remove the implant.

Moreover, in the early days of implant dentistry, and with these kinds of restorations, healing abutments were disconnected and reconnected several times, such as during impression taking and the fitting of the restoration and its placement. As it was considered inevitable, little attention was paid to this [1]. The repeated disconnection and connection of these abutments result in a negative bone response that manifests as bone loss at the marginal ridge level, accompanied by apical soft tissue migration [12]. In order to overcome these problems, the prosthetic procedures were modified and the "one-abutment one-time" protocol was introduced. This protocol included the placement of the permanent abutment immediately after implant placement, thereby eliminating the need for multiple implant–abutment disconnections [1,12]. As a result, the fragile soft tissue seal around an implant is not disrupted, the stability of the soft tissue is obtained [12,13], and the marginal bone is expected to be maximally preserved [1,12].

Consequently, the use of a multiunit abutment (also known as a transmucosal or transepithelial abutment) between the implant and abutment, as an alternative to the two-piece restorations, has become an increasingly common practice that brings numerous advantages from a clinical point of view. Firstly, it allows for the possibility of mounting the transepithelial abutment immediately after implant placement, avoiding the need to remove it later. This allows for working at the tissue level rather than the bone level during the next visit to the clinic [14]. Secondly, it allows for a decision regarding prosthetic emergence and height even after the implant is inserted; thus, it allows for the selection of the transepithelial component that best suits each case [15]. Thirdly, the transepithelial component favors sealing at the level of the implant platform [15].

As a consequence of some of the advantages mentioned above, several studies claim that there is lower crestal bone loss for restorations using a transepithelial component compared to direct-to-implant restorations [1,15–17]. With this kind of transepithelial abutment, the "one-abutment one-time" protocol can be followed, which is an advantage. Although its use has shown great advantages when used for the rehabilitation of multiple implants, its use in single-unit implants is less common, and few studies have been found that consider this topic.

However, evaluating stress distribution clinically in implant-supported prostheses is problematic. Therefore, finite element analysis (FEA) has been widely used for the mechanical testing of dental implants [18,19]. This method reflects the complexity of clinical conditions and has advantages over many analysis methods [18,20,21]. The data from FEA studies can be carefully extrapolated to daily clinical practice to improve the understanding of different scenarios, offering a suitable degree of reliability and accuracy without the risk and expense of implantation [22]. This method has been used as a tool to predict, for example, stresses in the peri-implant region and in the components of implant-supported restorations. Mathematically, FEA depends on the use of numerical techniques to solve the partial differential equations that govern the simulation problem. With FEA, the structures are to be converted into meshes using computer software. The resulting models consist of elements, nodes, and pre-defined boundary conditions. During the simulation, the loads are applied to specific nodes or elements specified by the user; then, the resulting displacement and stresses are evaluated using simulation analysis. FEA has been applied in many aspects of implant dentistry, such as the shape and design of restorations, crowns, or dental implants [19,23].

It is necessary to understand the biomechanics of implant-supported restorations in order to correctly design a FEA. In implant dentistry, as we have already mentioned, the abutment is usually connected to the implant by a retention screw. A tightening preload must be applied through this screw to prevent the loosening of this implant–abutment connection [19,23]. This preload is positively correlated with the screw-tightening torque, but only 10% of the torque is converted into the preload; the remaining 90% is used to overcome the friction between the surfaces of the joints of the components. The preload is the tension generated in the screw and the complementary clamping force between the head of the screw and the abutment. It is maintained by friction between the abutment-screw thread and the internal thread of the implant. When the abutment screw is tightened, a compressive force is generated along the interface between the abutment screw thread and the internal thread of the implant. Increasing the torque can increase the stability of the abutment-to-screw joint. The preload needs to be higher than the occlusal force to achieve a stable screw joint and to avoid screw loosening. Optimum preload should induce a force in the screw joint that is 75% of the yield strength of the screw [19]. However, excessive preloads can create screw stresses that exceed the yield limit of the material, resulting in plastic deformation of the screw threads and, hence, screw loosening or even fracture of the screw. The higher stresses created by excessive preloads can accelerate fatigue failure [23].

Screw loosening or the fracture of screws is a significant concern in implant-supported restorations. Achieving the right preload is crucial to prevent these complications and to ensure long-term success [19]. Following the torque specifications set by the manufacturer is critical to avoid problems and to ensure optimal operation; these vary, ranging from 18 Ncm to 45 Ncm. Inadequate tightening may result in joint separation and screw failure through fatigue, loosening, or even fracture [23]. The comprehensive relationship between the direction of load, the center of rotation, and the simultaneous stress distribution on the implant restoration simultaneously has rarely been investigated. Understanding the loading point and the direction of the load is critical for the design of implant prostheses that can withstand functional forces while minimizing stress concentration. Some research, such as a study by Kim et al. [24], has studied the correlation between the stress level and the various directions of the load on the occlusal surface using FE. The stress level was increased as the direction of the vector changed from the center of the implant connection [24].

Therefore, to complement the clinical studies that evaluate the clinical performance of single-implant restorations and to begin to gather evidence regarding the mechanical behavior of this type of restoration when we incorporate a transepithelial component, this study assesses the impact of using a transepithelial component versus a direct implant-supported restoration on the fatigue behavior of single dental implants. Our null hypothesis is that the force to which the screws are subjected is similar with or without an intermediate pillar. This study was carried out using the finite element method (FEM) and was experimentally validated. To the best of our knowledge, this is the first time an experiment has been conducted to analyze the biomechanical advantages of using intermediate abutments in the rehabilitation of single-unit implants.

2. Materials and Methods

An IIPSCA4513 Interna Plus implant (BTI Biotechnology Institute, Miñano, Spain) with a diameter of 4.5 mm and a length of 13 mm was selected. Two implants were inserted into the cortical bone juxtacrestally, that is, leaving the implant-abutment platform at bone level. In the first case, an INPPTU44 abutment (BTI Biotechnology Institute, Miñano, Spain) was mounted. The post was attached to the implant using an INTTUH screw (BTI®, Miñano, Spain) tightened to 35 Ncm. In the second case studied, an INTMIPU20 transepithelial abutment was mounted on the implant and tightened to 35 Ncm. A CPMIUPU abutment (BTI Biotechnology Institute, Miñano, Spain) was mounted on top and was joined to the assembly using the TTMIR prosthetic screw (BTI Biotechnology Institute, Miñano, Spain), tightened to 20 Ncm.

The implants, abutments, and transepithelial bodies were made of pure grade 4 titanium (Ti CP4), while the screws (including the one that comes with the transepithelial) were made of Ti6Al4V ELI (extra-low interstitials) (Ti Gr 5); the chemical composition is shown in Table 1.

Table 1. Chemical composition of titanium grade 5 (Ti Gr 5) in screws and titanium grade 4 (Ti CP 4) in implants, abutments, and transepithelial bodies.

Ti 6Al 4V ELI (Ti Gr 5)		Ti CP 4	
Composition	**Wt.%**	**Composition**	**Wt.%**
Al	5.5–6.5	N (max)	0.05
V	3.5–4.5	C (max)	0.08
Fe (max)	0.25	Fe (max)	0.5
O (max)	0.13	O (max)	0.4
C (max)	0.08	H (max)	0.0125
N (max)	0.05	-	-
H (max)	0.012	-	-

Wt.%: weight percent; Ti: titanium; Al: aluminum; V: vanadium; Fe (max): maximum allowable concentration of iron; O (max): maximum allowable concentration of oxygen; C (max): maximum allowable concentration of carbon; N (max): maximum allowable concentration of nitrogen; H (max): maximum allowable concentration of hydrogen.

Figure 1A shows the 3D models of the two dental restorations used in this study (direct-to-implant vs. using an intermediate abutment). The FE analyses were performed using Ansys Workbench® 19R1 (Ansys Iberia S.L., Madrid, Spain). A (cyclic) chewing force was simulated with an inclination of 30° with respect to the vertical, as indicated in ISO 14801 [25].

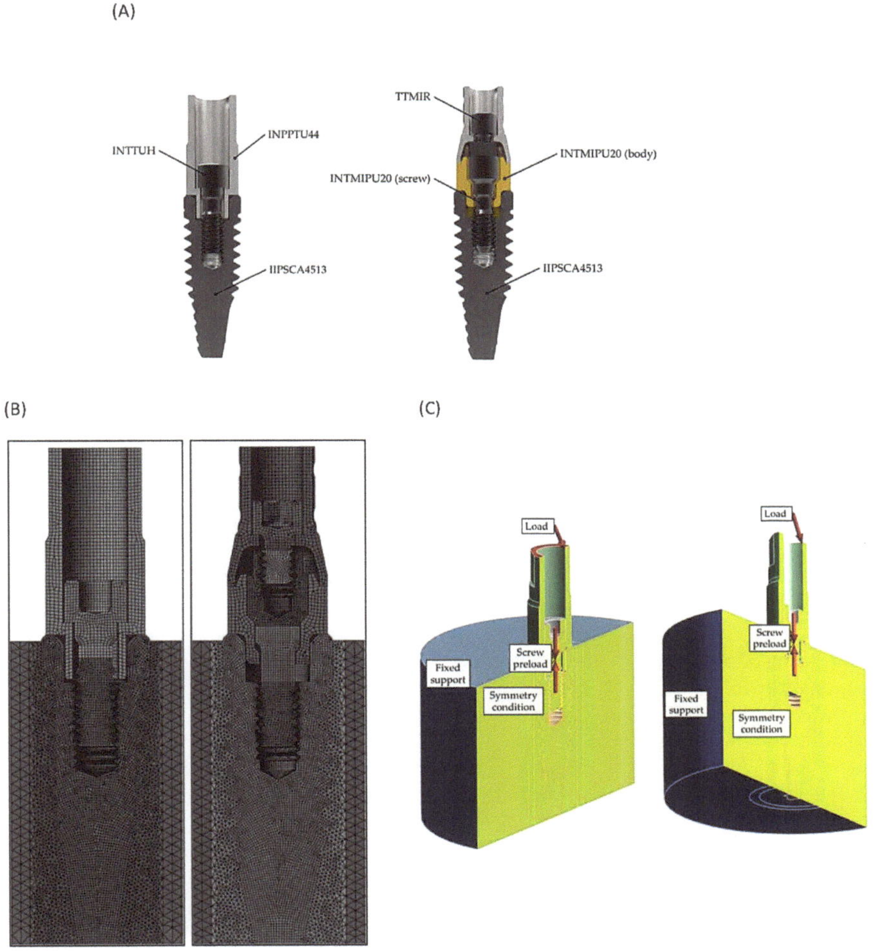

Figure 1. (**A**) Three-dimensional models of the two dental restorations. (**B**) The mesh of the two models under study. (**C**) Load, preload, fixed support, and symmetry condition.

In general, two types of loads have been applied in many studies—the vertical load and the 30° oblique load [24]. In order to consider a more severe case, one which is possibly also more realistic, the load was applied at a 10.5 mm height from the implant–abutment connection (IAC) [26]. The stress level of the implant increases with the increasing crown height, which appears to be a more critical factor than the crown–implant ratio. This is related to the class I lever effect. The fulcrum, load arm, and effort arm should be considered carefully to understand the stress distribution of the implant restoration [24].

Taking advantage of the symmetry of both the load and the model geometry, only half of the restoration was modeled; consequently, half of the preload and external load was applied. The threads of the screws and the internal thread of the implant were modeled as cylindrical threads instead of helical threads since this simplification does not imply an error greater than 3.5% [27]. Figure 1B shows the mesh of the two models under study (direct-to-implant vs. using an intermediate abutment), with a total of 2.7 million degrees of freedom (DoF). Both of the titanium materials were modeled as linearly elastic with

Young's modulus (E) = 103 GPa, Poisson's ratio (v) = 0.35 for CP 4 and v = 0.31 for Gr 5, while the cortical bone was also modeled as linearly elastic, with E = 13 GPa and v = 0.37. The contacts were defined as frictional contacts with friction coefficients of 0.17 for the screw–implant, screw–post, and screw–transepithelial contacts, and 0.21 for the implant–post, implant–transepithelial, and transepithelial–cylinder contacts [28]. Figure 1C shows the load, preload, fixed support, and symmetry condition of the system. However, the contacts are not shown, as there are so many that the authors could not find a way to show them all clearly.

The FEA performed consisted of two or three load steps, depending on the case under study. First, the screw preload corresponding to the recommended tightening torque was applied by means of a pretension section. In the case of the direct-to-implant restoration, this was performed in the first load step. In the case of the transepithelial-supported restoration, a first load step was necessary to preload the implant screw, and a second load step was needed to preload the prosthetic screw on the transepithelial component. This preload was calculated using the Motosh formula [27] and resulted in 814 N (direct-to-implant), 688 N (transepithelial), and 572 N (prosthetic screw over transepithelial). In a similar study, the preload of 825 N as a body force was applied to the upper part of the shank of the abutment screw, where the elongation of the screw was expected with tightening. Once the screw preloads were applied, a final loading step was used to apply the masticatory load with values from 0 to 400 N. The simulated bite forces employed in our finite element study were up to 400 N, as in other similar studies [29].

From the FEA, the contact reactions at the screw head were obtained: axial force, transverse force, and bending moment. These forces, once transferred to the critical section, the first thread in contact were used to calculate the nominal stress in this section. As the load cycle determined by ISO14801 [25] varies sinusoidally from the maximum load, and from 10% of this load, the nominal stress value at these two values of the load cycle were recorded. With both nominal stress values obtained, the effective fatigue stress was determined in order to later compare its behavior under a succession of masticatory loads. The finite element models used in this study are based on previously published studies [27,28,30] and a PhD thesis [31]. The results obtained through these finite element analyses were validated experimentally, providing very accurate results.

In this work, force reactions in screw contacts were considered. The structural behavior of a component does not need as high of a mesh refinement as for obtaining an accurate peak stress. Nevertheless, a minimum refinement must be performed to ensure proper contact behavior among components. In the following table, three mesh refinement grades are compared to determine the proper mesh. Force reactions were compared for the same instant during the analysis. The analyses were named from A to C, A being the least refined and C the most refined mesh. Table 2 shows the DoF used for each mesh.

Table 2. DoF used for each mesh.

	DoF
A	580 K
B	2.6 M
C	6.0 M

Figure 2A,B shows the axial force (2A) and bending moment reactions (2B) for a 0–400 N masticatory load range. As it can be appreciated, the axial force showed almost identical values regardless of the mesh. Regarding bending moment, only if a very coarse mesh is applied do the obtained values change. If a mesh refinement between B and C is performed, the moment values will not vary. In this study, the B mesh was used since it has been proven to be sufficient for accurate results.

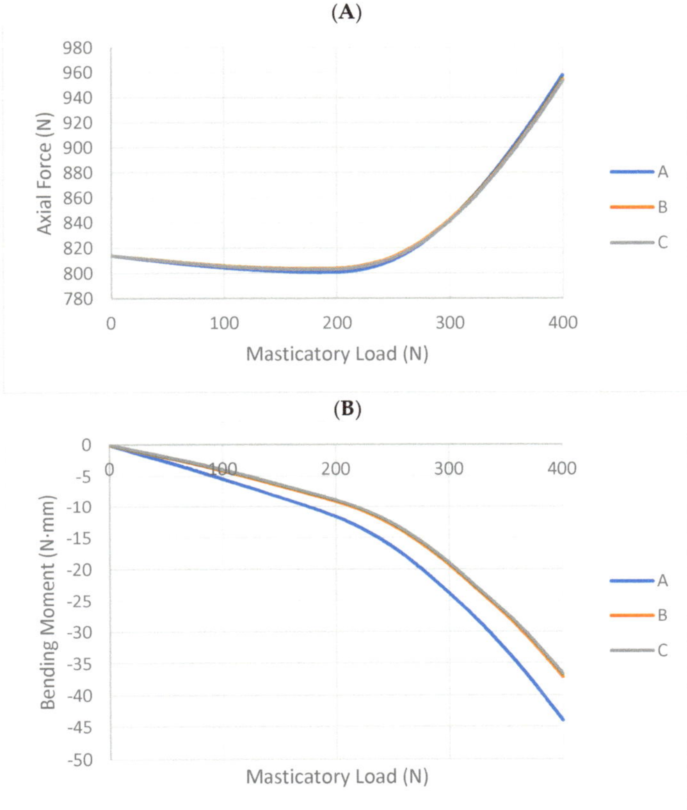

Figure 2. (**A**) Axial force and (**B**) bending moment reactions for a 0–400 N masticatory load range.

Finally, in order to verify that the breakages were produced by the site given by the FEM, a cyclic load was applied to three samples of each of the two dental restorations under study: the direct-to-implant and the transepithelial-supported restorations. The load was applied until a breakage of the dental restoration was detected and the component on which the breakage occurred was identified. The tests were performed on an INSTRON E 3000 Electropuls fatigue bench (Instron, Barcelona, Spain) mounted with a DYNACELLTM 2527-153 load cell (Instron, Barcelona, Spain) with a load range of ±5-kN. The setup was the same as that described for the FE analysis modeling, except for the specimen holder, which, in this case, was made of steel rather than cortical bone. Moreover, in order to facilitate a correct load application, a hemispherical device was added over the abutment to ensure that the load was applied at 10.5 mm from the height of the implant platform, using FEA.

Figure 3A shows a direct tension fatigue machine with standard fixtures for testing material specimens, and Figure 3B shows a specific set-up for conducting tests on dental restorations.

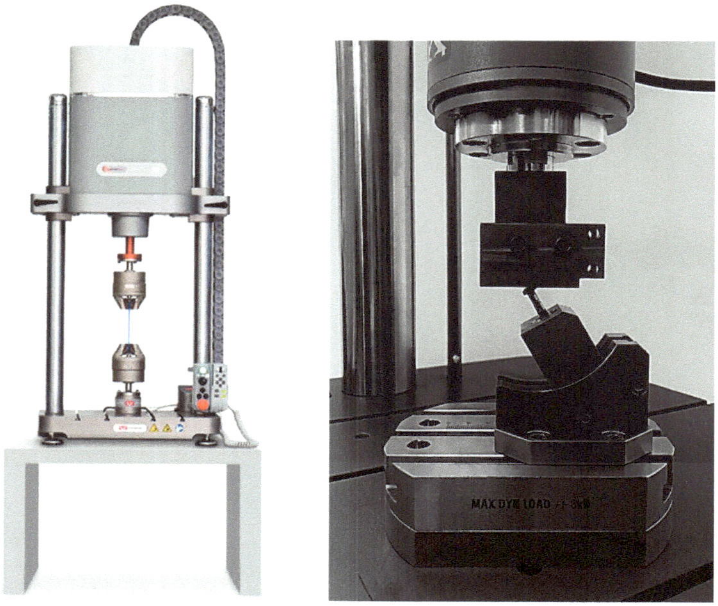

(A) (B)

Figure 3. Fatigue machine. (**A**) Direct tension fatigue machine with standard fixtures for testing material specimens. (**B**) Set-up for conducting tests on dental restorations.

3. Results

3.1. Stress Map

Figure 4 shows the stress map of the screws of the two dental restorations under study under a load of 300 N. As can be seen, when a transepithelial component is introduced, the screw into the implant undergoes a slightly lower stress status than in the case of a direct-to-implant restoration. Moreover, in the case of transepithelial-supported restoration, the prosthetic screw mounted over the transepithelial component suffers a higher stress status, which means that, in this case, the prosthetic screw is the fuse of the restoration.

3.2. Effective Fatigue Stress for Load Cycles

Figure 5 shows the effective fatigue stress for load cycles ranging from 100 to 400 N. S–N curves, also known as Wöhler curves, are graphical representations used in the field of materials science to depict the relationship between stress (S) and the number of cycles to failure (N) of material under cyclic loading conditions. These curves are derived from fatigue tests, where a material is subjected to repeated loading and unloading, and the number of cycles that cause failure is recorded. The curve helps engineers determine the fatigue life of a material, which is the number of stress cycles a material can withstand before failure occurs [32]. The primary objective of this investigation is not to precisely determine the fatigue lifespan of various dental restorations. Rather, the study adopts a broader methodology. Initially, it seeks to pinpoint the critical element in each dental restoration, and subsequently, it aims to discern which component endures the greatest/least amount of stress. Consequently, the research indirectly infers the longevity of different types of restorations, given that effective fatigue stress is intrinsically linked to the lifecycle of the component. The graph shows that the screw in the direct-to-implant restoration undergoes an effective stress that is almost identical to that of the screw attached to the implant in the restoration using a transepithelial component. In this case, it is the prosthetic screw (the upper screw) that suffers a slightly higher stress status. Therefore, the latter would fail in the event of a sufficiently high fatigue load.

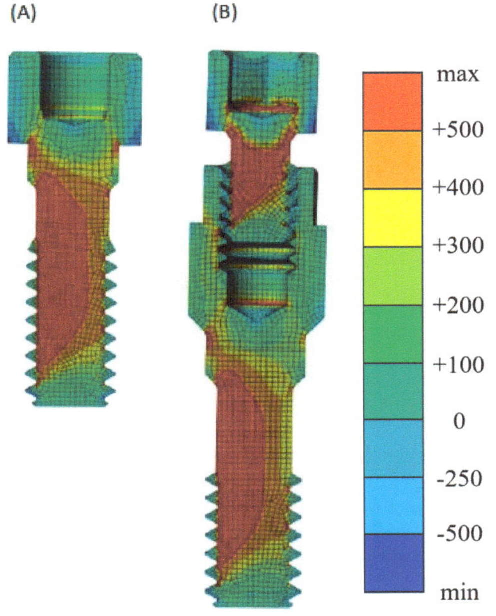

Figure 4. Stress map of the screws of the two dental restorations. (**A**) Direct to implant. (**B**) With an intermediate abutment. Signed normal stress in the vertical axis. + for tension and - for compression.

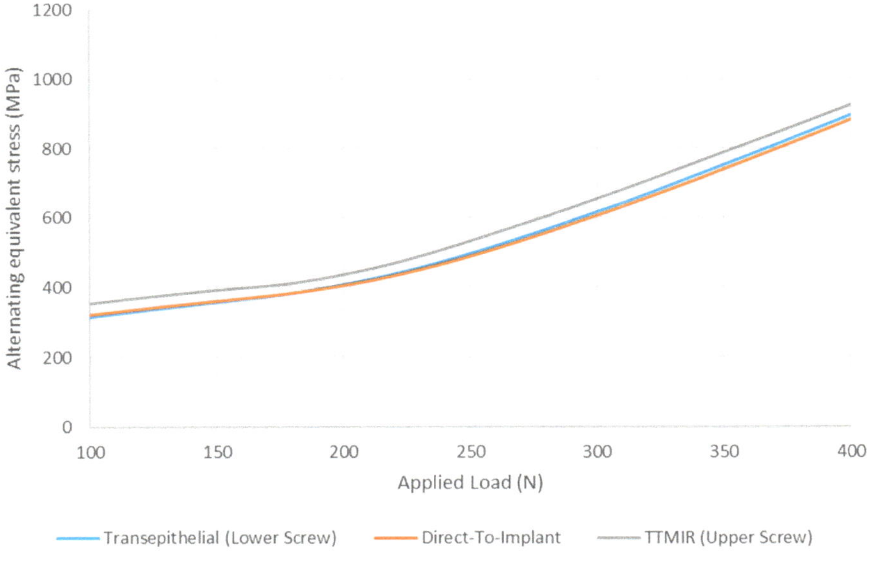

Figure 5. Effective fatigue stress.

3.3. Failing Component

Figure 6 shows the failing component of both dental restorations after the application of a cyclic fatigue load. It is therefore experimentally proven that, in transepithelial-supported restorations, the fuse changes from being the screw that goes into the implant (the lower one) to being the prosthetic screw (the upper one).

(A) (B)

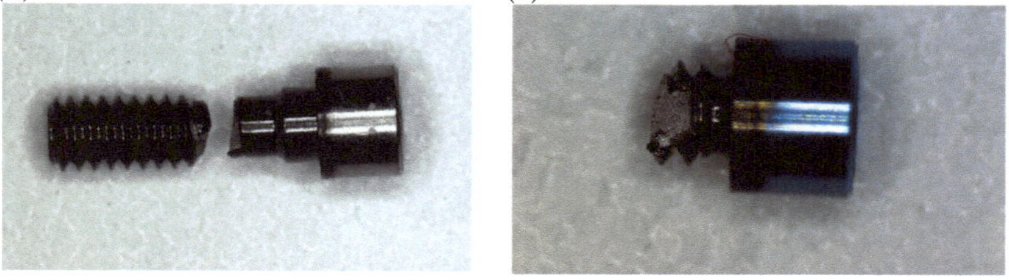

Figure 6. Failing component after cyclic fatigue load. (**A**) INTTUH screw. (**B**) TTMIR screw.

4. Discussion

In this study, we investigated how the use of a transepithelial component compares to a direct implant-supported restoration in terms of fatigue behavior. Given that one of the most common complications in single-unit restorations of implants is screw loosening or even fracture, it is intriguing to analyze the force distribution when adding an intermediate component. Additionally, investigating whether using a transepithelial abutment can prevent the unwanted fracture of the direct implant-to-screw connection, which could render the implant nonfunctional, is of significant interest.

We demonstrated that the component most likely to fracture is the screw of the intermediate abutment. In such cases, removing the fractured fragment or, in the worst-case scenario, the intermediate abutment itself is sufficient. Importantly, the implant remains uncompromised.

Considering the significance of the preload in preserving the structural and interfacial integrity of the implant abutment assembly, Honório et al. [23] conducted a study to evaluate the impact of varying preloads on the stress experienced by the retention screw and the microgap width of the internal conical connection. They employed finite element analysis (FEA) with a well-established model for preload assessment. This research shed light on the optimization of the preload to enhance the performance and longevity of dental implant components. When the screw-tightening torque was increased from 20 Ncm to 30 Ncm, it led to higher stresses in the abutment screw throughout various phases: before, during, and after occlusal loads. However, this higher torque also had some interesting effects, such as smaller micrograps. The higher torque also increased the occlusal load required to bridge the internal implant space. This could potentially help reduce bacterial leakage. Interestingly, the study found that the maximum stress in the abutment screw occurred at its neck on the distal surface. This suggests that the screw is more likely to fail in this specific region [23], which is consistent with the results obtained in our study.

Furthermore, FEA can be used for sensitivity analysis to study the effects of varying the material and geometrical parameters, such as the coefficient of friction, the screw diameter, the screw design, and the length of the implant fixture. Within the range of loads (10 N–280 N) tested in this FEA study, the gap sizes, especially those within the bridging zone, may not be large enough for the invasion of bacteria, the size of which can reach 6 μm. Therefore, further studies are needed to correlate the interfacial gap opening and bacterial microleakage [23].

In a FE study, Jung et al. [19] investigated the effects of the abutment screw preload on two different implant connection systems. Six three-dimensional finite element models were created based on various conditions: EO, an external hexagonal connection system with preload only; EN, an external hexagonal connection system with occlusal load only (no preload); EP, an external hexagonal system with both preload and occlusal load; IO, an internal hexagonal system with preload only; IN, an internal hexagonal system with occlusal load only; and IP, an internal hexagonal system with both preload and occlusal load. An 11.3-degree oblique load (100 N) was applied to the crown's occlusal surface for models with an occlusal load, and a preload of 825 N was applied to the abutment screw in the models. The abutment screw experienced the greatest increase in von Mises stress values under the occlusal load. The stress values ranged from 104.5 MPa (model EN) to 850 MPa (model EP) and from 37 MPa (model IN) to 674 MPa (model IP). Following the implant, the abutment showed the next highest stress levels. This study highlights the importance of considering preload conditions and connection system designs when evaluating the biomechanical behavior of dental implants. Understanding stress distribution patterns can contribute to the long-term success of implant-supported restorations [19]. Regardless of the conditions of the occlusal load, the models with preloads showed higher stress values than the models without preloads in both the external and internal connection systems. It seems essential to include the preload condition in finite element analysis, as a preload applied to the abutment screw influences the stress level in the implant system and bone.

Verri et al. [18] carried out a similar FE study to analyze the stress distributions of single implant-supported prostheses with different connections (external hexagon EH, internal hexagon IH, or Morse taper MT) in the anterior region of the maxilla, while varying the inclination of the applied load (0, 30, and 60 degrees) and the surgical technique for implant placement (monocortical/conventional, bicortical, and bicortical with nasal floor elevation) [18].

In this study conducted by Verri et al. [18], the EH implants exhibited higher stress levels on the fixation screw and implant, ranging from 100 to 600 MPa. However, these stress values did not necessarily affect the implant viability. The stress primarily indicated a mechanical tendency to result in issues such as screw loosening or failure, rather than a biological risk. The study found that the worst situations occurred under 60-degree loading. It is important to note that implants in clinical practice are typically not subjected to such large oblique forces. The internal connection implants showed a tendency toward implant-related problems. However, the amount of stress required to loosen the fixation screw of an EH implant should be less than the stress needed to fracture titanium. This observation might explain why EH implants are sometimes associated with more biomechanical issues than IC implants [18].

4.1. Impact of the Design of the Restoration

Whether prosthetic restorations supported by implants of different sizes and diameters placed adjacently should be separate or splinted is unclear. While splinted restorations may offer advantages in terms of stress distribution, individual patient factors and clinical realities play a significant role. The restoration designs can change the stress levels in adjacent implants of different lengths and diameters. Kul et al. [21] analyzed the stress and strain distribution around short and standard implants in the posterior mandible with splinted and separate crowns. The practical clinical considerations resulting from this study include the importance of an optimal and stable implant–abutment connection that plays a crucial role in the long-term success of dental implants because the stress is reduced. Oblique loads (loads applied at an angle) have a greater moment effect than purely vertical loads. As the angle between the direction of the oblique load and the implant axis increases, the moment effect becomes more pronounced. These oblique loads can significantly influence the stress distribution in implant-supported restorations. If a standard implant and a short implant are placed adjacently and restored with splinted crowns, the implants, abutments, and screws may be damaged; therefore, adjacent splinted implants should be of similar size. However, the perfect fit of the implant–abutment junction reduces stress, even in these adverse situations [21].

Other variables that should be taken into account during the manufacture of our restorations are the retention system (usually either screw- or cement-retained) and the type of restorative material used, which can also affect the stress distribution. For many years, metal–ceramic prostheses have been considered by some clinicians to be the gold standard of rehabilitation with an implant-supported prosthesis. Currently, there is increasing demand for metal-free restorations, such as those consisting of zirconia, as a more aesthetic rehabilitative treatment option. Lemos et al. [33] evaluated different implant–abutment connections, retention systems, and restorative materials in single crowns using 3D FEA. There was a higher concentration of stress in the fixation screw for the cemented prostheses in the external hexagon implants independently from the restorative materials used (increasing the risk of screw loosening/fracture). Furthermore, it should be acknowledged that in the event of screw loosening, the solution is more complex than in the screwed retention systems. Therefore, the combination of an external hexagon implant and a cement-retained prosthesis should be avoided [33].

Metal–ceramic and zirconia monolithic implant-supported single crowns had similar biomechanical behaviors in bone tissue and implants and their components. For Lemos et al., the similarities between the metal–ceramic and zirconia monolithic prostheses may be attributable to the similar mechanical properties, which may contribute to the sharing of stress across structures [33].

However, other researchers such as Pumnil et al. [34] present different evidence for the influence of the material on the transmission of loads. A comparative 3D FEA study of the stress distribution in the implant, screw, Ti-base, abutment, and restorative crown between the different customized abutment types has not been evaluated, and the proper abutment type selection for the implant-supported single crown is still an ambiguous issue. Therefore, Pumnil et al. [34] introduced a study to investigate the stress distribution using 3D FEA on the implant, screw, Ti-base, abutment, hybrid–abutment crown, and restorative crown among different abutment types, as follows: customized titanium abutment, customized titanium hybrid–abutment crown, customized zirconia abutment with Ti-base, and customized zirconia hybrid abutment crown with Ti-base. For all groups, oblique loading tended to generate higher stress values compared to purely vertical loads. Clinicians should avoid excessive oblique forces to prevent stress-related complications. Pumnil et al. [34] concluded that the choice of abutment type significantly impacts the stress distribution in implant-supported restorations. The presence of a titanium base within a zirconia abutment improved the stress distribution. Titanium has the ability to absorb stress, contributing to overall stability. This combination is a favorable option for implant-supported crowns. A customized titanium hybrid–abutment crown created stress concentration at the screw; thus, this abutment type should be used cautiously and maintained regularly. In addition, a customized zirconia hybrid–abutment crown with a titanium base caused stress concentration at the implant, and this abutment type should be maintained regularly. A thoughtful selection of abutment materials and diligent maintenance are crucial for successful implant restorations [34].

4.2. Impact of Intermediate Abutments

With regard to the use of intermediate abutments, Zincir et al. [20] compared the stress and strain values of the direct-to-implant system with the conventional angled multiunit abutment–implant connection system used in "all on four" rehabilitations in the implant parts and the surrounding bone using FEM. In the context of axial and oblique forces, they found that the direct-to-implant systems exhibited greater stress accumulation in the bone, prosthesis screws, and implants when compared to multiunit abutment–implant connection systems [20]. These results are in line with those reported in this study.

As seen in the stress map of the screws of the two dental restorations (Figure 5) in the case of transepithelial-supported restorations, the top screw is the one that would break in the event of an overload. This prevents the rest of the restoration from being damaged. This study, based on the previous literature, was performed with the assumption that the maximum functional force was 400 N [19,23]. Occlusal loading of 100 N is considered to correspond with light clenching while loading of 200 N is considered to correspond with middle clenching [35]. We wanted to understand how stress is distributed within the system when it faces extreme loads, such as those experienced by bruxism patients. Our goal was to identify the most vulnerable parts of the system (Figure 4). Additionally, we observed that in S–N curves (Figure 5), stress on the screws exponentially increases with load. This observation aligns with the finding that bruxism patients tend to experience greater mechanical complications compared to non-bruxism individuals [36].

It is worth mentioning that the stress level of the upper screw (Figure 5) indicates that the fatigue life of the screw of the intermediate abutment would be slightly lower than in the case of direct-to-implant restoration. However, the difference is minimal; so, this reduction in fatigue life may not be relevant in the patient's mouth. In any case, as mentioned above, the fact that the upper prosthetic screw acts as a fuse serves to protect the rest of the dental restoration. It is beneficial that the fracture is usually of the abutment screw, as we have seen in this study. Furthermore, in the case when a repair is needed, this would be conducted at the tissue level, that is, with the replacement of the upper screw and possibly the abutment, without touching the implant–abutment connection or the bone surrounding it. Therefore, these advantages far outweigh the minimal reduction in fatigue life.

However, there are additional potential benefits. Even if the screw does not break, when screw loosening occurs, there are biological implications. The microgap at the implant interface permits fluid passage independently of the implant system. Functional rocking effects and screw loosening may contribute to increased leakage. Moreover, the clinical phenomenon of bleeding and malodor, which are attributable to anaerobic bacteria on the removal of abutments or healing screws, may partly be the result of the effects of microleakage [37].

When the implant–abutment interface is positioned at the alveolar bone level, it leads to persistent peri-implant inflammation and significant bone loss. This suggests that the inflammatory stimulus originates precisely at the implant–abutment interface, and there is a direct relationship between the extent of the inflammation and the magnitude of the alveolar bone loss [38].

Abutment screw loosening, as mentioned above, is a common mechanical complication in dental implants. It occurs primarily because the abutment screw is the weakest part of the implant system. Stable connections between implant components are crucial for treatment success. A review by Goodacre et al. [39] revealed that screw loosening occurs in 8% of cases, and this figure can rise to 45% in single crowns. Additionally, abutment screw loosening may lead to other complications, including screw fracture, marginal gaps, peri-implantitis, microbial leakage, crown loosening, and patient discomfort [40].

To mitigate this complication, it is essential to ensure an optimal component fit, minimize the abutment micromovement, reduce the prosthetic misfit, optimize the prosthetic design and occlusion, and maintain a sustained preload [37].

The use of an intermediate abutment in single–implant restorations offers several advantages. Firstly, the implant–abutment interface remains better sealed, avoiding disconnection during prosthesis fabrication. This inherently reduces contamination and minimizes microleakage around the implant platform. Additionally, as mechanical overload primarily affects the abutment screw, it is more likely to loosen. Importantly, this loosening occurs away from the bone crest, mitigating potential biological complications and preventing marginal bone loss.

Although the use of intermediate abutments has been more commonly accepted in multiple implants, it is not as common in single implants. Not all commercial implant manufacturers offer this possibility yet. However, following the results of this study, we can emphasize that, among the other advantages already described regarding the use of an intermediate abutment, in the event of a fracture of the abutment screw, it would be possible to remove it and put in a new one. In the worst-case scenario, when trying to rescue the fragment of the screw, the abutment could be damaged but not the implant. Additionally, in the event of screw loosening, it would not happen at the implant–abutment interface. Consequently, there may be a reduction in biological complications related to microfiltrations, although research and further investigation are essential to advance our understanding and to address unanswered questions.

Limitations of the study: Finite element analysis (FEA) has its limitations, and critical considerations are necessary when interpreting its results. FEA relies on input material properties, which may not perfectly mimic real-world conditions. Variations in material behavior can impact simulation outcomes. Clinicians should be cautious when directly applying FEA findings to clinical practice because it is a virtual model and cannot fully replicate the complexities of biological tissues. Clinical validation is essential to confirm the observed biomechanical effects. However, FEA provides valuable insights, and its findings should be complemented by empirical evidence from clinical studies. Moreover, our investigation was restricted to a particular brand of implant and a specific material. However, it is essential to recognize that contemporary clinicians have an array of options available, including various brands and materials for implants and implant components. Notably, zirconia and diverse metal combinations are among these alternatives. By acknowledging these limitations, we demonstrate both our awareness of this study's scope and our commitment to transparent reporting. Furthermore, we encourage future research to explore broader material choices and their implications.

Practical clinical applications: With the results obtained in this study, we can affirm that the use of intermediate abutments may be beneficial when rehabilitating single-unit implants. The use of these abutments has been relatively common in multiple implant rehabilitation; however, it has been less common in single-unit implants. The biomechanical advantages that their use can provide appear to be proven.

Future research directions: Future studies that incorporate even more sophisticated models are essential. These studies would allow for a comprehensive evaluation of, for example, the impact of simulating the preload condition in the abutment screw during advanced finite element analysis (FEA) applications. By considering these factors, researchers can gain deeper insights into the behavior of implant-supported restorations and enhance clinical outcomes.

It should be taken into account that this study was limited to the analysis of the mechanical behavior of a specific transepithelial abutment model. There are transepithelial abutments with different prosthetic emergences and it is possible that these different prosthetic platforms would have an influence on the behavior of the prosthetic screw (the upper screw). This study is proposed as a future line of the current research.

The authors' objective is to conduct a clinical trial that applies the insights derived from this study, thereby substantiating its clinical validity.

5. Conclusions

Introducing a transepithelial component into the dental restoration causes the fuse or critical component to be the upper screw, that is, the one that is mounted on the transepithelial component, rather than the screw that is mounted on the implant, as is the case with direct-to-implant restorations. This means that the rest of the dental restoration is better protected; thus, in the case of the need for restoration, it may be performed at the tissue level (at the level of the prosthetic platform) instead of the bone level (at the level of the implant platform). This benefit is achieved without compromising the fatigue behavior of the dental restoration.

Author Contributions: Conceptualization, E.A., P.T.D. and J.S.-C.; Data curation, N.L. and M.A.; Formal analysis, P.T.D., N.L. and M.A.; Funding acquisition, E.A.; Investigation, P.T.D.; Methodology, P.T.D., N.L., M.A. and J.S.-C.; Project administration, P.T.D.; Resources, E.A.; Software, M.A.; Supervision, E.A., J.P.V. and J.S.-C.; Validation, E.A., J.P.V. and J.S.-C.; Visualization, J.P.V. and J.S.-C.; Writing—original draft, P.T.D. and M.A.; Writing—review and editing, P.T.D., J.P.V. and J.S.-C. All authors have read and agreed to the published version of the manuscript.

Funding: This research received no external funding.

Institutional Review Board Statement: Not applicable.

Informed Consent Statement: Not applicable.

Data Availability Statement: Data are contained within the article. If any raw data supporting the conclusions are needed, the dataset can be made available upon request from the authors.

Acknowledgments: The support given by "Biotechnology Institute I mas" for their donations in kind (some of the materials and technology used) is acknowledged.

Conflicts of Interest: The authors declare no conflicts of interest.

References

1. Lambrechts, T.; Doornewaard, R.; De Bruyckere, T.; Matthijs, L.; Deschepper, E.; Cosyn, J. A Multicenter Cohort Study on the Association of the One-abutment One-time Concept with Marginal Bone Loss around Bone Level Implants. *Clin. Oral Implants Res.* **2021**, *32*, 192–202. [CrossRef] [PubMed]
2. Mat-Baharin, N.H.; Patel, A. Peri-Implantitis and Practical Management: A Review. *Arch. Orofac. Sci.* **2022**, *17*, 11–19. [CrossRef]
3. Hamed, M.T.; Abdullah Mously, H.; Khalid Alamoudi, S.; Hossam Hashem, A.B.; Hussein Naguib, G. A Systematic Review of Screw versus Cement-Retained Fixed Implant Supported Reconstructions. *Clin. Cosmet. Investig. Dent.* **2020**, *12*, 9–16. [CrossRef] [PubMed]
4. Al-Thobity, A.M. Titanium Base Abutments in Implant Prosthodontics: A Literature Review. *Eur. J. Dent.* **2022**, *16*, 49–55. [CrossRef] [PubMed]
5. Shemtov-Yona, K.; Rittel, D. Fatigue of Dental Implants: Facts and Fallacies. *Dent. J.* **2016**, *4*, 16. [CrossRef] [PubMed]
6. Pjetursson, B.; Asgeirsson, A.; Zwahlen, M.; Sailer, I. Improvements in Implant Dentistry over the Last Decade: Comparison of Survival and Complication Rates in Older and Newer Publications. *Int. J. Oral Maxillofac. Implants* **2014**, *29*, 308–324. [CrossRef] [PubMed]
7. Dhima, M.; Paulusova, V.; Lohse, C.; Salinas, T.J.; Carr, A.B. Practice-Based Evidence from 29-Year Outcome Analysis of Management of the Edentulous Jaw Using Osseointegrated Dental Implants. *J. Prosthodont.* **2014**, *23*, 173–181. [CrossRef] [PubMed]
8. Pow, E.H.-N.; Dent, P.; Leung, K.C.-M.; Dent, P. Prosthodontic Complications in Dental Implant Therapy. *Hong Kong Dent. J.* **2008**, *5*, 79–83.
9. Katsavochristou, A.; Koumoulis, D. Incidence of Abutment Screw Failure of Single or Splinted Implant Prostheses: A Review and Update on Current Clinical Status. *J. Oral Rehabil.* **2019**, *46*, 776–786. [CrossRef] [PubMed]
10. Di Francesco, F.; De Marco, G.; Cristache, C.; Vernal, R.; Cafferata, E.; Lanza, A. Survival and Mechanical Complications of Posterior Single Implant–Supported Restorations Using Prefabricated Titanium Abutments: A Medium- and Long-Term Retrospective Analysis with up to 10 Years Follow-Up. *Int. J. Prosthodont.* **2022**, *35*, 278–286. [CrossRef]
11. Lee, B.-A.; Kim, B.-H.; Kweon, H.H.I.; Kim, Y.-T. The Prosthetic Abutment Height Can Affect Marginal Bone Loss around Dental Implants. *Clin. Implant Dent. Relat. Res.* **2018**, *20*, 799–805. [CrossRef] [PubMed]
12. Nicolás-Silvente, A.I.; Rivas-Pérez, A.; García-López, R.; Alemán-Marín, J.; Chiva-García, F.; Sánchez-Pérez, A. La Utilización de Pilares Transmucosos Definitivos de Colocación Inmediata: The Use of Immediate Placement of Definitive Transmucosal Abutments. *Av. Odontoestomatol.* **2020**, *36*, 99–106. [CrossRef]
13. Canullo, L.; Pesce, P.; Tronchi, M.; Fiorellini, J.; Amari, Y.; Penarrocha, D. Marginal Soft Tissue Stability around Conical Abutments Inserted with the One Abutment-one Time Protocol after 5 Years of Prosthetic Loading. *Clin. Implant Dent. Relat. Res.* **2018**, *20*, 976–982. [CrossRef]
14. Atieh, M.A.; Tawse-Smith, A.; Alsabeeha, N.H.M.; Ma, S.; Duncan, W.J. The One Abutment–One Time Protocol: A Systematic Review and Meta-Analysis. *J. Periodontol.* **2017**, *88*, 1173–1185. [CrossRef] [PubMed]
15. Henao, P.; Queija, L.; Linder, S.; Dard, M.; González, A.; Carrión, J. Clinical and Radiographic Outcomes of a New Fully Tapered Implant with the One-Abutment One-Time Approach: In-Line Clinical Case Series with a 1-Year Follow-Up. *Int. J. Oral Maxillofac Implants* **2023**, *38*, 943–953. [CrossRef] [PubMed]
16. Canullo, L.; Omori, Y.; Amari, Y.; Iannello, G.; Pesce, P. Five-year Cohort Prospective Study on Single Implants in the Esthetic Area Restored Using One-abutment/One-time Prosthetic Approach. *Clin. Implant Dent. Relat. Res.* **2018**, *20*, 668–673. [CrossRef] [PubMed]
17. Bressan, E.; Grusovin, M.G.; D'Avenia, F.; Neumann, K.; Sbricoli, L.; Luongo, G.; Esposito, M. The Influence of Repeated Abutment Changes on Peri-Implant Tissue Stability: 3-Year Post-Loading Results from a Multicentre Randomised Controlled Trial. *Eur. J. Oral Implant.* **2017**, *10*, 373–390.
18. Verri, F.; Santiago, J.; Almeida, D.; De Souza Batista, V.; Aparecido Araujo Lemos, C.; Mello, C.; Pellizzer, E. Biomechanical Three-Dimensional Finite Element Analysis of Single Implant–Supported Prostheses in the Anterior Maxilla, with Different Surgical Techniques and Implant Types. *Int. J. Oral Maxillofac. Implants* **2017**, *32*, e191–e198. [CrossRef] [PubMed]

19. Jung, W.; Lee, W.; Kwon, H.-B. Effects of Abutment Screw Preload in Two Implant Connection Systems: A 3D Finite Element Study. *J. Prosthet. Dent.* **2019**, *122*, 474.e1–474.e8. [CrossRef]
20. Zincir, Ö.Ö.; Parlar, A. Comparison of Stresses in Monoblock Tilted Implants and Conventional Angled Multiunit Abutment-Implant Connection Systems in the All-on-Four Procedure. *BMC Oral Health* **2021**, *21*, 646. [CrossRef]
21. Kul, E.; Korkmaz, İ.H. Effect of Different Design of Abutment and Implant on Stress Distribution in 2 Implants and Peripheral Bone: A Finite Element Analysis Study. *J. Prosthet. Dent.* **2021**, *126*, 664.e1–664.e9. [CrossRef] [PubMed]
22. Sabri, L.; Hussein, F.; AL-Zahawi, A.; Abdulrahman, B.; Salloomi, K. Biomechanical Finite Element Analysis of a Single Implant Threaded in Anterior and Posterior Regions of Maxilla Bone. *Indian J. Dent. Res.* **2020**, *31*, 203–208. [CrossRef] [PubMed]
23. Honório Tonin, B.S.; He, Y.; Ye, N.; Chew, H.P.; Fok, A. Effects of Tightening Torque on Screw Stress and Formation of Implant-Abutment Micrograps: A Finite Element Analysis. *J. Prosthet. Dent.* **2022**, *127*, 882–889. [CrossRef] [PubMed]
24. Kim, J.-H.; Noh, G.; Hong, S.-J.; Lee, H. Biomechanical Stress and Microgap Analysis of Bone-Level and Tissue-Level Implant Abutment Structure According to the Five Different Directions of Occlusal Loads. *J. Adv. Prosthodont.* **2020**, *12*, 316. [CrossRef] [PubMed]
25. *ISO 14801*; 2016-Dynamic Fatigue Test for Endosseous Dental Implants. ISO: Geneva, Switzerland, 2016.
26. Khan, M.; Khan, M.A.; Hussain, U. Clinical Crown Lenght, Width and the Width/Length Ratio in the Maxillary Anterior Region in a Sample of Mardan Population. *Pak. Oral Dent. J.* **2015**, *35*, 738–741.
27. Armentia, M.; Abasolo, M.; Coria, I.; Albizuri, J. Fatigue Design of Dental Implant Assemblies: A Nominal Stress Approach. *Metals* **2020**, *10*, 744. [CrossRef]
28. Armentia, M.; Abasolo, M.; Coria, I.; Bouzid, A.-H. On the Use of a Simplified Slip Limit Equation to Predict Screw Self-Loosening of Dental Implants Subjected to External Cycling Loading. *Appl. Sci.* **2020**, *10*, 6748. [CrossRef]
29. Hollanders, A.C.C.; Kuper, N.K.; Huysmans, M.C.D.N.J.M.; Versluis, A. The Effect of Occlusal Loading on Cervical Gap Deformation: A 3D Finite Element Analysis. *Dent. Mater.* **2020**, *36*, 681–686. [CrossRef]
30. Armentia, M.; Abasolo, M.; Coria, I.; Sainitier, N. Effect of the Geometry of Butt-Joint Implant-Supported Restorations on the Fatigue Life of Prosthetic Screws. *J. Prosthet. Dent.* **2022**, *127*, 477.e1–477.e9. [CrossRef]
31. Armentia Sánchez, M. *Mechanical Analysis and Design of Dental Restorations: Fatigue, Microgap and Screw Loosening*; Universidad del País Vasco: Leioa, Bizkaia, 2023.
32. Connor, N. What Is Fatigue Life—S-N Curve—Woehler Curve—Definition. Available online: https://material-properties.org/what-is-fatigue-life-s-n-curve-woehler-curve-definition/ (accessed on 27 March 2024).
33. Lemos, C.A.A.; Verri, F.R.; Noritomi, P.Y.; De Souza Batista, V.E.; Cruz, R.S.; De Luna Gomes, J.M.; De Oliveira Limírio, J.P.J.; Pellizzer, E.P. Biomechanical Evaluation of Different Implant-Abutment Connections, Retention Systems, and Restorative Materials in the Implant-Supported Single Crowns Using 3D Finite Element Analysis. *J. Oral Implantol.* **2022**, *48*, 194–201. [CrossRef]
34. Pumnil, S.; Rungsiyakull, P.; Rungsiyakull, C.; Elsaka, S. Effect of Different Customized Abutment Types on Stress Distribution in Implant-Supported Single Crown: A 3D Finite Element Analysis. *J. Prosthodont.* **2022**, *31*, e2–e11. [CrossRef] [PubMed]
35. Yoshitani, M.; Takayama, Y.; Yokoyama, A. Significance of Mandibular Molar Replacement with a Dental Implant: A Theoretical Study with Nonlinear Finite Element Analysis. *Int. J. Implant Dent.* **2018**, *4*, 4. [CrossRef] [PubMed]
36. Al-Ani, Z.; Maghaireh, H. Occlusion on a Single Implant-Supported Crown: Any Differences? *Prim. Dent. J.* **2022**, *11*, 32–38. [CrossRef]
37. Gross, M.; Abramovich, I.; Weiss, E.I. Microleakage at the Abutment-Implant Interface of Osseointegrated Implants: A Comparative Study. *Int J Oral Maxillofac Implants* **2000**, *14*, 94–100.
38. Broggini, N.; McManus, L.M.; Hermann, J.S.; Medina, R.; Schenk, R.K.; Buser, D.; Cochran, D.L. Peri-Implant Inflammation Defined by the Implant-Abutment Interface. *J. Dent. Res.* **2006**, *85*, 473–478. [CrossRef] [PubMed]
39. Goodacre, C.J.; Bernal, G.; Rungcharassaeng, K.; Kan, J.Y.K. Clinical Complications with Implants and Implant Prostheses. *J. Prosthet. Dent.* **2003**, *90*, 121–132. [CrossRef]
40. Alsubaiy, E. Abutment Screw Loosening in Implants: A Literature Review. *J. Fam. Med. Prim. Care* **2020**, *9*, 5490–5494. [CrossRef]

prosthesis

MDPI

Article

Implant-Prosthetic Rehabilitation of the Agenesis of Maxillary Lateral Incisors: A 2-Year Prospective Clinical Study with Full Digital Workflow

Roberto Sorrentino [1], Maria Irene Di Mauro [1,*], Gennaro Ruggiero [1], Renato Leone [1], Edoardo Ferrari Cagidiaco [2], Marco Annunziata [3], Marco Ferrari [2] and Fernando Zarone [1]

1 Scientific Unit of Digital Dentistry, Department of Neurosciences, Reproductive and Odontostomatological Sciences, University "Federico II" of Napoli, 80138 Napoli, Italy; roberto.sorrentino@unina.it (R.S.); gennaro_ruggiero@hotmail.it (G.R.); natoreleone@gmail.com (R.L.); fernando.zarone@mac.com (F.Z.)

2 Department of Medical Biotechnologies, University of Siena, 53100 Siena, Italy; edoardo.ferrari.cagidiaco@gmail.com (E.F.C.); ferrarimar@unisi.it (M.F.)

3 Multidisciplinary Department of Medical-Surgical and Dental Specialties, University of Campania "Luigi Vanvitelli", 81100 Caserta, Italy; marco.annunziata@unicampania.it

* Correspondence: mariadimauro94@gmail.com

Abstract: The main objectives of the present prospective clinical study were to evaluate the survival and success rates of implant-supported zirconia single crowns fabricated with a full digital workflow for the rehabilitation of mono- and bilateral agenesis of maxillary lateral incisors after 2 years of clinical function; biological and technical parameters affecting the prosthetic restorations were recorded, as well as the patient-satisfaction score. Twenty-two patients showing mono- or bilateral agenesis of the maxillary lateral incisors were included in this study, and a total of 30 narrow-diameter implants were inserted. Thirty screw-retained monolithic cubic zirconia single crowns with internal connections were fabricated. Objective outcome evaluations were performed by means of the Functional Implant Prosthodontic Score, whereas the patient-satisfaction score was evaluated using Visual Analog Scales. Descriptive statistics were performed and the Kaplan–Meier analysis was run to analyze time-to-event data. After 2 years of clinical function, the overall FIPS found in the present study was 9.2, whereas the average patient-satisfaction score was 8.7. The Kaplan–Meier analysis at the 2-year follow-up reported a cumulative survival rate of 100% and a cumulative success rate of 93.3%. The implant-prosthetic rehabilitation with a full digital workflow proved to be an effective and reliable procedure for the functional and aesthetic treatment of the agenesis of maxillary lateral incisors in the short-term. Clinical investigations with wider sample populations and longer observational follow-ups could be useful to validate, in the long-term, the clinical outcomes of the present prospective clinical study.

Keywords: implant-prosthesis; prosthodontics; agenesis; zirconia; dental implant

1. Introduction

Dental agenesis is defined as the absence or failure of formation of a tooth, and permanent maxillary lateral incisors have been reported to be the teeth most likely to be missing [1,2]. This condition can affect the Oral Health Related Quality of Life (OHRQoL) of patients, as it represents peoples' subjective perspectives regarding various experiences and symptoms related to oral functions, aesthetic perceptions, and psychological comfort and self-esteem.

In the literature, the prevalences of maxillary lateral incisor agenesis varied across population on the basis of race and sex [2,3].

This anomaly was reported to be frequently bilateral and often associated with tooth ectopias and/or other abnormal dental conditions, such as smaller or peg-shaped teeth on the contralateral side [2–4].

In particular, smiles showing agenesis of maxillary lateral incisors were ranked as less attractive by patients and laypeople, probably because of anatomical differences between the lateral incisors and canines [5]. The presence of canines, which are more conical, and the absence of lateral incisors, which are smaller and flat-faced, were considered disharmonious aspects that were seen as less pleasant in a smile by evaluators [5]. The lack of maxillary lateral incisors was referred to as a reason of concern to patients for both functional and aesthetic reasons; therefore, several options were proposed for the rehabilitation of this condition [5]. The chosen treatment should be the less invasive option that could satisfy both the functional issues and the aesthetic expectations of patients. Careful interdisciplinary treatment planning is always advisable, keeping in mind that improper patient selection could result in unsatisfactory clinical outcomes [4].

If the deciduous maxillary lateral incisors are present in the arch, then a short-term conservative approach consists of only an esthetic reshaping of the deciduous teeth with composite resins [6]. Alternatively, canine substitution can be performed by carrying out a coronal reshaping and resin composite camouflage of the canine to be transformed in a lateral incisor; however, such a solution may not be completely satisfactory from both the functional and esthetic point of views. Removable partial dentures (RPDs) are usually considered interim restorations whereas more invasive prosthetic approaches are based on adhesive bridges (i.e., Maryland or Rochette bridge), cantilevered restorations, or fixed dental prostheses (metal-ceramic or all-ceramic FDPs) sustained by the central incisor and the canine so as to replace the missing maxillary lateral incisor [7].

Nevertheless, according to the literature, patients seem to prefer an interdisciplinary treatment based on an orthodontic approach to close the edentulous space or conversely to open it and carry out an implant-prosthetic treatment, as no tooth preparation is required [8,9]; these can be considered the most conservative and widespread treatments [10], and the present study focused on this specific treatment option.

As regards implant-prosthetic treatment, narrow-diameter implants were found to be comparable to standard fixtures in the anterior zone, with users reporting satisfactory cumulative success rates ranging from 84.2% to 100% (mean: 95.2%) [11–13]. Implant-prosthetic rehabilitations can be performed by means of either conventional or digital workflows.

The use of a digital workflow in daily dental practice is increasing, allowing clinicians to optimize chair time and, simultaneously, improving the patients' comfort and compliance [14–17]. A full digital workflow involves several aspects such as 3D radiographic acquisition, optical impressions by means of intraoral scanners (IOSs), digital smile planning and CAD-CAM fabrication of the prostheses by means of milling or 3D printing. Nowadays, the advantages of digital technologies are well known, including offering a 3D pre-visualization and planning of the region of interest and reducing working time, according to the skill and experience of clinical operators [14–16]. Patients seem to prefer optical impressions, in terms of anxiety, nausea, taste, and discomfort related to the conventional impression-taking procedure. Furthermore, the digital workflow allows the avoidance of possible distortions associated with conventional impression materials, offering the possibility to re-scan a defective area with better acceptance by patients [14–17].

Recent investigations reported satisfactory clinical outcomes in esthetic areas when implant-prosthetic rehabilitations were carried out following a digital workflow, although more clinical prospective studies are needed to establish valid protocols [18–20].

The present prospective clinical study was designed to evaluate the 2-year survival and success rates of implant-supported zirconia single crowns produced with a digital workflow for the rehabilitation of mono- and bilateral agenesis of maxillary lateral incisors. Biological (i.e., marginal bone levels and peri-implant soft-tissue conditions) and technical parameters (i.e., mechanical complications) possibly affecting the prosthetic restorations were recorded, as well as the patients' satisfaction scores.

2. Materials and Methods

The present study was designed as a 2-year prospective clinical study, following the international guidelines STROBE (STrengthening the Reporting of OBservational studies in Epidemiology) and respecting the Declaration of Helsinki (2013) or comparable ethical standards reviewed and approved by the hosting institution. Digital data collected from patients were protected by means of a password and the access was limited to clinicians who performed the present study. This study was performed by expert prosthodontists from the Scientific Unit of Digital Dentistry (SUDD) at the Department of Prosthodontics of the University "Federico II" of Napoli (Italy) and was authorized by the Institutional Review Board of the University "Federico II".

The recruitment of patients was carried out between April and June 2020 according to the following inclusion criteria established by the literature for implant-supported prostheses:

- minimum age: 18 years (proven completion of facial growth);
- single edentulous space (mono- or bilateral agenesis of maxillary lateral incisors);
- presence of at least 10 pairs of opposing teeth;
- intact adjacent teeth, restored with functionally and esthetically congruous reconstructions or restored with prostheses precluding the possibility of adding missing teeth;
- refusal of alternative treatments (i.e., canine replacement, removable prosthesis, adhesive prosthesis, conventional or cantilevered fixed dental prostheses).

In addition, the following exclusion criteria were used:

- symptomatic temporo-mandibular dysfunctions;
- inability to undergo surgical procedures;
- pregnancy or breastfeeding;
- abuse of medication and/or drugs;
- psychosis and/or dysmorphophobia;
- unachievable esthetic expectations;
- poor bone quantity and/or quality (i.e., D3 or D4) or unsatisfactory conditions of the implant site (as highlighted by clinical and X-ray examinations);
- bone volume in the implant site not sufficient to position a 3.3 mm × 10 mm narrow-diameter implant;
- mouth opening and/or space between the dental arches insufficient for implant components (>4 cm);
- incomplete facial growth and/or tooth eruption.

Subjects recruited for this study had to meet all of the inclusion criteria; the meeting of one or more exclusion criteria made the subject not suitable for this study population. The included subjects received exhaustive explanations about treatment risks, therapeutic alternatives, and study aims and design; they expressed their willingness to participate by signing a written informed consent form.

This study was performed according to the following timeline:

- 0–3 months: patient recruitment;
- 4–10 months: periodontal and orthodontic preparation (if necessary) and implant surgery;
- 11–12 months: prosthetic finalization and baseline control (T0);
- 24 months: 1-year follow-up;
- 36 months: 2-year follow-up.

Based on the inclusion and exclusion criteria, 22 patients were recruited for the present study (15 women and 7 men), aged between 18 and 37 years.

Of the subjects recruited for this study, 14 presented monolateral agenesis whereas 8 showed bilateral agenesis of the maxillary lateral incisors (Figure 1).

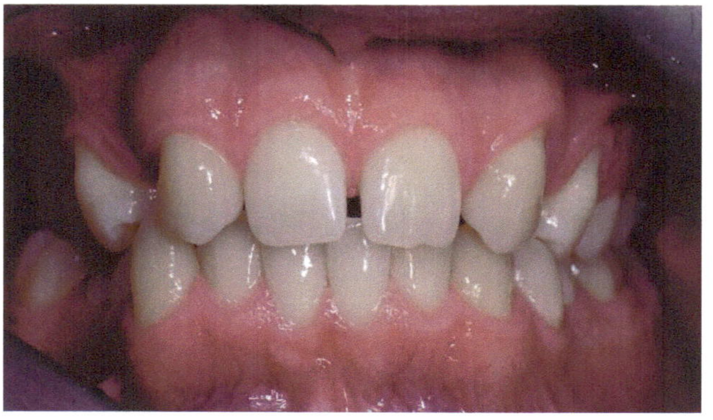

Figure 1. Pre-operative intraoral view.

Once recruited for this study, the patients underwent periodontal preparation through professional oral hygiene and motivation to maintain correct oral hygiene at home with the help of a dental hygienist. According to the interdisciplinary treatment plan, 9 patients underwent orthodontic therapy preparatory to implant surgery. At the end of the periodontal and/or orthodontic preparation, the local anatomical conditions were carefully re-evaluated before proceeding with the surgical placement of the implants.

The surgical and prosthetic planning of the cases were carried out after acquiring the volumes of the loco-regional anatomy by means of 3D CBCT radiographs and detecting the morphology of the dental and mucous tissues adjacent to the implant site by means of intraoral digital optical scans with an IOS system (Trios 4, 3Shape, Copenhagen, Denmark). The relative DICOM and STL files were imported into specific software that allowed us to superimpose the digital images, obtaining high fidelity 3D models. These models were used to create printed surgical templates for the guided surgical positioning of non-submerged implants.

Thirty non-submerged implants with a narrow diameter of 3.3 mm and a length of 10 mm (NC Bone Level, Straumann, Basel, Switzerland) were inserted by the same experienced oral surgeon. Peri-implant-tissue profile designers (Iphysio, LYRA ETK, Sal-lanches, France) were used as healing abutments to provide an initial peripheral conditioning of the transmucosal path (Figures 2 and 3).

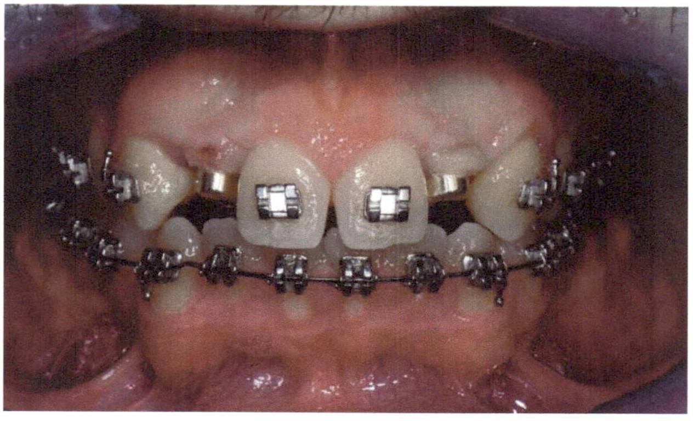

Figure 2. Buccal view of profile designers used as healing abutments and scan bodies for digital impression making with IOS to fabricate the temporary prostheses.

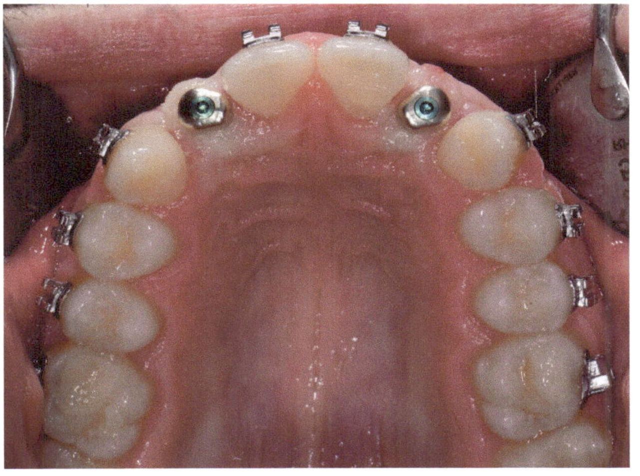

Figure 3. Occlusal view of profile designers used as healing abutments and scan bodies for digital impression making with IOS to fabricate the temporary prostheses.

Depending on the surgical procedures performed and the local conditions of each case, the healing and osseointegration period lasted from 3 to 6 months before proceeding to the prosthetic rehabilitation.

After the healing period, proper osseointegration was checked by means of either clinical and radiographic examinations; individual X-ray trays were made for each implant site to standardize radiographic examinations, and they were used in the same position at each follow-up appointment.

The same experienced prosthodontist performed all of the prosthetic procedures. The same profile designers that were employed as healing abutments were used as scan bodies to make digital impressions by means of an IOS system (Figure 4).

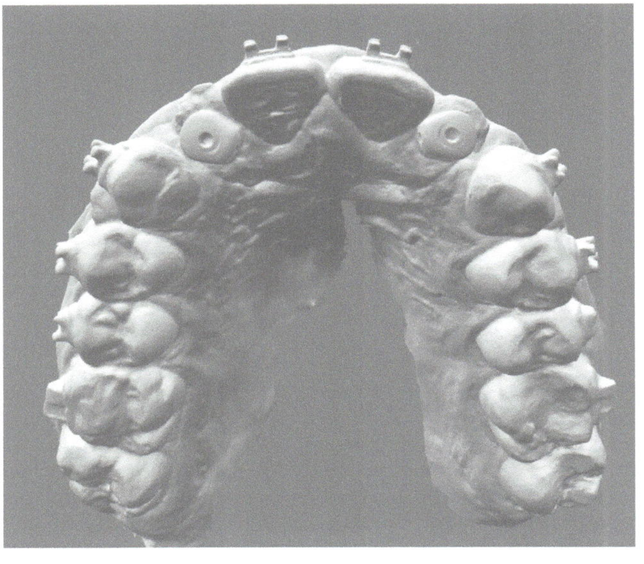

Figure 4. Digital scanning performed using profile designers.

By means of CAD-CAM manufacturing, temporary screw-retained single crowns in polymetylmetacrylate (PMMA) were fabricated to test occlusion, esthetics, and phonetics and to customize the peri-implant emergence profiles. When necessary, the temporary restorations were modified by relining with composite resin, in order to optimize the 3D morphology of the transmucosal path and obtain an optimal emergence profile of the restorations. After achieving proper shape and volume of the transmucosal paths and waiting for the maturation and stabilization of peri-implant soft tissues, the triple scan technique (i.e., temporary in situ, temporary extraoral, scanbody) was used for final digital impression making with an IOS system as previously described, in order to detect both the 3D position of the implant and the architecture of the soft tissues as conditioned by the morphology of the temporary prosthesis (Figures 5 and 6).

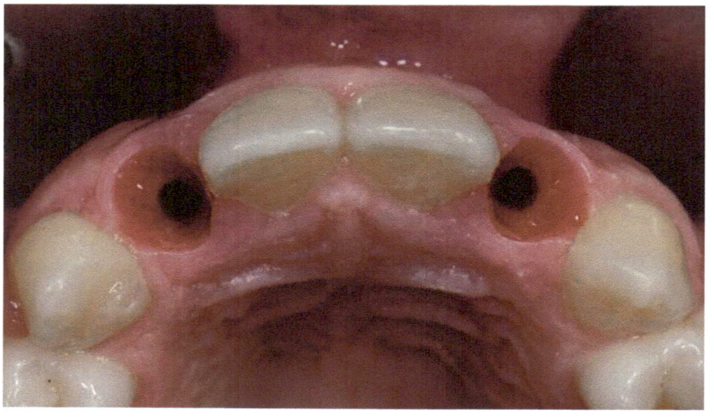

Figure 5. Occlusal view of peri-implant emergence profiles conditioned by means of screw-retained temporary single crowns.

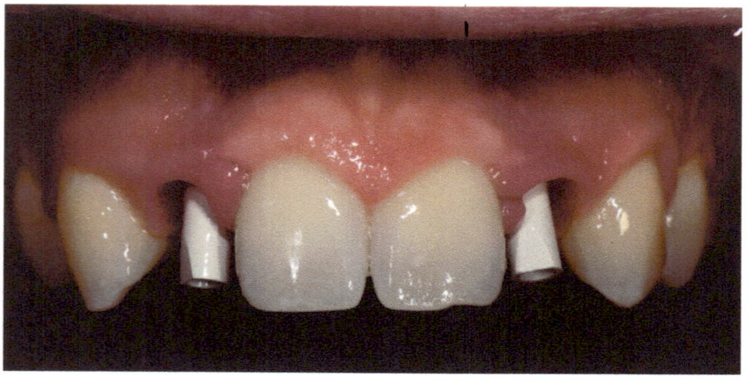

Figure 6. Buccal view of scan bodies for digital impression making with IOS to fabricate the final prostheses.

As for the temporary prostheses, CAD-CAM manufacturing was used to fabricate 30 monolithic cubic zirconia crowns (5Y-TZP; GC Initial Zirconia Disks, GC Co., Tokyo, Japan) that were cemented onto screw-retained implant Ti-bases with internal connection. Micro-layering with veneering ceramics was made only onto the buccal surfaces, leaving all of the functional areas (i.e., transmucosal, interproximal, palatal, and incisal) in polished zirconia, in order to promote epithelial attachment and avoid any possible chipping (Figure 7).

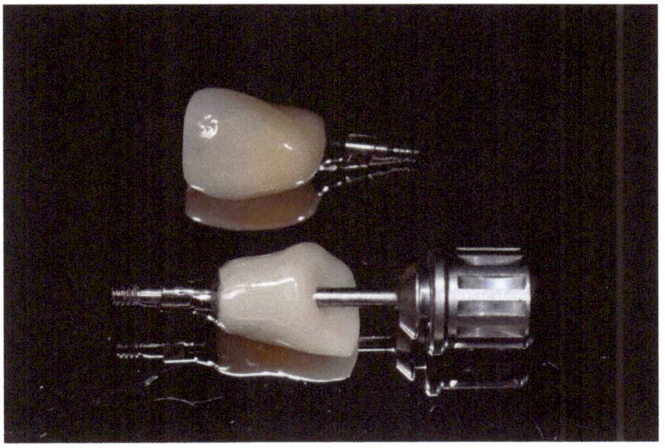

Figure 7. Final screw-retained zirconia single crowns.

The restorations were tried on intraorally, carefully verifying the occlusal and interproximal contacts as well as the coupling of the implant-prosthetic components using standardized intraoral radiographs. After possible occlusal adjustments, the final crowns were screwed onto the implants with a torque wrench at 25 Ncm and the screw channels were sealed with teflon tapes and resin composites (Figures 8 and 9).

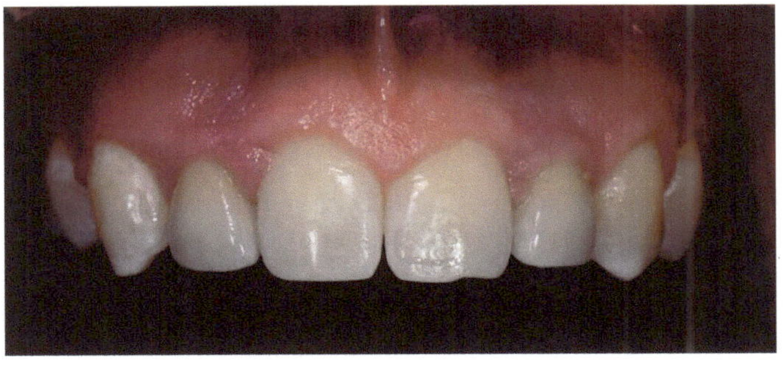

Figure 8. Post-operative intraoral view.

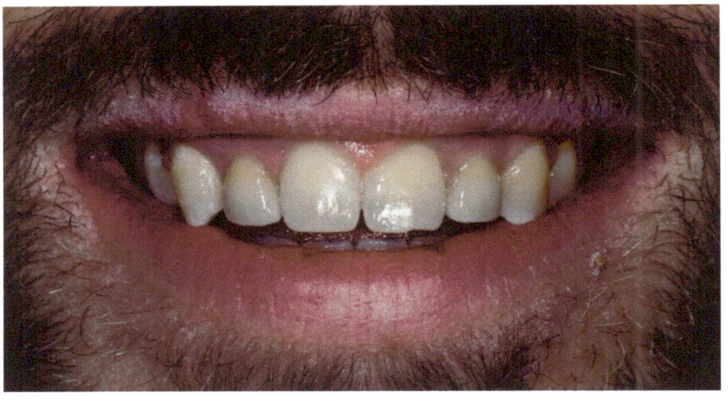

Figure 9. Post-operative extraoral view.

At the baseline and follow-up assessments, the levels of the marginal bone tissues were recorded clinically by means of peri-implant probing with plastic periodontal probes in order to damage neither the zirconia of the prosthetic crowns nor the titanium of the implant necks; moreover, standardized periapical radiographs were taken as previously described to record the marginal bone levels radiographically (Figure 10) and to use software allowing overlapping of the radiographic images and collection of the relative measurements over time.

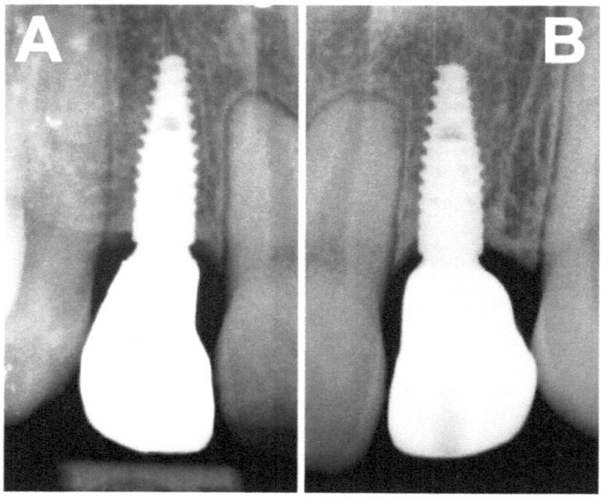

Figure 10. Standardized periapical radiographs in a case of bilateral agenesis at the baseline. (**A**) tooth 12 and (**B**) tooth 22.

The conditions of the peri-implant soft tissues were evaluated qualitatively and quantitatively from a clinical point of view by the same expert periodontist.

Any possible mechanical or biological complications affecting implants, crowns, or peri-implant tissues were recorded. The clinical variables affecting the outcome of restorations were subjected to an objective evaluation by means of the Functional Implant Prosthodontic Score (FIPS) [21]. This score can vary from 0 to 10, attributing a numerical value from 0 (worst) to 2 (best) to 5 clinical parameters as follows (Table 1):

- interproximal conditions (contact areas and papillae);
- occlusion (static and dynamic);
- appearance of the crown (margin and color);
- peri-implant soft tissues (quantity and quality);
- marginal bone tissue (radiographic evaluation).

Table 1. Variables for the objective evaluation of FIPS (Functional Implant Prosthodontic Score) [21].

Variable	**0**	**1**	**2**
Interproximal conditions (contact areas and papillae)	Major discrepancies (2x incomplete)	Minor discrepancies (1x incomplete)	No discrepancy (2x complete)
Occlusion (static and dynamic)	Major discrepancies (precontact)	Minor discrepancies (infraocclusion)	No discrepancy
Appearance of the crown (margin and color)	Major discrepancies (margin)	Minor discrepancies (color)	No discrepancy
Peri-implant soft tissues (quantity and quality)	Non-keratinized, non-adherent	Non-keratinized, adherent	Keratinized, adherent
Marginal bone tissue (RX)	Marginal resorption > 1.5 mm	Marginal resorption < 1.5 mm	No marginal resorption

Furthermore, the above-mentioned clinical variables were subjectively evaluated by patients using Visual Analog Scales (VASs) to rank the degree of patients' satisfaction [22]; such scales allowed patients to express an opinion on the clinical experience and satisfaction with the restorations received, expressing a vote from 0 (worst) to 10 (best) (Table 2).

Table 2. VAS scale for the subjective assessment of the degree of patients' satisfaction.

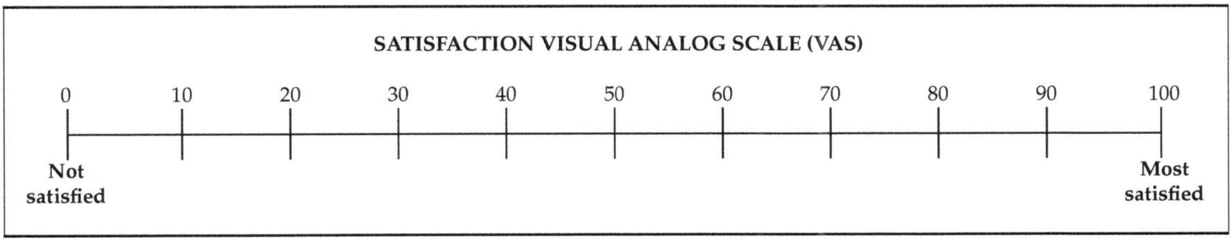

Data produced using the VAS and FIPS scales were collected by the same expert prosthodontist and periodontist who performed this study.

Patients were monitored for a minimum follow-up period of 24 months; controls were performed at T0, 7 days, 1 month, 3 months, 6 months, 1 year and 2 years.

The recordings of the study variables using FIPS and VASs were carried out at T0 and at the periodic follow-up controls at 1 and 2 years. The values obtained were statistically analyzed and cumulative 2-year survival and success rates were calculated according to the Kaplan–Meier analysis. Two independent curves were analyzed separately. Dedicated software (SPSS 17, SPSS Inc., Chicago, IL, USA) was used to perform statistical analyses.

3. Results

Twenty-two patients were recruited and a total of 30 narrow-diameter implants were inserted. Osseointegration was achieved for all of the implants. At 1- and 2-year follow-ups, non-significant values of marginal bone resorption were found for the implants and optimal qualitative and quantitative conditions of the peri-implant soft tissues were reported.

After 2 years of clinical function, the average recorded FIPS was 9.2 in bilateral agenesis (Table 3) and 9.3 in monolateral agenesis (Table 4), respectively. The overall average FIPS found in the present study was 9.2, showing an optimal functional and esthetic integration of the prosthetic restorations as well as a fully satisfactory short-term stability.

As regards the subjective evaluation of patients, both the function and esthetics of the restorations were considered fully satisfactory; in particular, the following scores were reported according to the VASs evaluation, with an average patient-satisfaction score of 8.7:

- score from 0 to 5: 0 restorations
- score 6: 1 restoration
- score 7: 3 restorations
- score 8: 8 restorations
- score 9: 9 restorations
- score 10: 9 restorations

As regards the survival (i.e., permanence in the oral cavity even in the presence of minor complications that do not compromise function) and success (i.e., permanence as delivered in T0) rates of both implants and prosthetic restorations, the Kaplan–Meier analysis at the 2-year follow-up reported a cumulative survival rate of 100% and a cumulative success rate of 93.3% (Figure 11).

In particular, after 1 year of clinical function, 1 event of mucositis, and 1 unscrewing of a crown were observed in 2 patients showing bilateral agenesis. No event affected clinical function; the mucositis was resolved, motivating the patient to increase oral hygiene at home, whereas the unscrewing was treated by tightening the restoration again at 25 Ncm.

Both of these drawbacks occurred in patients who did not undergo any orthodontic preparation. Consequently, cumulative survival and success rates of 100% were recorded in orthodontic patients, whereas cumulative survival and success rates of 100% and 93.3%, respectively were recorded in non-orthodontic patients.

Table 3. FIPS of restorations of patients affected by bilateral agenesis.

#	INTER PROXIMAL	OCCLUSION	DESIGN	MUCOSA	BONE	TOTAL
1a	2	2	2	2	2	10
1b	2	2	2	2	2	10
2a	1	1	2	2	2	8
2b	2	1	2	2	2	9
3a	2	2	2	2	2	10
3b	2	2	1	2	1	8
4a	2	2	2	1	2	9
4b	2	2	2	2	2	10
5a	2	1	2	2	2	9
5b	2	2	2	2	2	10
6a	2	2	2	1	1	8
6b	1	2	2	1	1	7
7a	2	2	2	2	2	10
7b	2	2	2	2	2	10
8a	2	2	1	2	2	9
8b	2	2	2	2	2	10

Table 4. FIPS of restorations of patients affected by monolateral agenesis.

#	INTER PROXIMAL	OCCLUSION	DESIGN	MUCOSA	BONE	TOTAL
9	2	2	2	2	2	10
10	1	2	2	2	2	9
11	2	1	2	2	2	9
12	2	2	1	2	2	9
13	2	2	2	2	2	10
14	2	2	2	0	1	7
15	2	2	2	2	2	10
16	2	1	2	1	2	8
17	2	2	2	2	2	10
18	2	2	2	2	2	10
19	2	2	2	2	2	10
20	2	2	1	2	2	9
21	2	2	2	2	2	10
22	2	2	2	2	1	9

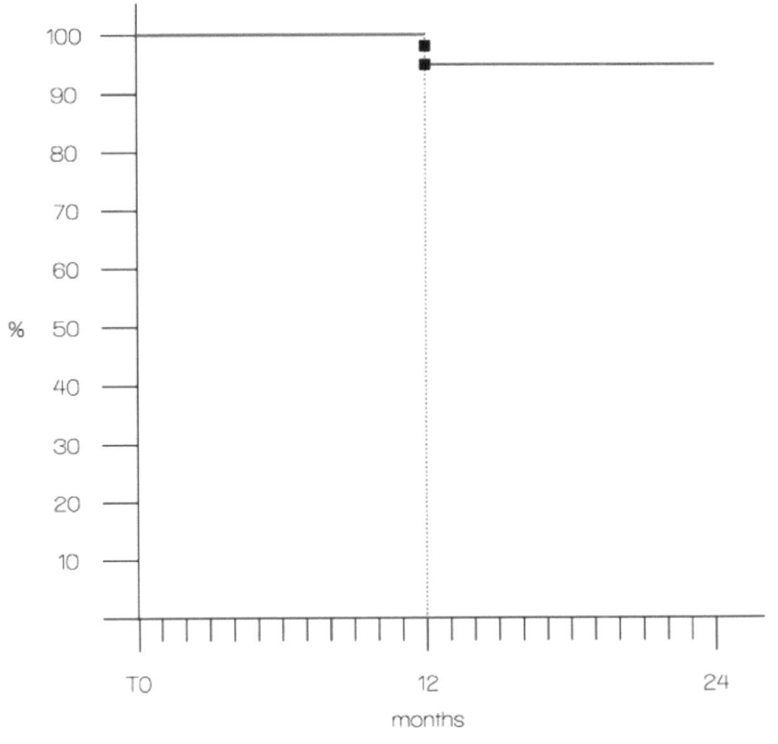

Figure 11. Kaplan–Meier graph showing the cumulative success rate in relation to time.

4. Discussion

The treatment of maxillary lateral incisor agenesis often requires an interdisciplinary approach. In particular, the implant-prosthetic rehabilitation of this condition is sometimes preceded by an orthodontic treatment that allows patients to achieve the proper surgical space to place an implant and obtain the best functional and esthetic outcomes. Furthermore, the implant-prosthetic approach offers a good cost/benefit ratio and is considered biologically conservative towards the adjacent teeth.

In the present prospective study, all of the implants achieved osseointegration, showing good marginal bone stability and peri-implant soft-tissue response in the short-term.

Previous studies demonstrated satisfactory results in the medium- and long-term for the implant-prosthetic treatment of maxillary lateral incisor agenesis [23–25]. In particular, high cumulative survival (95.7%) and success rates (87.1%) were reported after 16 years of clinical function for implant-supported, all-ceramic cemented crowns used to restore missing maxillary lateral incisors [24].

Although the implant-prosthetic approach is a well-known treatment option, evidence of screw-retained implant-prosthetic restorations used to rehabilitate maxillary lateral incisor agenesis with a full digital workflow are quite scant in the literature.

In the present prospective clinical study, implant-supported, screw-retained cubic zirconia crowns were used to restore maxillary lateral incisor agenesis, showing optimal function, esthetic integration and, at the same time, reducing the risk of periimplantitis related to cementation.

The data collected in the present short-term prospective study confirmed the feasibility of this treatment option as pointed out by previous investigations [23–25], reporting herein high cumulative survival (100%) and success rates (93.3%).

In addition, the high overall FIPS value of 9.2 supported the results of the descriptive statistics, showing the optimal functional and esthetic integration of the implant-supported zirconia crowns in the short-term. Patients enrolled in the present prospective study evaluated as highly satisfactory the clinical experience and the restorations received, reporting an average patient-satisfaction score of 8.7.

These data are consistent with the good objective evaluations obtained using the FIPS and with the findings from other investigations that demonstrated high levels of patient satisfaction when treating maxillary lateral incisor agenesis with implant-prosthetic restorations [26,27]. The authors expect that several factors, such as proper 3D implant positioning, a conservative design of the surgical flaps, and the correct application of temporary acrylic resin restorations may have played an important role in obtaining this result. Furthermore, the use of a digital workflow, in particular for the optical impression, contributed to the reduction of the discomfort of patients in terms of anxiety, nausea, and taste, which are usually related to conventional impression.

As reported by previous studies, a full digital workflow could reduce treatment time, improving each patient's clinical experience and perception of quality as well as their psychological comfort and compliance with the treatment. Nowadays, patients require convenience-oriented treatment timing with reduced chair time and a shortened number of appointments [28].

From a technical point of view, digital workflows help in simplifying the production process, reducing human intervention and overcoming different manual fabrication steps; furthermore the standardization offered by CAD-CAM technologies could contribute to producing high quality and precise prosthetic restorations [29].

Nevertheless, the present prospective clinical study presented some limitations that have to be considered in the interpretation of the obtained clinical data; in particular, the observational period was limited to the short-term (i.e., 2 years), the study population had a limited number of implant-prosthetic study units (i.e., 30), the study lacked a control group, and there were implicit limitations in the scales used (VAS and FIPS). Furthermore, the literature on this topic is quite scant: the present study aimed to provide preliminary short-term data, in the hope of increasing the number of enrolled patients in the future and providing long-term data. Further long-term clinical studies and a larger sample size would be advisable to corroborate the findings of the present clinical investigation and establish validated protocols.

5. Conclusions

Given the limitations of the present prospective clinical study, in accordance with the obtained results, the following conclusions can be drawn:

- in the 2-year short-term, both implants and zirconia crowns did not show significant technical or biological complications, achieving high survival and success rates;
- in standard clinical conditions, the implant-prosthetic rehabilitation can be considered a viable choice option for the treatment of the agenesis of maxillary lateral incisors.
- as regards the subjective evaluation of patients, both the function and esthetics of the restorations were considered fully satisfactory.
- an interdisciplinary approach to the treatment plan represents an essential pre-requisite for achieving functional and esthetic success.

Author Contributions: Conceptualization, R.S., M.F. and F.Z.; methodology M.I.D.M.; software M.I.D.M. and M.A.; validation, R.L., G.R. and M.A.; formal analysis, F.Z.; investigation, R.S. and M.I.D.M.; data curation, E.F.C., M.A. and M.F.; writing—original draft preparation, M.I.D.M. and G.R.; writing—review and editing, R.S., E.F.C. and M.A.; supervision, M.F. and F.Z. All authors have read and agreed to the published version of the manuscript.

Funding: This research received no external funding.

Institutional Review Board Statement: As structured clinical-assistance personnel at the AOU Federico II of Naples, clinical studies based on validated and non-experimental procedures (in terms of testing materials and/or techniques that are already validated and accredited) do not require any specific authorization from our university hospital, apart from the informed consent signed by the patients. The processing of sensitive data is ensured by the recruitment procedures carried out at the aforementioned institution.

Informed Consent Statement: Informed consent was obtained from all subjects involved in this study.

Data Availability Statement: Data available on request due to privacy and ethical restrictions.

Acknowledgments: The authors would like to thank the following parties for their clinical, technical and manufacturing support: Anna Mariniello (orthodontics), Fabio Cozzolino (implant surgery), Iorioinlab (dental laboratory), and LYRA ETK (Iphysio profile designers).

Conflicts of Interest: The authors declare no conflicts of interest.

References

1. The Glossary of Prosthodontic Terms: Ninth Edition. *J. Prosthet. Dent.* **2017**, *117*, e1–e105. [CrossRef] [PubMed]
2. Stamatiou, J.; Symons, A.L. Agenesis of the permanent lateral incisor: Distribution, number and sites. *J. Clin. Pediatr. Dent.* **1991**, *15*, 244–246. [PubMed]
3. Pandey, P.; Ansari, A.A.; Choudhary, K.; Saxena, A. Familial aggregation of maxillary lateral incisor agenesis (MLIA). *BMJ Case Rep.* **2013**, *2013*, bcr2012007846. [CrossRef] [PubMed]
4. Hua, F.; He, H.; Ngan, P.; Bouzid, W. Prevalence of peg-shaped maxillary permanent lateral incisors: A meta-analysis. *Am. J. Orthod. Dentofac. Orthop.* **2013**, *144*, 97–109. [CrossRef]
5. Rosa, M.; Olimpo, A.; Fastuca, R.; Caprioglio, A. Perceptions of dental professionals and laypeople to altered dental esthetics in cases with congenitally missing maxillary lateral incisors. *Prog. Orthod.* **2013**, *14*, 34. [CrossRef]
6. Laverty, D.P.; Thomas, M.B. The restorative management of microdontia. *Br. Dent. J.* **2016**, *221*, 160–166. [CrossRef]
7. Kinzer, G.A.; Kokich, V.O., Jr. Managing congenitally missing lateral incisors. Part II: Tooth-supported restorations. *J. Esthet. Restor. Dent.* **2005**, *17*, 76–84. [CrossRef] [PubMed]
8. Pithon, M.M.; Vargas, E.O.A.; da Silva Coqueiro, R.; Lacerda-Santos, R.; Tanaka, O.M.; Maia, L.C. Impact of oral-health-related quality of life and self-esteem on patients with missing maxillary lateral incisor after orthodontic space closure: A single-blinded, randomized, controlled trial. *Eur. J. Orthod.* **2021**, *43*, 208–214. [CrossRef]
9. Priest, G. The treatment dilemma of missing maxillary lateral incisors-Part II: Implant restoration. *J. Esthet. Restor. Dent.* **2019**, *31*, 319–326. [CrossRef]
10. Šikšnelytė, J.; Guntulytė, R.; Lopatienė, K. Orthodontic canine substitution vs. implant-supported prosthetic replacement for maxillary permanent lateral incisor agenesis: A systematic review. *Stomatologija* **2021**, *23*, 106–113.
11. Momberger, N.; Mukaddam, K.; Zitzmann, N.U.; Bornstein, M.A.; Filippi, A.; Kühl, S. Esthetic and functional outcomes of narrowdiameter implants compared in a cohort study to standard diameter implants in the anterior zone of the maxilla. *Quintessence Int.* **2022**, *53*, 502–509. [PubMed]
12. Parize, H.N.; Bohner, L.O.L.; Gama, L.T.; Porporatti, A.L.; Mezzomo, L.A.M.; Martin, W.C.; Gonçalves, T.M.S.V. Narrow-diameter implants in the anterior region: A meta-analysis. *Int. J. Oral Maxillofac. Implant.* **2019**, *34*, 1347–1358. [CrossRef] [PubMed]
13. Telles, L.H.; Portella, F.F.; Rivaldo, E.G. Longevity and marginal bone loss of narrow-diameter implants supporting single crowns: A systematic review. *PLoS ONE* **2019**, *14*, e0225046. [CrossRef] [PubMed]
14. Siqueira, R.; Galli, M.; Chen, Z.; Mendonça, G.; Meirelles, L.; Wang, H.L.; Chan, H.L. Intraoral scanning reduces procedure time and improves patient comfort in fixed prosthodontics and implant dentistry: A systematic review. *Clin. Oral Investig.* **2021**, *25*, 6517–6531. [CrossRef] [PubMed]
15. Haddadi, Y.; Bahrami, G.; Isidor, F. Evaluation of Operating Time and Patient Perception Using Conventional Impression Taking and Intraoral Scanning for Crown Manufacture: A Split-mouth, Randomized Clinical Study. *Int. J. Prosthodont.* **2018**, *31*, 55–59. [CrossRef]
16. Resende, C.C.D.; Barbosa, T.A.Q.; Moura, G.F.; Tavares, L.D.N.; Rizzante, F.A.P.; George, F.M.; Neves, F.D.D.; Mendonça, G. Influence of operator experience, scanner type, and scan size on 3D scans. *J. Prosthet. Dent.* **2021**, *125*, 294–299. [CrossRef] [PubMed]
17. Bishti, S.; Tuna, T.; Rittich, A.; Wolfart, S. Patient-reported outcome measures (PROMs) of implant-supported reconstructions using digital workflows: A systematic review and meta-analysis. *Clin. Oral Implant. Res.* **2021**, *32* (Suppl. S21), 318–335. [CrossRef]
18. Barros Vde, M.; Costa, N.R.; Martins, P.H.; Vasconcellos, W.A.; Discacciati, J.A.; Moreira, A.N. Definitive Presurgical CAD/CAM-Guided Implant-Supported Crown in an Esthetic Area. *Braz. Dent. J.* **2015**, *26*, 695–700. [CrossRef]
19. da Silva Salomão, G.V.; Chun, E.P.; Panegaci, R.D.S.; Santos, F.T. Analysis of Digital Workflow in Implantology. *Case Rep. Dent.* **2021**, *2021*, 6655908. [CrossRef]

20. Gianfreda, F.; Pesce, P.; Marcano, E.; Pistilli, V.; Bollero, P.; Canullo, L. Clinical Outcome of Fully Digital Workflow for Single-Implant-Supported Crowns: A Retrospective Clinical Study. *Dent. J.* **2022**, *10*, 139. [CrossRef]
21. Joda, T.; Zarone, F.; Zitzmann, N.U.; Ferrari, M. The Functional Implant Prosthodontic Score (FIPS): Assessment of reproducibility and observer variability. *Clin. Oral Investig.* **2018**, *22*, 2319–2324. [CrossRef] [PubMed]
22. Wang, Y.; Bäumer, D.; Ozga, A.K.; Körner, G.; Bäumer, A. Patient satisfaction and oral health-related quality of life 10 years after implant placement. *BMC Oral Health* **2021**, *21*, 30. [CrossRef] [PubMed]
23. King, P.; Maiorana, C.; Luthardt, R.G.; Sondell, K.; Øland, J.; Galindo-Moreno, P.; Nilsson, P. Clinical and radiographic evaluation of a small-diameter dental implant used for the restoration of patients with permanent tooth agenesis (hypodontia) in the maxillary lateral incisor and mandibular incisor regions: A 36-month follow-up. *Int. J. Prosthodont.* **2016**, *29*, 147–153. [CrossRef] [PubMed]
24. Sorrentino, R.; Di Mauro, M.I.; Leone, R.; Ruggiero, G.; Annunziata, M.; Zarone, F. Implant–Prosthetic Rehabilitation of Maxillary Lateral Incisor Agenesis with Narrow Diameter Implants and Metal–Ceramic vs. All-Ceramic Single Crowns: A 16-Year Prospective Clinical Study. *Appl. Sci.* **2023**, *13*, 964. [CrossRef]
25. Branzén, M.; Eliasson, A.; Arnrup, K.; Bazargani, F. Implant-Supported Single Crowns Replacing Congenitally Missing Maxillary Lateral Incisors: A 5-Year Follow-Up. *Clin. Implant Dent. Relat. Res.* **2015**, *17*, 1134–1140. [CrossRef]
26. De-Marchi, L.M.; Pini, N.I.; Ramos, A.L.; Pascotto, R.C. Smile attractiveness of patients treated for congenitally missing maxillary lateral incisors as rated by dentists, laypersons, and the patients themselves. *J. Prosthet. Dent.* **2014**, *112*, 540–546. [CrossRef] [PubMed]
27. Jamilian, A.; Perillo, L.; Rosa, M. Missing upper incisors: A retrospective study of orthodontic space closure versus implant. *Prog. Orthod.* **2015**, *16*, 2. [CrossRef] [PubMed]
28. Joda, T.; Bragger, U. Digital vs. conventional implant prosthetic workflows: A cost/time analysis. *Clin. Oral Implant. Res.* **2015**, *26*, 1430–1435. [CrossRef]
29. Joda, T.; Gintaute, A.; Brägger, U.; Ferrari, M.; Weber, K.; Zitzmann, N.U. Time-efficiency and cost-analysis comparing three digital workflows for treatment with monolithic zirconia implant fixed dental prostheses: A double-blinded RCT. *J. Dent.* **2021**, *113*, 103779. [CrossRef]

prosthesis

MDPI

Article

Evaluation of Bone Turnover around Short Finned Implants in Atrophic Posterior Maxilla: A Finite Element Study

Andrii Kondratiev [1,2,*], Vladislav Demenko [3], Igor Linetskiy [4], Hans-Werner Weisskircher [5] and Larysa Linetska [6]

1 Department of Materials Science and Engineering of Composite Structures, O.M. Beketov National University of Urban Economy in Kharkiv, Marshal Bazhanov Str. 17, 61002 Kharkiv, Ukraine
2 Department of Engineering, University of Cambridge, Trumpington Str., Cambridge CB2 1PZ, UK
3 Department of Aircraft Strength, National Aerospace University "Kharkiv Aviation Institute", Chkalova Str. 17, 61070 Kharkiv, Ukraine; v.demenko@khai.edu
4 Department of Oral and Maxillofacial Surgery, General University Hospital in Prague, U Nemocnice 499/2, 128 08 Prague, Czech Republic; igor.linetskiy@gmail.com
5 Private Practice, Bahnhofstrasse 7, 54298 Igel, Germany; weisskircher@t-online.de
6 Department of Rehabilitation Medicine, National Technical University "Kharkiv Polytechnic Institute", Kyrpychova Str. 2, 61002 Kharkiv, Ukraine; larisa.linetska@gmail.com
* Correspondence: andrii.kondratiev@kname.edu.ua

Abstract: Background/Objectives: Dental implants have emerged as a modern solution for edentulous jaws, showing high success rates. However, the implant's success often hinges on the patient's bone quality and quantity, leading to higher failure rates in poor bone sites. To address this issue, short implants have become a viable alternative to traditional approaches like bone sinus lifting. Among these, Bicon® short implants with a plateau design are popular for their increased surface area, offering potential advantages over threaded implants. Despite their promise, the variability in patient-specific bone quality remains a critical factor influencing implant success and bone turnover regulated by bone strains. Excessive strains can lead to bone loss and implant failure according to Frost's "Mechanostat" theory. To better understand the implant biomechanical environment, numerical simulation (FEA) is invaluable for correlating implant and bone parameters with strain fields in adjacent bone. The goal was to establish key relationships between short implant geometry, bone quality and quantity, and strain levels in the adjacent bone of patient-dependent elasticity to mitigate the risk of implant failure by avoiding pathological strains. **Methods**: Nine Bicon Integra-CP™ implants were chosen. Using CT scans, three-dimensional models of the posterior maxilla were created in Solidworks 2022 software to represent the most challenging scenario with minimal available bone, and the implant models were positioned in the jaw with the implant apex supported by the sinus cortical bone. Outer dimensions of the maxilla segment models were determined based on a prior convergence test. Implants and abutments were considered as a single unit made of titanium alloy. The bone segments simulated types III/IV bone by different cancellous bone elasticities and by variable cortical bone elasticity moduli selected based on an experimental data range. Both implants and bone were treated as linearly elastic and isotropic materials. Boundary conditions were restraining the disto-mesial and cranial surfaces of the bone segments. The bone–implant assemblies were subjected to oblique loads, and the bone's first principal strain fields were analyzed. Maximum strain values were compared with the "minimum effective strain pathological" threshold of 3000 microstrain to assess the implant prognosis. **Results**: Physiological strains ranging from 490 to 3000 microstrain were observed in the crestal cortical bone, with no excessive strains detected at the implant neck area across different implant dimensions and cortical bone elasticity. In cancellous bone, maximum strains were observed at the first fin tip and were influenced by the implant diameter and length, as well as bone quality and cortical bone elasticity. In the spectrum of modeled bone elasticity and implant dimensions, increasing implant diameter from 4.5 to 6.0 mm resulted in a reduction in maximum strains by 34% to 52%, depending on bone type and cortical bone elasticity. Similarly, increasing implant length from 5.0 to 8.0 mm led to a reduction in maximum strains by 15% to 37%. Additionally, a two-fold reduction in cancellous bone elasticity modulus (type IV vs. III) corresponded to an increase in maximum strains by 16% to 59%. Also, maximum strains increased by 86% to 129% due to a decrease in patient-dependent cortical bone elasticity from the softest to the most rigid bone. **Conclusions**: The findings have practical implications for dental practitioners planning short finned implants in the posterior maxilla. In cases where the quality of cortical bone is uncertain and bone height is insufficient, wider 6.0 mm diameter implants should be preferred to mitigate the risk of pathological strains. Further investigations of cortical bone architecture and elasticity in the posterior maxilla are recommended to develop comprehensive clinical recommendations considering bone volume and quality limitations. Such research can potentially enable the placement of narrower implants in cases of insufficient bone.

Keywords: Frost's "mechanostat" theory; plateau implant; bone quality; FEA

1. Introduction

The success of dental implants relies on maintaining a stable attachment to the bone tissue, which is influenced by various factors including bone availability and quality, implant design, and dimensions. Effective remodeling of the bone is essential for supporting secure anchoring, with bone restructuring being particularly important [1,2].

Although dental implant treatments often have high success rates, long-term success can be challenged by different biomechanical factors [3]. The posterior maxillary region, in particular, is at greater risk of failure due to insufficient bone density, reduced bone volume, and increased masticatory forces, indicating poor bone quality [4,5].

Short implants have emerged as a practical solution in compromised conditions, especially in the maxillary molar region, eliminating the need for bone grafting and traditional implant placement [6–8]. Studies suggest that short implants can achieve comparable success rates to longer ones [9,10]. However, their smaller surface area can lead to increased stress and strain concentrations in the crestal bone compared to conventional implants [11,12]. To mitigate this issue, the use of wide short implants has been proposed in cases of insufficient bone height and significantly higher occlusal loads in molar sites. This approach aims to increase surface area, thereby improving stress and strain distribution in the adjacent bone, particularly at the critical area of the bone–implant interface.

Plateau implants, introduced in 1985, stand out as a unique type of dental implant characterized by multiple parallel circular threads known as plateau or fins. Among these, the Bicon® screwless design, featuring a plateau root-formed body, is widely utilized [13,14]. Notably, it provides 30% more surface area compared to threaded implants of similar size. These implants are especially recommended for patients with inadequate bone height, and their short lengths (<8 mm) help eliminate the need for preoperative procedures like grafting or sinus lifting. The increased surface area of bone–implant contact in plateau implants reduces stresses and strains by enhancing load transfer along the bone–implant interface, which is particularly advantageous when bone height is limited.

Plateau implants demonstrate significant efficacy in preventing bone loss, facilitated by the creation of 'healing chambers'—hollow spaces between the implant and bone. These chambers, formed due to the interaction between implant design and drilling dimensions, promote the development of intramembranous-like woven bone formation [14–16]. These spaces are initially filled by the blood clot immediately after implantation and gradually filled by new bone apposition over time.

Numerous studies have investigated the behavior of short implants, examining factors such as diameter, length, and macrostructure, as well as the bone healing response to different implant root shapes and the cumulative survival rates of short implants. The refinement of plateau root form designs has significantly increased the cumulative survival rate to over 90%, optimizing biological responses during early endosseous peri-implant healing.

Successful osseointegration of the implant with marginal bone significantly increases the implant's load-bearing capacity and bone turnover regulation. However, predicting the mechanical behavior of the bone–implant interface remains challenging due to variations in patient-specific cortical bone elasticity and strength. Bone strains are recognized as mechanical stimuli that influence bone turnover and maintain mechanical strength through primary cilia in bone-forming cells [16,17]. This process ensures the adaptation of bone morphology to functional loads throughout an individual's life. Biomechanical feedback, in line with Frost's "bone mechanostat" hypothesis, regulates the relationship between bone density and load magnitude, optimizing bone structure through modeling and remodeling. According to Frost, when bone strains surpass the "minimum effective strain pathological" threshold (MESp = 3000 microstrain), microdamages accumulate, leading to bone failure [18]. To promote peri-implant bone mass, strains should be maintained above the minimum effective strain modeling threshold (MESm) of 1000–1500 microstrain [19,20].

Predicting treatment success and the longevity of implants hinges on selecting dimensions that align with a safe strain spectrum and the available bone quantity. This process typically involves an initial evaluation of the patient's jaw bone properties, taking into account factors like placement site, bone shape, and dimensions. However, it often overlooks the physical and mechanical properties of bone tissues, which are crucial for preventing bone resorption. In cases of thin, atrophic bone where the cancellous bone core lacks reliability as a load-bearing element, understanding the impact of adjacent bone elasticity on neck area strains becomes paramount. Selecting suitable implants becomes particularly challenging when dealing with poor bone quality, insufficient volume, and variations in cortical bone mechanical properties [21].

To address the complex issue of correlating implant geometry and dimensions with bone properties and strains, computer modeling, specifically, finite element analysis (FEA), proves invaluable [4,22]. FEA allows for the assessment of strain concentrations in peri-implant bone, considering factors such as implant shape, dimensions, bone quality, and quantity [23]. This approach provides a comprehensive understanding of implant biomechanics under functional loads, aiding in the selection of appropriate implants for each unique patient scenario [24].

Several studies have explored finite element (FE) strain analysis in adjacent bone, examining the biomechanical effects of implant dimensions and bone quality in osseointegrated and immediate implants [22,23,25]. However, these studies present a wide range of strain magnitudes that do not directly correlate with the pathological strain threshold at the critical area of the bone–implant interface, where maximal strains may trigger bone loss. As a result, these findings cannot reliably predict implant success or failure or provide recommendations for implant sizing selection, especially in compromised maxillary cases.

The objective of this study was to use FEA simulation to evaluate the impact of short finned implants and crestal bone quality/quantity on strain magnitudes in adjacent cortical bone with patient-variable elasticity. The aim was to characterize bone turnover in the posterior maxilla and recommend the placement of Bicon Integra-CP™ short plateau implants based on their post-osseointegration perspective. This assessment sought to ensure their ability to withstand functional loading and prevent cortical bone loss.

2. Materials and Methods

Nine different geometric designs of Bicon Integra-CP™ implants (Bicon, Boston, MA, USA) were investigated, varying in length (5.0 mm for S, 6.0 mm for I, and 8.0 mm for L) and diameter (4.5 mm for N, 5.0 mm for M, and 6.0 mm for W). To model these implants and abutments, their dimensions and designs were obtained using digital calipers, photographs, and images captured with a light optic microscope.

To create solid models of posterior maxilla alveolar bone segments, computed tomographic (CT) images from unidentified patients were selected from the authors' CT database to delineate cortical and cancellous bone contours. Simplified 3D geometry segments with a length of 40 mm were reconstructed using these CT images in DICOM format with Solidworks 2022 software (Dassault Systemes SolidWorks Corporation, Waltham, MA, USA).

The 3D models of the implants were crestally placed into nine posterior maxilla segment models including 0.5 mm of crestal cortical bone. They represented type III/IV bone (classified according to Lekholm and Zarb) simulated by varying the cancellous bone modulus of elasticity (1.37/0.69 GPa) (see Figure 1).

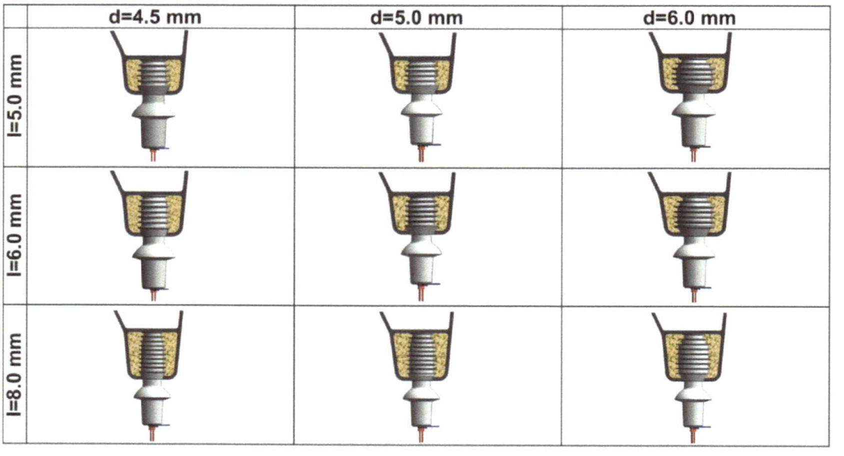

Figure 1. Maxillary bone segments with nine inserted implants. Oblique loading is applied to the center of abutment upper surface at 7.0 mm distance from the upper bone margin.

The size of the maxilla segment model was 40 × 13 × 9 mm (length × height × width), determined based on a previous convergence test. The dimensions were chosen to replicate the most critical scenario of minimal available bone, necessitating crestal placement as a necessary compromise.

Implant and bone tissues were considered to be linearly elastic and isotropic, with homogeneous material volumes. Implants and abutments were treated as a continuous unit and were assumed to be made of titanium alloy, with a modulus of elasticity of 114 GPa and a Poisson's ratio of 0.34 [26]. The Poisson's ratio for both cortical and cancellous bone tissues was assumed to be 0.3 [27].

For computer simulation, a spectrum of elasticity moduli representing different levels of human cortical bone elasticity was selected based on experimental investigations, ranging from 4.0 to 13.7 GPa [21]. The elasticity moduli were assigned as follows: E_1 = 13.7 GPa, E_2 = 12.0 GPa, E_3 = 10.0 GPa, E_4 = 8.0 GPa, E_5 = 6.0 GPa, and E_6 = 4.0 GPa.

The boundary conditions involved restraining the disto-mesial surfaces of the bone segment as well as the cranial surfaces in all models (see Figure 2). Functional loading of the implant was simulated at the center of a 7 Series Low 0° abutment in three dimensions. A mean maximal functional load of 120.9 N was applied obliquely at an angle of approximately 75° to the abutment top surface [28]. The loading components were determined as 116.3 N in the axial direction, 17.4 N lingually, and 23.8 N disto-mesially. The last two components represented the resultant vector of a 29.5 N horizontal functional load acting in the plane of the critical bone–implant interface [24,28,29]. The implants were assumed to be fully osseointegrated [30].

A numerical analysis of bone–implant models was conducted using FE software Solidworks Simulation (Dassault Systemes SolidWorks Corporation, Waltham, MA, USA). A mesh convergence analysis was performed to determine the optimal element size.

The mesh refinement process involved gradually reducing the element sizing from 2.0 mm to 0.010 mm, and the change in the maximum first principal strain in the bone–implant interface was studied. The maximum strain converged toward a finite value as the mesh density increased, with a convergence criterion set to less than 2% for the changes in the maximal strain of all the elements [31,32]. The analysis was stopped at the FEA size of 0.020 mm. The total number of 3D finite elements ranged from 1,659,134 to 2,000,652, and the nodes ranged from 2,238,052 to 2,695,410. An example of FE meshing for a 5.0 × 5.0 mm implant is provided in Figure 3.

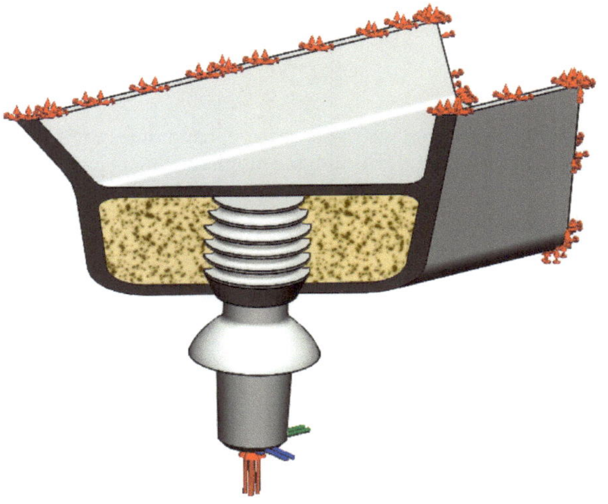

Figure 2. Illustration of 3D view of 5.0 × 5.0 mm implant placed in maxillary bone segment with 0.5 mm crestal and sinus cortical bone thickness. Oblique loading is applied at the center of 7 Series Low 0° abutment upper surface at 7.0 mm distance from the upper bone margin. Disto-mesial and cranial surfaces of the bone segment are restrained.

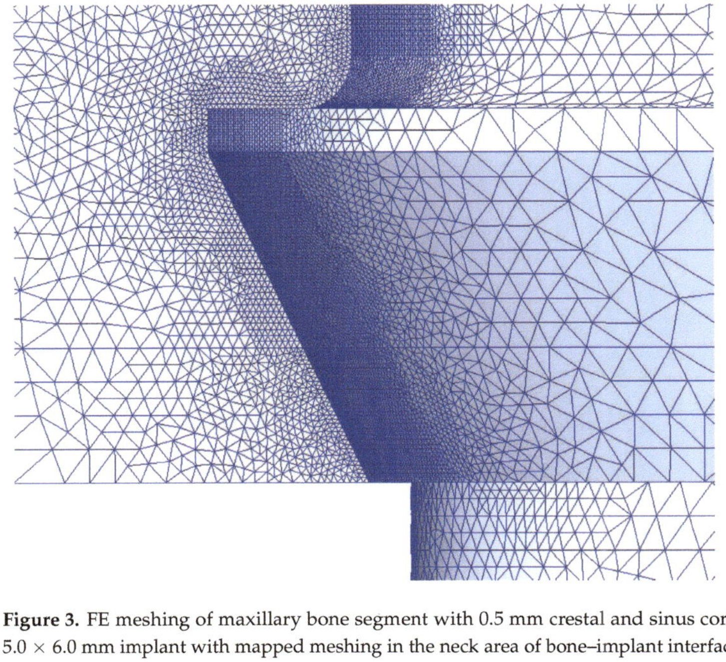

Figure 3. FE meshing of maxillary bone segment with 0.5 mm crestal and sinus cortical bone and 5.0 × 6.0 mm implant with mapped meshing in the neck area of bone–implant interface. Minimal FE size is 0.020 mm.

In the bone–implant assembly strain analysis, first principal strains (FPSs) were chosen as the measure of bone turnover. FPS localizations in the peri-implant area of the critical bone–implant interface were studied for 108 bone–implant combinations (9 bone–implant models × 2 bone types × 6 cortical bone elasticity moduli) to determine maximal FPS (MFPS). These values were then correlated with 3000 microstrain threshold (MFPSp) to evaluate implant lifetime prognosis in terms of physiological/pathological bone strains in the anchorage area [18–20].

3. Results

The distributions of FPS at the bone–implant interface is depicted in Figure 4, while Figure 5 displays the variation of FPS along the critical bone–implant contact length.

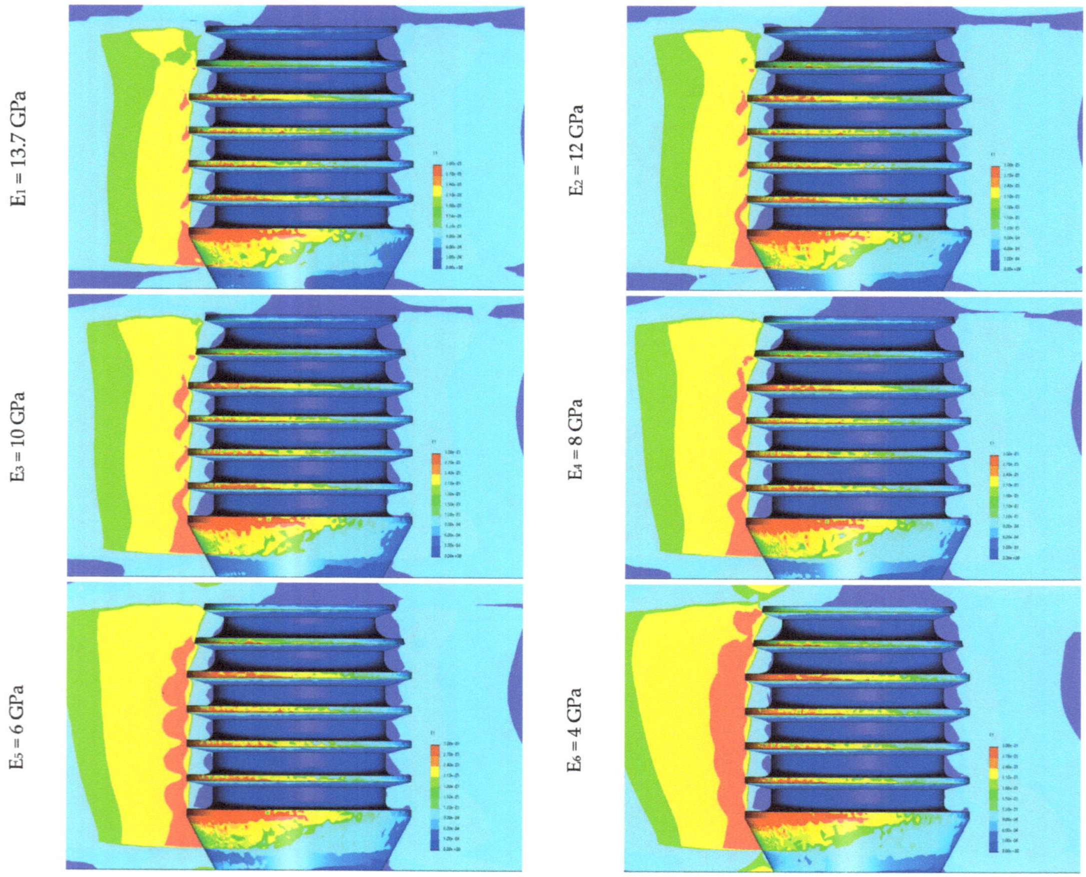

Figure 4. First principal strain localization in the plane of the critical bone–implant interface for the studied 5.0 × 6.0 mm Bicon SHORT® implant, type IV bone, and six degrees of cortical bone elasticity corresponding to E_1–E_6 moduli of elasticity.

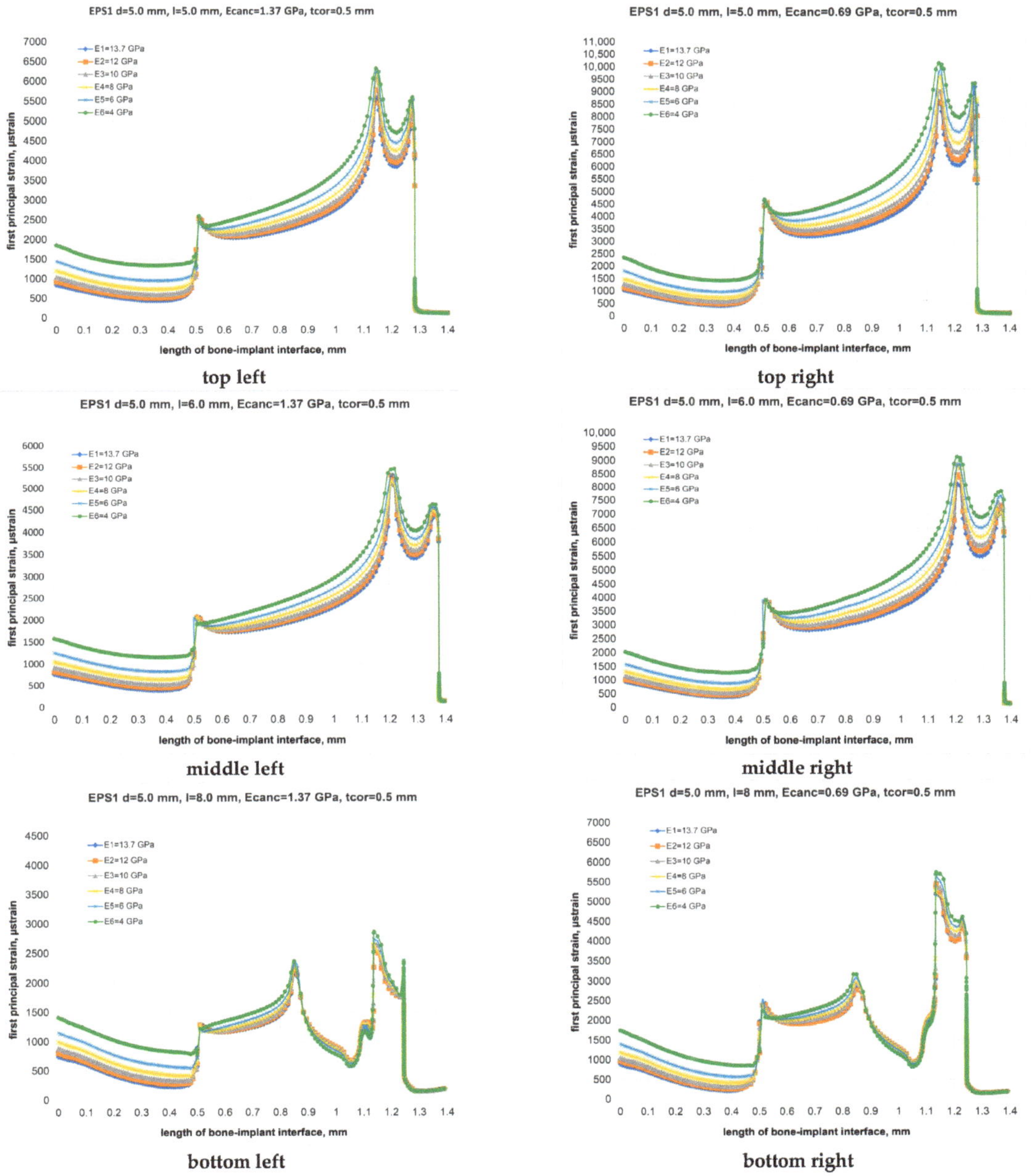

Figure 5. Illustration of first principal strain distribution along the critical line of bone–implant interface for 5.0 × 5.0 mm (**top**), 5.0 × 6.0 mm (**middle**), 5.0 × 8.0 mm (**bottom**) implants placed into bone segments of types III (**left**) and IV (**right**) bone (E_{III} = 1.37 GPa and E_{IV} = 0.69 GPa) at E_1–E_6 degrees of cortical bone elasticity.

Maximal magnitudes of FPS (MFPSs) were sought, particularly on the surface of cortical bone, influenced by factors such as implant dimensions, cortical bone elasticity, and bone quality (Figures 4–7).

The analysis revealed that there were no instances of overstrains (MFPS > 3000 µstrain) at the implant neck area. Instead, a spectrum of safe maximal MFPS (490–3000 µstrain) was observed in the crestal cortical bone, with W implants causing strains ranging from 490 to 1860 µstrain and N implants inducing strains from 860 to 3000 µstrain. In cancellous bone, MFPSs were located at the first fin position, similarly influenced by implant dimensions, bone quality, and cortical bone elasticity.

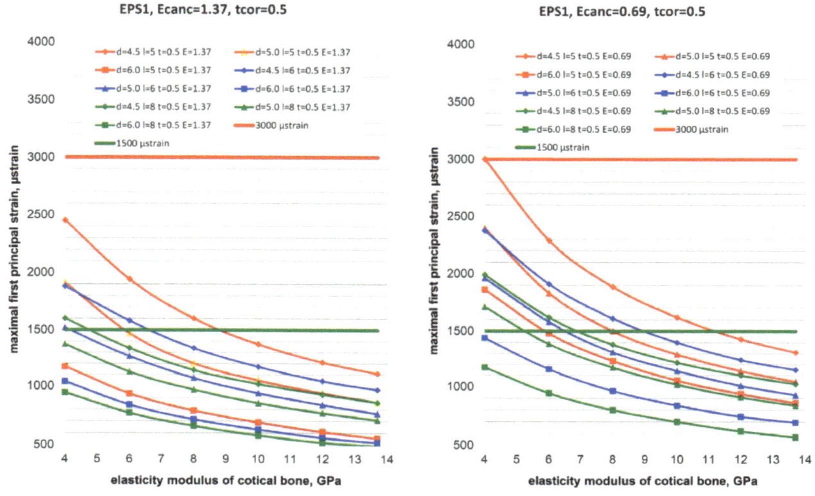

Figure 6. Dependence of maximal first principal strains (MFPSs) in cortical bone on its modulus of elasticity for the spectrum of implants placed into bone segments with 0.5 mm cortical bone thickness for type III (**left**) and IV (**right**) bone and the studied degrees of patient-specific cortical bone elasticity E_1–E_6. Red line corresponds to 3000 microstrain of Frost "minimum effective strain pathological" (MESp).

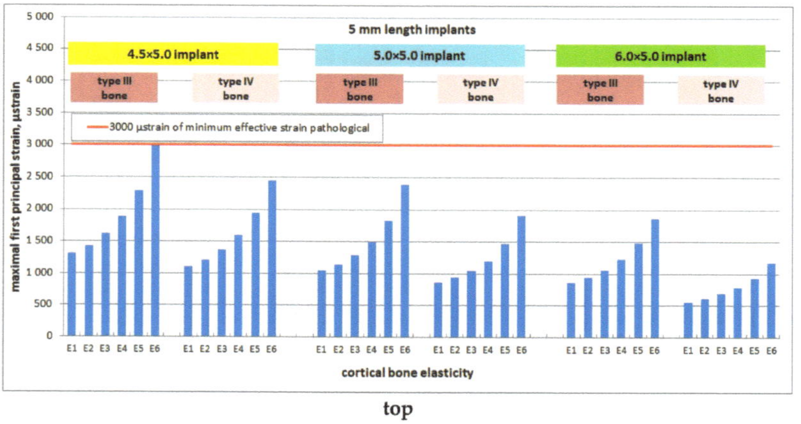

top

Figure 7. *Cont.*

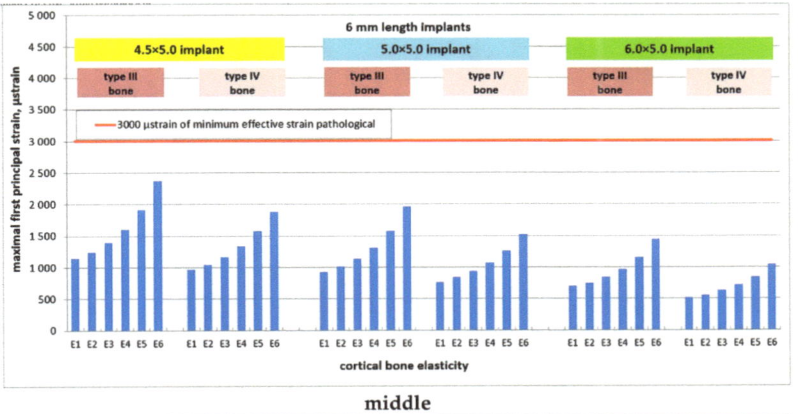

middle

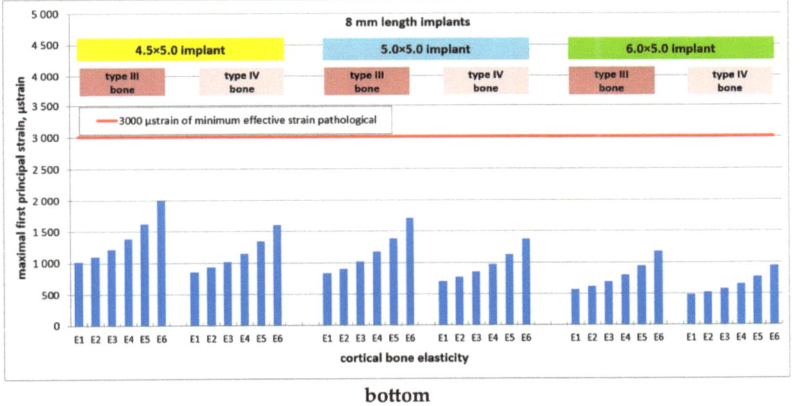

bottom

Figure 7. Dependence of maximal first principal strains (MFPSs) in cortical bone on its modulus of elasticity for implants of length 5.0 mm (**top**), 6.0 mm (**middle**), 8.0 mm (**bottom**) placed into bone segments of type III and IV bone and the studied E1–E6 degrees of patient-specific cortical bone elasticity. Red line corresponds to 3000 microstrain of Frost "minimum effective strain pathological" (MESP).

For a range of cortical bone elasticity (E_1–E_6) and type III bone, increasing implant diameter from 4.5 to 6.0 mm resulted in FPS reductions of approximately 50% for S-implants, 47% for I-implants, and 43% for L-implants. In type IV bone under similar conditions, FPS reductions were around 36%, 40%, and 42% for S-, I-, and L-implants, respectively. These trends are depicted in Figures 8 and 9.

For the E1–E6 spectrum of cortical bone elasticity and type III bone, MFPS reduction due to increase in length from 5.0 to 8.0 mm was (23–35) % for N-implants, (17–28)% for M-implants, and (12–19) % for W-implants. For type IV bone and otherwise same conditions, the corresponding FPS reduction was (22–34) % for N-implants, (20–29) % for M-implants, and (34–37) % for W-implants. These findings are illustrated in Figures 10 and 11.

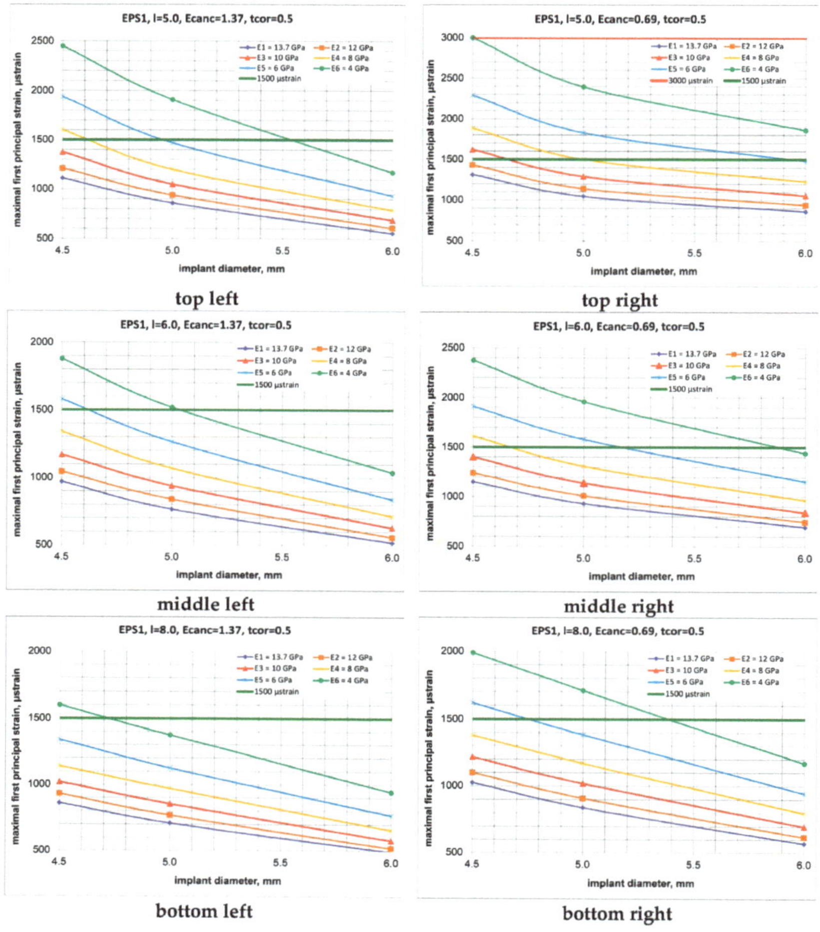

Figure 8. Maximal first principal strains' (MFPSs') dependence on the implants' diameter increase for implants of length 5.0 mm (**top**), 6.0 mm (**middle**), 8.0 mm (**bottom**) placed into bone segments of type III (**left**) and IV (**right**) bone and the studied E_1–E_6 degrees of patient-specific cortical bone elasticity. Red line corresponds to 3000 microstrain of Frost "minimum effective strain pathological" (MESP).

Bone quality was found to have a substantial impact on the biomechanical state of cortical bone: for N, M, and W implants, a two-fold reduction in elasticity modulus (0.69 against 1.37 GPa) corresponded to an 18, 22, and 55% MFPS rise for S-implants, 19, 22, and 34% MFPS rise for I-implants, and 19, 18, and 16% MFPS rise for L-implants. These data correspond to the E1 level of cortical bone elasticity. For the E2, E3, E4, E5, and E6 levels, the corresponding MFPS rise was as follows: E2 level—18, 21, and 54% (S-implants), 19, 20, and 33% (I-implants), and 19, 19, and 21% (L-implants); E3 level—18, 23, and 53% (S-implants), 20, 21, and 33% (I-implants), and 19, 19, and 21% (L-implants); E4 level—18, 25, and 56% (S-implants), 20, 22, and 35% (I-implants), and 21, 21, and 22% (L-implants)); E5 level—18, 24, and 58% (S-implants), 21, 25, and 38% (I-implants), and 21, 22, and 23% (L-implants)); E6 level—22, 25, and 59% (S-implants), 26, 29, and 38% (I-implants), and 24, 24, and 25% (L-implants). The percentage of MFPS rise is illustrated in Figure 12.

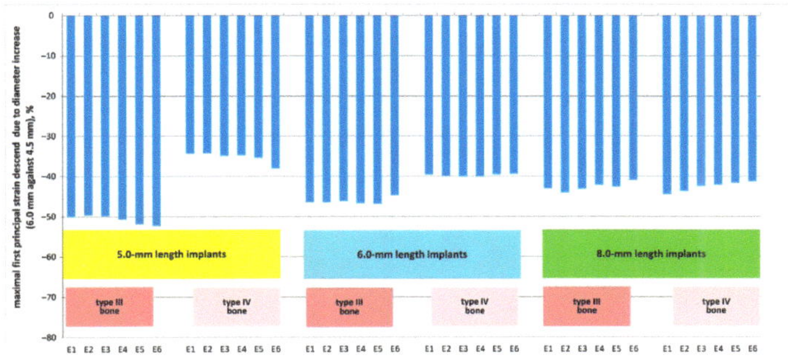

Figure 9. Maximal first principal strain (MFPS) reduction due to the implants' diameter increase from 4.5 mm to 6.0 mm for the spectrum of implants placed into bone segments of types III and IV bone.

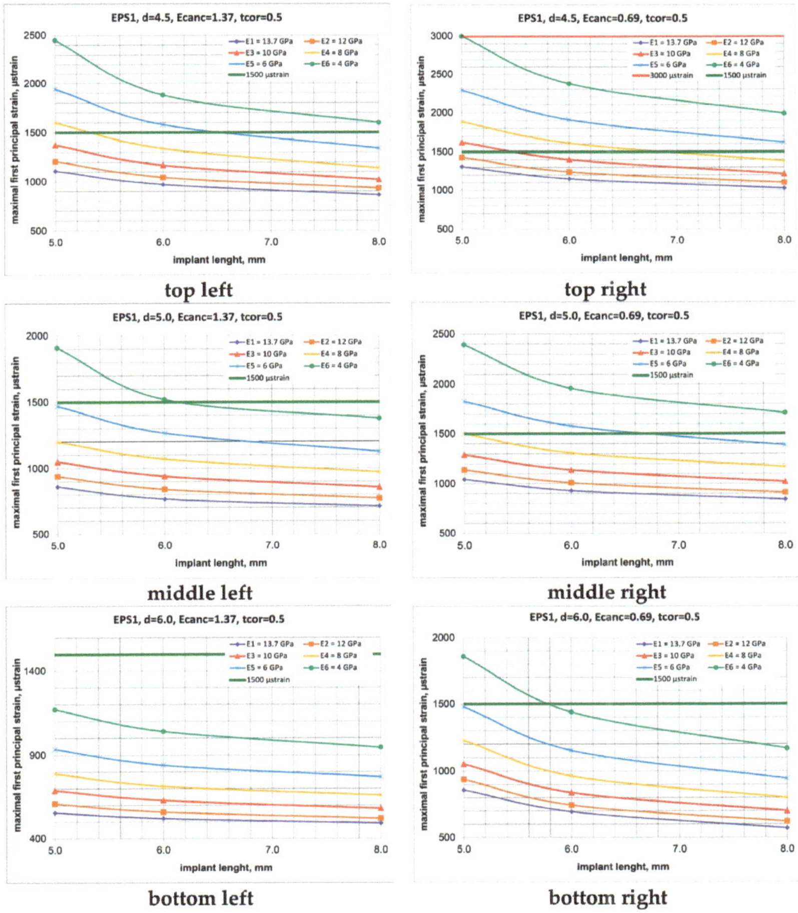

Figure 10. Maximal first principal strain (MFPS) dependence on the implants' length increase for implants of diameter 4.5 mm (**top**), 5.0 mm (**middle**), 6.0 mm (**bottom**) placed into bone segments of

type III (**left**) and IV (**right**) bone and the studied E_1–E_6 degrees of patient-specific cortical bone elasticity. Red line corresponds to 3000 microstrain of Frost "minimum effective strain pathological" (MESp).

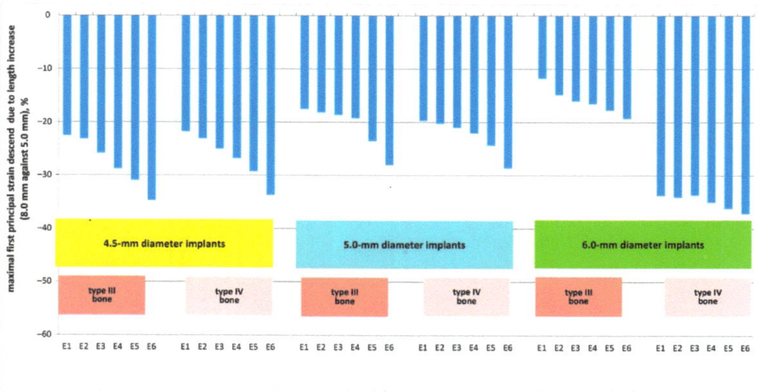

Figure 11. Maximal first principal strain (MFPS) reduction due to the implants' length increase from 4.5 mm to 6.0 mm for the spectrum of implants placed into bone segments of types III and IV bone.

A reduction in cortical bone elasticity caused its significant overstrain: for N-, M-, W-implants, the MFPS rise due to decrease in cortical bone rigidity (E6 against E1) for type III bone was 121/122/111, 94/99/101, and 86/94/93% (S-/I-/L-implants). For type IV bone, cortical bone and implant parameters, the MFPS rise was 129/129/116, 107/110/107, and 94/104/105% (S-/I-/L-implants). The results of an MFPS rise (percentage) are illustrated in Figure 13.

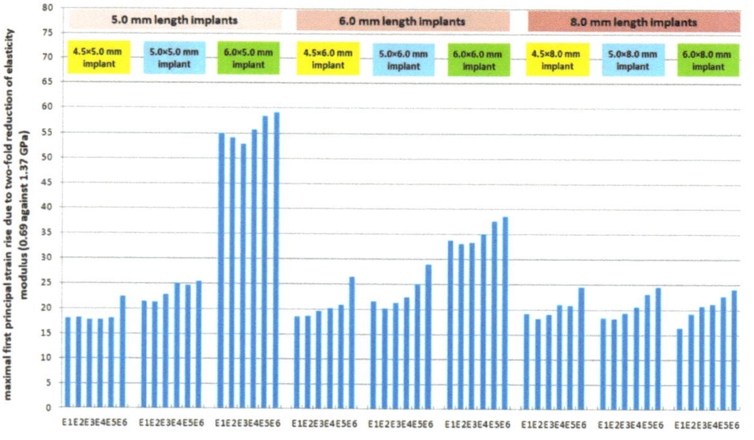

Figure 12. Impact of bone quality on maximal first principal strain (MFPS) rise in terms of two-fold reduction in the cancellous bone elasticity modulus (0.69 against 1.37 GPa) for the spectrum of implants placed into bone segments at E_1–E_6 degrees of cortical bone elasticity.

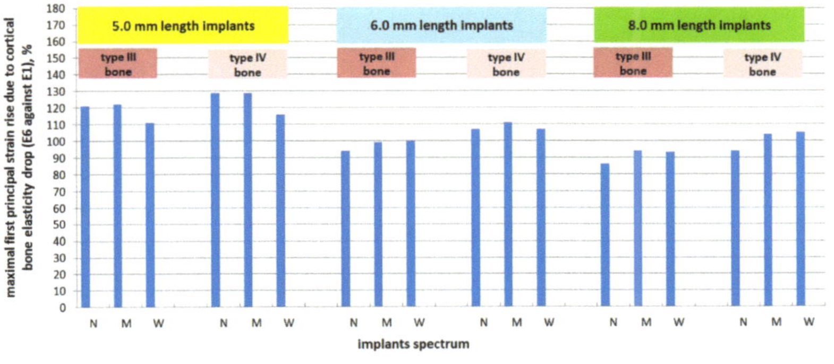

Figure 13. Maximal first principal strain (MFPS) rise due to cortical bone elasticity reduction (E_6 against E_1).

4. Discussion

Studying the strain fields induced by osseointegrated implants of varying sizes in the anchorage area of the maxilla, often characterized by insufficient volume and low quality (types III and IV), is crucial for achieving optimal transfer of functional loads to marginal bone and ensuring physiological turnover. In compromised conditions like these, understanding the impact of cortical bone elasticity on neck area strains becomes a priority, particularly in terms of bone turnover. However, relying solely on clinical and experimental studies may not fully address the biomechanical response, especially considering patient-variable bone elasticity on loading transfer. Establishing quantitative relationships between key influencing parameters such as mechanical stresses and strains in bone, maxillofacial system geometry and dimensions, material properties, and bone quality, and subsequently selecting patient-personified implants, requires advanced methods of analysis.

Computer modeling serves as a contemporary in silico approach to comprehending the biomechanical behavior of implants and implant-supported restorations, as well as predicting implant lifetime by maintaining tolerable strain concentrations in peri-implant bone to facilitate adequate bone turnover through viable implant selection [22,23,25,33–37].

Although several finite element (FE) studies have analyzed strain distribution in adjacent bone concerning implant dimensions and bone quality [25,35,38], their findings primarily help in refining the parameters of a bone–implant system rather than assessing the risk of structural failure. While some studies have considered strains in evaluating bone's mechanical behavior, they have struggled to establish clear patterns linking specific implant dimensions, adjacent bone quality, and the origins of pathological bone strains (see, for example, [16]).

The issues outlined above underscore the objective of this study, wherein we aimed to establish a methodological framework for selecting viable implants based on finite element analysis and Frost's "Mechanostat" theory. This method enables the prediction of safe strains in adjacent bone by appropriately sizing the implant, considering the quality and quantity of patient-specific bone. This approach appears more effective than the conventional clinical procedure, which typically involves a preliminary examination of the patient's jaw bone properties, such as placement site, bone shape, and dimensions. However, this conventional method overlooks the physical and mechanical properties of bone tissues, which are crucial indicators of potential bone failure.

Therefore, the proposed method offers an advantage in addressing the clinically significant issue of implant selection, as it theoretically predicts the success of implant treatment by preventing pathological bone strains. This methodology allows for the identification of specific requirements regarding both the quality/rigidity of the bone and the size of the implant necessary for successful placement in a particular patient's bone. Consequently, this lays the groundwork for the development of tailored clinical recommendations for implant selection.

Despite the abundance of finite element analysis (FEA) studies on bone–implant interaction over the past five decades, the reliability and precision of these in silico investigations are not always beyond dispute. It has been emphasized [39] that FEA studies of biological structures should ideally undergo experimental validation whenever feasible. Additionally, it has been suggested [40] that FEA studies should meet minimum requirements, including comparisons with data from other studies or real-world observations. According to the American Society of Mechanical Engineers Committee on verification and validation in computational solid mechanics, verification is defined as "the process of determining that a computational model accurately represents the underlying mathematical model and its solution", while validation is defined as "the process of determining the degree to which a model is an accurate representation of the real world from the perspective of the intended uses of the model". In simple terms, verification is the process of "solving the equations right", whereas validation is the process of "solving the right equations" [41]. Given these contentious circumstances, our focus was on ensuring the high accuracy of our computations. We conducted an adequate mesh convergence test to determine the optimal element size, thus enhancing the reliability of our results.

Despite the high precision of finite element (FE) analysis, its limitations stem from the assumptions and simplifications adopted, which directly affect the accuracy of strain calculations. One major limitation is the design of an adequate 3D bone segment model. On one hand, it is necessary to eliminate the effects of patient-specific anatomical variations to ensure the comparability of results across different models [42]. On the other hand, the difference between the actual anatomic models of maxilla and the 3D computer model should be minimal [43]. To strike a balance, we first developed 3D models of the maxilla surrounding Bicon plateau implants by importing CT images to simulate the shape and dimensions of bone components. Then, minor simplifications were made to create models that are more comparable and universally applicable in a biomechanical context, consistent with similar studies in the field [35,42,44,45].

In most studies examining the biomechanical behavior of dental implant systems, the implant is assumed to be 100% osseointegrated [22,30,35,43,44], with no sliding allowed between the implant and the bone. In reality, osseointegration does not occur clinically [30,46]. It is highly dependent on bone quality, healing process, and loads applied to the implant during function [4]. Therefore, deviating from the concept of complete osseointegration in the present comparative study would have introduced significant uncertainties into the results, making it impossible to compare them with those of other authors. Hence, the concept of complete osseointegration was incorporated into the designed algorithm of numerical analysis.

In our designed models, all structures were assumed to be homogeneous, isotropic, and linearly elastic, consistent with many contemporary studies in the field [35,44]. However, clinical reality and material properties differ from these assumptions. Cortical bone, for example, is transversely isotropic and inhomogeneous; bone anisotropy better reflects actual clinical conditions [46,47], and it is more appropriate in simulating actual clinical conditions. Undoubtedly, detailed architecture and morphology of each bone layer, if included in the model, may provide more realistic strain fields and could result in significant local variations in bone strains. While including detailed architecture and morphology of each bone layer in the model could provide more realistic strain fields and lead to significant local variations in bone strains, studies such as Limbert et al.'s micro-CT-based work [48] have shown that bone strain levels remain within the homeostatic range even when these variations, partly caused by the trabecular nature of bone, are modeled in detail.

It is worth noting that Frost's experimental work also does not differentiate for these effects but provides a general guideline for bone remodeling [18,49]. Considering the challenges in obtaining adequate anisotropic elastic data and the comparative nature of our study, we opted for the widely accepted isotropic approach. Like many well-recognized studies in the field [35,37,44], our investigation aimed to assess the biomechanical conditions of peri-implant bone behavior. However, it is important to acknowledge that future finite element models should address these issues, and bone should be analyzed as anisotropic and nonhomogeneous when more sophisticated approaches are planned to analyze living tissue behavior under mechanical loading.

To predict potential bone failure and enhance our understanding of the impact of bone quality and implant dimensions on lifetime prognosis, information on elasticity modulus is crucial. This parameter plays a fundamental role in assigning bone material properties to finite element (FE) meshes during model generation and subsequent strain analysis. While calculating heterogeneous elasticity modulus magnitudes from Digital Imaging and Communications in Medicine (DICOM) files, containing comprehensive tissue structure data, would be more appropriate, this capability is currently not widely available to most clinicians [50]. Therefore, in our study, we opted for an averaged phenomenological approach based on experimental tests of bone specimens using ultrasonic transmission techniques or nanoindentation [21]. This approach involves equalizing data across cortical and cancellous bone volumes and allows for correlation with similar research in the field, facilitating the generalization of relationships across a wide spectrum of human bone elasticity modulus variations.

In our study, we improved critical strain calculation by determining the plane of the critical bone–implant interface based on a preliminary evaluation of strain field contours. This detail enabled us to assess implant success by comparing maximal principal strain values with Frost's MESp threshold for each bone–implant system studied. This approach seemed more appropriate compared to the study [16], where "overstrain fractions" are analyzed instead of exact strain data.

Our study focuses solely on static loadings because their duration of application is brief, classifying them as static forces according to the generally accepted classification in the mechanics of deformable solids [39]. This approach aligns with most finite element studies, where static loads simulate not only vertical loads and horizontal forces but also combined loads (e.g., oblique occlusal forces) to reflect more realistic conditions and obtain a more accurate mechanical response.

Although dental implants are subjected to both static and repeated (cyclic) loads [22,44,45,51], our study does not address cyclic loading. In fact, dental implants are subjected not only to static loads but also to repeated (cyclic) loads [52]. It has been suggested that bone strength decreases when subjected to cyclic loading, potentially leading to different stress/strain calculations [53]. However, this aspect falls beyond the scope of our present study.

Despite controversial limitations, the results obtained in terms of critical stress–strain state location are in agreement with foregoing studies, in which the authors try to understand the biomechanical behavior of implants in living surroundings [25,33–36]. Same as in our study, the works mentioned above have the limitations of being a finite element simulation. Similar to their content, we have considered specific situations, like a simplified 3D bone model, standardized bone elasticity and variable quality, average cortical bone thicknesses, non-axial static loading, completely osseointegrated implants, ranked bone elasticity moduli, that do not exactly reflect clinical situations. Thus, the values obtained may not correspond to the clinical behavior of considered implant systems. It is necessary to contrast these results with those obtained in in vitro and in vivo studies where possible.

The significance of our study lies in our attempt to enhance the understanding of the fundamental processes involved in load transmission to both bone and implants and to evaluate the conditions leading to eventual failure due to pathological turnover. Additionally, we aimed to explore the relationships between principal biomechanical parameters of bone structures and dental system geometry. While our research currently offers modest clinical recommendations, future systematic studies on the ordered testing of the elastic and strength properties of cortical and cancellous bone could pave the way for establishing more robust relationships between the aforementioned parameters and the properties of the studied structures. This, in turn, will likely reduce clinical discrepancies between the results of computer modeling and real clinical practice, reinforcing the significance of following clinical recommendations for practitioners.

5. Conclusions

According to Frost's "Mechanostat" theory, strains exceeding MESp (3000 μstrain) will cause bone failure. In order to produce a peri-implant bone mass increase after the healing period and avoid bone mass loss, strains should be kept above the MESm threshold (1000–1500 μstrain). In the present study, Frost's "Mechanostat" hypothesis was applied to establish basic relationships between the factors which predetermine implant success in terms of physiological/pathological bone turnover.

The studied implants were found to be sensitive to the spectrum of the influence factors, but the analysis revealed that there were no instances of overstrains (MFPS > 3000 μstrain) at the implant neck area of Bicon Integra-CP™ implants.

The largest effect of implant diameter increase on strain reduction in the type III bone was found for short implants for all studied levels of cortical bone elasticity and the smallest for long implants. In Type IV bone, such an effect was nondependent on the implant length.

The effect of increase in implant length on strain reduction was significantly dependent on bone quality in the case of wide implants where type IV bone provided two times more effect of strain reduction than type III bone.

A wide implant was the most sensitive to the deterioration of bone quality. It induced the most rise of bone strains than N and M implants. This trend occurred at all studied cortical bone elasticity levels, especially for most soft bone.

A significant variation of the cortical bone elasticity induced the most significant increase in cortical bone strains for all implants and bone types. The larger strain increase corresponded to narrow implants in both bone types.

The findings are useful for dental practitioners planning short finned implants in the posterior maxilla. In cases of uncertain cortical bone quality and insufficient bone height, wider implants from the Bicon Integra-CP™ catalogue should be preferred to mitigate the risk of pathological strains. Further investigations of cortical bone architecture and elasticity in the posterior maxilla site are recommended to develop comprehensive clinical recommendations considering bone volume and quality limitations. Such research can potentially enable the placement of narrower implants in cases of insufficient bone.

Author Contributions: Conceptualization, H.-W.W. and V.D.; methodology, A.K.; validation, I.L., L.L. and H.-W.W.; formal analysis, H.-W.W. and A.K.; investigation, I.L. and V.D.; resources, V.D. and L.L.; data curation, V.D., I.L. and A.K.; writing—original draft preparation, V.D. and A.K.; writing—review and editing, L.L. and H.-W.W.; visualization, H.-W.W.; supervision, A.K.; project administration, H.-W.W. and V.D. All authors have read and agreed to the published version of the manuscript.

Funding: This research received no external funding.

Institutional Review Board Statement: Not applicable.

Informed Consent Statement: Not applicable.

Data Availability Statement: The original contributions presented in the study are included in the article, further inquiries can be directed to the corresponding author.

Acknowledgments: The authors thank the University of Cambridge for their support.

Conflicts of Interest: The authors declare no conflicts of interest.

References

1. Overmann, A.L.; Aparicio, C.; Richards, J.T.; Mutreja, I.; Fischer, N.G.; Wade, S.M.; Potter, B.K.; Davis, T.A.; Bechtold, J.E.; Forsberg, J.A.; et al. Orthopaedic osseointegration: Implantology and future directions. *J. Orthopaed. Res.* **2020**, *38*, 1445–1454. [CrossRef] [PubMed]
2. Verma, A.; Singh, S.V.; Arya, D.; Shivakumar, S.; Chand, P. Mechanical failures of dental implants and supported prostheses: A systematic review. *J. Oral Biol. Craniofac. Res.* **2023**, *13*, 306–314. [CrossRef] [PubMed]

3. Segal, P.; Makhoul, A.; Eger, M.; Lucchina, A.G.; Winocur, E.; Mijiritsky, E. Preliminary Study to Evaluate Marginal Bone Loss in Cases of 2- and 3-Implant-Supported Fixed Partial Prostheses of the Posterior Mandible. *J. Craniofac. Surg.* **2019**, *30*, 1068–1072. [CrossRef] [PubMed]
4. Hingsammer, L.; Pommer, B.; Hunger, S.; Stehrer, R.; Watzek, G.; Insua, A. Influence of Implant Length and Associated Parameters Upon Biomechanical Forces in Finite Element Analyses: A Systematic Review. *Implant Dent.* **2019**, *28*, 296–305. [CrossRef] [PubMed]
5. Fernandes, G.V.O.; Ferreira, N.D.R.N.; Heboyan, A.; Nassani, L.M.; Pereira, R.M.A.; Fernandes, J.C.H. Clinical assessment of short (>6 mm and ≤8.5 mm) implants in posterior sites with an average follow-up of 74 Months: A Retrospective Study. *Int. J. Oral Maxillofac. Implant* **2023**, *38*, 915–926. [CrossRef]
6. Cenkoglu, B.G.; Balcioglu, N.B.; Ozdemir, T.; Mijiritsky, E. The Effect of the Length and Distribution of Implants for Fixed Prosthetic Reconstructions in the Atrophic Posterior Maxilla: A Finite Element Analysis. *Materials* **2019**, *12*, 2556. [CrossRef]
7. Fernandes, G.; Costa, B.; Trindade, H.F.; Castilho, R.M.; Fernandes, J. Comparative analysis between extra-short implants (≤6 mm) and 6 mm-longer implants: A meta-analysis of randomized controlled trial. *Aust. Dent. J.* **2022**, *67*, 194–211. [CrossRef]
8. Markose, J.; Eshwar, S.; Srinivas, S.; Jain, V. Clinical outcomes of ultrashort sloping shoulder implant design: A survival analysis. *Clin. Implant Dent. Relat. Res.* **2018**, *20*, 646–652. [CrossRef]
9. Taschieri, S.; Lolato, A.; Testori, T.; Francetti, L.; Del Fabbro, M. Short dental implants as compared to maxillary sinus augmentation procedure for the rehabilitation of edentulous posterior maxilla: Three-year results of a randomized clinical study. *Clin. Implant Dent. Relat. Res.* **2018**, *20*, 9–20. [CrossRef]
10. Cruz, R.S.; Lemos, C.A.A.; Batista, V.E.S.; Oliveira, H.F.F.E.; Gomes, J.M.L.; Pellizzer, E.P.; Verri, F.R. Short implants versus longer implants with maxillary sinus lift. A systematic review and meta-analysis. *Braz. Oral Res.* **2018**, *32*, e86. [CrossRef]
11. Linetskiy, I.; Demenko, V.; Yefremov, O.; Linetska, L.; Kondratiev, A. Crestal versus subcrestal short plateau implant placement. In *Integrated Computer Technologies in Mechanical Engineering*; ICTM 2023 Lecture Notes in Networks and Systems; Springer International Publishing: Cham, Switzerland, 2024; Volume 1008, pp. 258–267. [CrossRef]
12. Alqahtani, A.R.; Desai, S.R.; Patel, J.R.; Alqhtani, N.R.; Alqahtani, A.S.; Heboyan, A.; Fernandes, G.V.O.; Mustafa, M.; Karobari, M.I. Investigating the impact of diameters and thread designs on the Biomechanics of short implants placed in D4 bone: A 3D finite element analysis. *BMC. Oral Health* **2023**, *23*, 686. [CrossRef] [PubMed]
13. Linetska, L.; Kipenskyi, A.; Demenko, V.; Linetskiy, I.; Kondratiev, A.; Yefremov, O. Finite element study of biomechanical behavior of short dental implants with bone loss effects–evaluation of bone turnover. In Proceedings of the 2023 IEEE 4th KhPI Week on Advanced Technology (KhPIWeek), Kharkiv, Ukraine, 2–6 October 2023; pp. 1–6. [CrossRef]
14. Calì, M.; Zanetti, E.M.; Oliveri, S.M.; Asero, R.; Ciaramella, S.; Martorelli, M.; Bignardi, C. Influence of thread shape and inclination on the biomechanical behaviour of plateau implant systems. *Dent. Mater.* **2018**, *34*, 460–469. [CrossRef] [PubMed]
15. McKenna, G.J.; Gjengedal, H.; Harkin, J.; Holland, N.; Moore, C.; Srinivasan, M. Effect of autogenous bone graft site on dental implant survival and donor site complications: A systematic review and meta-analysis. *J. Evid. Based Dent. Pract.* **2022**, *22*, 101731. [CrossRef] [PubMed]
16. Qiu, P.; Cao, R.; Li, Z.; Fan, Z. A comprehensive biomechanical evaluation of length and diameter of dental implants using finite element analyses: A systematic review. *Heliyon* **2024**, *10*, e26876. [CrossRef] [PubMed]
17. Leucht, P.; Monica, S.D.; Temiyasathit, S.; Lenton, K.; Manu, A.; Longaker, M.T.; Jacobs, C.R.; Spilker, R.L.; Guo, H.; Brunski, J.B.; et al. Primary cilia act as mechanosensors during bone healing around an implant. *Med. Eng. Phys.* **2013**, *35*, 392–402. [CrossRef]
18. Frost, H.M. A 2003 update of bone physiology and Wolff's Law for clinicians. *Angle Orthod.* **2004**, *74*, 3–15. [CrossRef]
19. Frost, H.M. Bone "mass" and the "mechanostat": A proposal. *Anat. Rec.* **1987**, *219*, 1–9. [CrossRef]
20. Misch, C.E. (Ed.) Chapter 32—Progressive Bone Loading: Increasing the Density of Bone with a Prosthetic Protocol. In *Dental Implant Prosthetics*; Mosby: Maryland Heights, MO, USA, 2015; pp. 913–937. [CrossRef]
21. Saab, X.E.; Griggs, J.A.; Powers, J.M.; Engelmeier, R.L. Effect of abutment angulation on the strain on the bone around an implant in the anterior maxilla: A finite element study. *J. Prosthet. Dent.* **2007**, *97*, 85–92. [CrossRef]
22. Falcinelli, C.; Valente, F.; Vasta, M.; Traini, T. Finite element analysis in implant dentistry: State of the art and future directions. *Dent. Mater.* **2023**, *39*, 539–556. [CrossRef]
23. Lisiak-Myszke, M.; Marciniak, D.; Bieliński, M.; Sobczak, H.; Garbacewicz, Ł.; Drogoszewska, B. Application of Finite Element Analysis in Oral and Maxillofacial Surgery-A Literature Review. *Materials* **2020**, *13*, 3063. [CrossRef]
24. Linetskiy, I.; Sutcliffe, M.; Kondratiev, A.; Demenko, V.; Linetska, L.; Yefremov, O. A novel method of load bearing ability analysis of short plateau implants placed in compromised bone. In Proceedings of the 2023 IEEE 4th KhPI Week on Advanced Technology (KhPIWeek), Kharkiv, Ukraine, 2–6 October 2023; pp. 1–6. [CrossRef]
25. Sugiura, T.; Yamamoto, K.; Horita, S.; Murakami, K.; Tsutsumi, S.; Kirita, T. The effects of bone density and crestal cortical bone thickness on micromotion and peri-implant bone strain distribution in an immediately loaded implant: A nonlinear finite element analysis. *J. Periodontal Implant Sci.* **2016**, *46*, 152–165. [CrossRef] [PubMed]
26. Bozkaya, D.; Muftu, S.; Muftu, A. Evaluation of load transfer characteristics of five different implants in compact bone at different load levels by finite elements analysis. *J. Prosthet. Dent.* **2004**, *92*, 523–530. [CrossRef] [PubMed]
27. Lemos, C.A.A.; Verri, F.R.; Noritomi, P.Y.; Kemmoku, D.T.; Souza Batista, V.E.; Cruz, R.S.; de Luna Gomes, J.M.; Pellizzer, E.P. Effect of bone quality and bone loss level around internal and external connection implants: A finite element analysis study. *J. Prosthet. Dent.* **2021**, *125*, 137.e1–137.e10. [CrossRef]

28. Mericske-Stern, R.; Zarb, G.A. In vivo measurements of some functional aspects with mandibular fixed prostheses supported by implants. *Clin. Oral Implants Res.* **1996**, *7*, 153–161. [CrossRef] [PubMed]
29. Sahin, S.; Cehreli, M.C.; Yalçin, E. The influence of functional forces on the biomechanics of implant-supported prostheses—A review. *J. Dent.* **2002**, *30*, 271–282. [CrossRef]
30. Li, J.; Jansen, J.A.; Walboomers, X.F.; van den Beucken, J.J. Mechanical aspects of dental implants and osseointegration: A narrative review. *J. Mech. Behav. Biomed. Mater.* **2020**, *103*, 103574. [CrossRef]
31. Gere, J.M.; Timoshenko, S.P. *Mechanics of Materials*; Chapman & Hall: London, UK, 1991.
32. Kondratiev, A.; Potapov, O.; Tsaritsynskyi, A.; Nabokina, T. Optimal design of composite shelled sandwich structures with a honeycomb filler. In *Advances in Design, Simulation and Manufacturing IV. DSMIE 2021*; Lecture Notes in Mechanical Engineering; Springer: Berlin/Heidelberg, Germany, 2021; pp. 546–555. [CrossRef]
33. Jafarian, M.; Mirhashemi, F.S.; Emadi, N. Finite element analysis of stress distribution around a dental implant with different amounts of bone loss: An in vitro study. *Dent. Med. Probl.* **2019**, *56*, 27–32. [CrossRef]
34. Yalçın, M.; Kaya, B.; Laçin, N.; Arı, E. Three-Dimensional Finite Element Analysis of the Effect of Endosteal Implants with Different Macro Designs on Stress Distribution in Different Bone Qualities. *Int. J. Oral Maxillofac. Implant* **2019**, *34*, e43–e50. [CrossRef]
35. Oliveira, H.; Brizuela Velasco, A.; Ríos-Santos, J.V.; Sánchez, L.F.; Lemos, B.F.; Gil, F.J.; Carvalho, A.; Herrero-Climent, M. Effect of Different Implant Designs on Strain and Stress Distribution under Non-Axial Loading: A Three-Dimensional Finite Element Analysis. *Int. J. Environ. Res. Public Health* **2020**, *17*, 4738. [CrossRef]
36. Chen, J.; Guo, J.; Yang, L.; Wang, L.; Zhang, X. Effect of different implant angulations on the biomechanical performance of prosthetic screws in two implant-supported, screw-retained prostheses: A numerical and experimental study. *J. Prosthet. Dent.* **2023**, *130*, 240.e1–240.e10. [CrossRef]
37. Demenko, V.; Linetskiy, I.; Nesvit, K.; Hubalkova, H.; Nesvit, V.; Shevchenko, A. Importance of diameter-to-length ratio in selecting dental implants: A methodological finite element study. *Comput. Methods Biomech. Biomed. Eng.* **2014**, *17*, 443–449. [CrossRef] [PubMed]
38. Yoda, N.; Zheng, K.; Chen, J.; Li, W.; Swain, M.; Sasaki, K.; Li, Q. Bone morphological effects on post-implantation remodeling of maxillary anterior buccal bone: A clinical and biomechanical study. *J. Prosthodont. Res.* **2017**, *61*, 393–402. [CrossRef] [PubMed]
39. De Matos, J.D.M.; Queiroz, D.A.; Nakano, L.J.N.; Andrade, V.C.; Ribeiro, N.D.R.; Borges, A.L.S.; Bottino, M.A.; Lopes, G.D.S. Bioengineering tools applied to dentistry: Validation methods for in vitro and in silico analysis. *Dent. J.* **2022**, *10*, 145. [CrossRef] [PubMed]
40. Hannam, A.G. Current computational modelling trends in craniomandibular biomechanics and their clinical implications. *J. Oral Rehabil.* **2011**, *38*, 217–234. [CrossRef]
41. Anderson, A.E.; Ellis, B.J.; Weiss, J.A. Verification, validation and sensitivity studies in computational biomechanics. *Comput. Methods Biomech. Biomed. Eng.* **2007**, *10*, 171–184. [CrossRef]
42. Yan, X.; Zhang, X.; Chi, W.; Ai, H.; Wu, L. Comparing the influence of crestal cortical bone and sinus floor cortical bone in posterior maxilla bi-cortical dental implantation: A three-dimensional finite element analysis. *Acta Odontol. Scand.* **2015**, *73*, 312–320. [CrossRef]
43. Okumura, N.; Stegaroiu, R.; Kitamura, E.; Kurokawa, K.; Nomura, S. Influence of maxillary cortical bone thickness, implant design and implant diameter on stress around implants: A three-dimensional finite element analysis. *J. Prosthodont. Res.* **2010**, *54*, 133–142. [CrossRef]
44. Barbosa, F.T.; Zanatta, L.C.S.; de Souza Rendohl, E.; Gehrke, S.A. Comparative analysis of stress distribution in one-piece and two-piece implants with narrow and extra-narrow diameters: A finite element study. *PLoS ONE* **2021**, *16*, e0245800. [CrossRef]
45. Chou, I.-C.; Lee, S.-Y.; Ming-Chang, W.; Sun, C.W.; Jiang, C.P. Finite element modelling of implant designs and cortical bone thickness on stress distribution in maxillary type IV bone. *Comput. Methods Biomech. Biomed. Eng.* **2014**, *17*, 516–526. [CrossRef]
46. Alper, B.; Gultekin, P.; Yalci, S. Application of Finite Element Analysis in Implant Dentistry. In *Finite Element Analysis-New Trends and Developments*; InTech: Rijeka, Croatia, 2012. [CrossRef]
47. Tretto, P.H.W.; Dos Santos, M.B.F.; Spazzin, A.O.; Pereira, G.K.R.; Bacchi, A. Assessment of stress/strain in dental implants and abutments of alternative materials compared to conventional titanium alloy-3D non-linear finite element analysis. *Comput. Methods Biomech. Biomed. Eng.* **2020**, *23*, 372–383. [CrossRef]
48. Limbert, G.; van Lierde, C.; Muraru, O.L.; Walboomers, X.F.; Frank, M.; Hansson, S.; Middleton, J.; Jaecques, S. Trabecular bone strains around a dental implant and associated micromotions--a micro-CT-based three-dimensional finite element study. *J. Biomech.* **2010**, *43*, 1251–1261. [CrossRef] [PubMed]
49. Frost, H.M. The mechanostat: A proposed pathogenic mechanism of osteoporoses and the bone mass effects of mechanical and nonmechanical agents. *Bone. Miner.* **1987**, *2*, 73–85. [PubMed]
50. Grauer, D.; Cevidanes, L.S.; Proffit, W.R. Working with DICOM craniofacial images. *Am. J. Orthod. Dentofac. Orthop.* **2009**, *136*, 460–470. [CrossRef] [PubMed]
51. Bordin, D.; Castro, M.B.; Carvalho, M.A.; Araujo, A.M.; Cury, A.A.D.B.; Lazari-Carvalho, P.C. Different Treatment Modalities Using Dental Implants in the Posterior Maxilla: A Finite Element Analysis. *Braz. Dent. J.* **2021**, *32*, 34–41. [CrossRef] [PubMed]

52. Khosravani, M.R. Mechanical behavior of restorative dental composites under various loading conditions. *J. Mech. Behav. Biomed. Mater.* **2019**, *93*, 151–157. [CrossRef]
53. Kurniawan, D.; Nor, F.M.; Lee, H.Y.; Lim, J.Y. Finite element analysis of bone-implant biomechanics: Refinement through featuring various osseointegration conditions. *Int. J. Oral Maxillofac. Surg.* **2012**, *41*, 1090–1096. [CrossRef]

prosthesis

MDPI

Article

The Effect of the Incorporation of a 3D-Printed Titanium Framework on the Mechanical Properties CAD/CAM Denture Base Materials

Rafael Delgado-Ruiz [1,*], Ido Brintouch [1], Aisha Ali [1], Yiwei Fang [2], Georgios Romanos [1] and Miriam Rafailovich [2]

[1] School of Dental Medicine, Stony Brook University, Stony Brook, NY 11733, USA; brintouchido@gmail.com (I.B.); aisha.m.ali@stonybrook.edu (A.A.); georgios.romanos@stonybrookmedicine.edu (G.R.)

[2] Department of Materials Science and Chemical Engineering, Stony Brook University, Stony Brook, NY 11733, USA; yiwei.fang@stonybrook.edu (Y.F.); miriam.rafailovich@stonybrook.edu (M.R.)

* Correspondence: rafael.delgado-ruiz@stonybrookmedicine.edu; Tel.: +1-631-632-6913

Abstract: Background: Complete dentures should withstand occlusal forces and wear. However, over time, dentures can suffer fatigue and develop cracks, chipping, and fractures. Conventional methods for the fabrication of complete dentures involve injection molding, thermal curing, and the use of microwaves with polymethyl methacrylate (PMMA)-based materials. These methods have served well for many years. More recently, the incorporation of computer-aided design and computer-aided manufacturing (CAD/CAM) to fabricate complete dentures has been shown to enhance the dentures' mechanical properties, including resistance to wear and impact strength. This study aims to investigate the mechanical properties and fracture types of CAD/CAM denture base materials (both milled and printed) as compared to a novel proprietary method that embeds a 3D-printed framework within PMMA-milled blocks. The null hypothesis is that incorporating a 3D-printed framework does not affect the mechanical properties of milled PMMA blocks. Methods: Three groups of bars were fabricated using CAD/CAM methods: printed (P), milled (M), and milled with a 3D-printed metallic framework reinforcement (M + F). A three-point bending test evaluated deformation, followed by an impact fracture test for fracture toughness. A descriptive fractographic analysis assessed the fracture characteristics. A statistical analysis using a paired t-test compared the differences between the groups. Results: The P group showed more elastic deformation than the M and M + F groups ($p < 0.05$). The M + F group achieved a higher fracture toughness as compared to the M and P groups ($p < 0.05$). Conclusions: Within the limitations of this experimental study, the null hypothesis can be rejected. Milled samples with an embedded 3D-printed titanium framework possess higher resistance to impact than milled samples without frameworks, and printed samples and milled samples with embedded 3d-printed titanium frameworks present increased flexural strength and lower elastic deformation as compared to milled samples without frameworks and printed samples.

Keywords: denture reinforcement; 3D-printed; milling; CAD/CAM; denture base; impact fracture test; three-point bending test; titanium framework

1. Introduction

Denture base resins should exhibit the necessary strength, fracture toughness, and dimensional stability to endure forces during function over many years [1,2]. Although there are different materials for denture base fabrication, PMMA (polymethyl methacrylate) remains the primary choice due to its aesthetics, ease of processing, cost-effectiveness, and easiness to repair [3]. However, PMMA based materials also present shortcomings including the presence of residual monomer, the tendency to exacerbate allergies, variable mechanical properties resulting from the method of fabrication, and shrinkage during setting [4].

A common concern associated with PMMA denture-based prostheses is the fracture of the denture base or denture teeth, which can result from fabrication defects, improper design, inadequate fit, bruxism, and the relatively low fracture resistance of the acrylic resins [5]. Specifically, maxillary fractures result from a combination of fatigue due to occlusal forces and impact, such as accidental drops on hard surfaces, while around 80% of mandibular fractures are primarily caused by impact [6]. Regarding the location of the fractures, maxillary dentures are more susceptible to midline fractures [7–10].

To increase the fracture resistance and flexural strength of denture bases, different approaches are used, including impact-resistant polymers (by chemical modification) and reinforcement of the PMMA (nanoparticles/nanotubes, fibers, and metal reinforcements) [11–18]. The use of nanoparticles aims to improve the mechanical properties of the denture base polymer (specifically resin hardness); however, if the nanoparticles are non-homogeneously distributed and produce nanoparticles agglomerates, the toughness, flexural strength, and tensile strength are impaired given their uneven dispersion within the polymer matrix [12,13]. Furthermore, the properties of nanoparticle-reinforced polymers are influenced by factors such as nanoparticle geometry, form, orientation, surface treatment, and interfacial adhesion with the polymer matrix, which are difficult to control [14,15]. Additionally, variations in standardization, polymerization cycles, and manipulation methods can further impair the flexural strength of the nanoparticle-reinforced polymer [16].

Another option is reinforcing the PMMA with different fibers, such as nylon, polyethylene, polyamide, and glass fibers, which in the laboratory show enhanced flexural strength, impact strength, and fatigue resistance of the denture bases [17]. However, the literature indicates that this method is technique-sensitive and presents inconsistent values in its reinforcing effects [17].

Metal framework reinforcements can be added to complete or partial dentures, offering certain advantages such as light weight (compared with thick denture bases), increased patient comfort (due to minimal thickness of the metal framework and less invasion of the intraoral spaces), high strength, excellent biocompatibility, and increased fracture toughness [18,19]; in addition, metal frameworks can be casted or digitally manufactured [20]. In general, frameworks can be fabricated by casting methods that use different metallic alloys, including cobalt–chromium (Co–Cr), titanium (Ti), and gold (Au) [21].

Computer-aided design/computer-assisted manufacturing (CAD/CAM) can be used for the fabrication of dentures, denture bases, and denture teeth, offering advantages like simplified workflows, improved patient experiences, and improved mechanical properties as compared to the conventional fabrication methods [11]. Two CAD/CAM manufacturing methods are available for fabricating denture bases and denture teeth: 3-D printing and milling. When comparing printed versus milled CAD/CAM denture base materials, the 3D-printed materials show lower flexural strength [22]. CAD/CAM milled materials possess better mechanical properties as compared to printed and conventional materials because their fabrication results in less internal porosity, minimal free monomers, and higher density per volume area. However, they are exposed to the same risks of fatigue and impacts experienced by conventional denture materials [22].

Furthermore, CAD/CAM denture materials (3D-printed or milled) exposed to thermocycling have shown impaired hardness, reduced fracture strength, and changes in surface roughness, thus demonstrating that the material can degrade over time [23], and if the CAD/CAM materials are exposed to denture cleansers, hardness and fracture toughness also decrease [24].

Furthermore, in patients with implant overdentures, the thickness of the denture base over the implants is thinner, and fractures are most found in those areas [25]. Thus, the insertion of metal frameworks in the denture base could decrease the stress concentration around the portion of the denture base that surrounds the implant housing [25].

Recently, AvaDent Digital Denture Solutions (AvaDent®, Scottsdale, AZ, USA) developed a proprietary method to embed inside the CAD/CAM resin a titanium 3D-printed framework before the resin is processed and milled. This method combines the possibility of customizing the 3D-printed framework to almost any ridge configuration. While metal reinforcement is a recognized strategy for enhancing the fracture toughness of conventional denture bases [18–20], the benefits of integrating a 3D-printed metal framework into CAD/CAM denture base materials remain unknown. The present study aimed to evaluate the deformation under the three-point bending test and determine the elastic portion of the stress/strain curves and the fracture toughness of three denture base materials: 3D-printed denture base, milled denture base, and milled denture base with an embedded 3D-printed titanium framework.

2. Materials and Methods

2.1. Experimental Design

This was an experimental, exploratory in vitro study. The sample size was calculated using the Statsmodels library in Python (ChatGPT4.0). The calculations were based on a significance level of 0.05, a power of 70%, and an effect size of 0.35. The sample size was determined as n = 22 samples per group for three experimental groups (Printed, Milled, and Milled + Reinforcement) for a total of N = 66 samples (Figure 1).

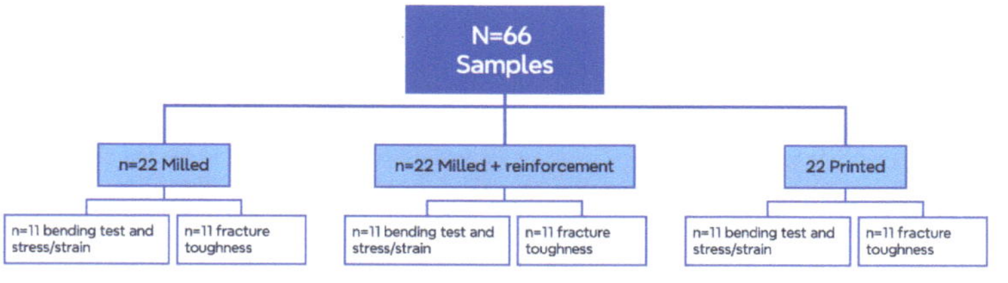

Figure 1. Scheme of the sample distribution. Three groups of 22 samples were fabricated by different CAD/CAM methods. Two experiments were carried out: a three-point bending test and an impact fracture test.

2.2. Sample Design

A bar design was created using Tinker CAD (Autodesk Inc., San Rafael, CA, USA) with the following dimensions: 60 mm-length, 10 mm-width, and 4 mm-thickness. The CAD file was exported as standard tessellation language (STL) to fabricate the three experimental groups of samples: Group 1: Milled; Group 2: Printed; and Group 3: Milled + Titanium framework reinforcement (Figure 2a,b).

To fabricate the milled samples (without and with reinforcement), the STL file was sent to Avadent (AvaDent®: Scottsdale, AZ, USA), who applied our design to mill the samples using their proprietary technology. A brief description was provided by the manufacturer as follows: a 3D-printed framework was designed to the desired geometry and fabricated by the laser printing of titanium powders. Afterward, the framework was embedded into liquid resin, and the resin was processed by heat and pressure, which resulted in the incorporation of the framework into the denture base material pucks. Afterward, the samples were milled to their final dimensions. Meanwhile, the printed samples were fabricated on-site using a Form3 3D-printer (Formlabs, Somerville, MA, USA) using OP (Original Pink) denture base material (Ref. PKG-RS-F2-DB) from Formlabs (Formlabs, Somerville, MA, USA). The samples were printed with a horizontal orientation of 0° parallel to the printing surface. After printing, the supports were removed and the samples were washed in isopropyl alcohol for 15 min (Form Wash, Formlabs, Somerville, MA, USA), and post-cured with UV light at 45 °C for 30 min (Form Cure, Formlabs, Somerville, MA, USA). The samples were maintained in a controlled environment at 21 degrees Celsius with a relative humidity of 30%. To preserve the materials' maximum strength prior to any aging or thermocycling, no immersion in water or conditioning was performed before the mechanical tests.

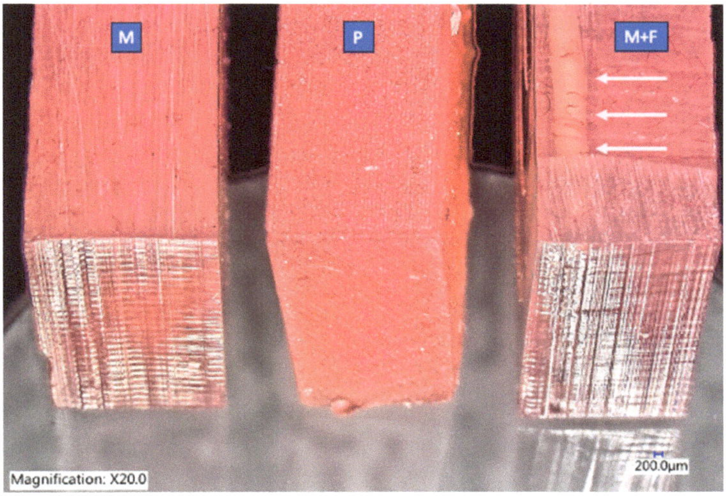

(**a**)

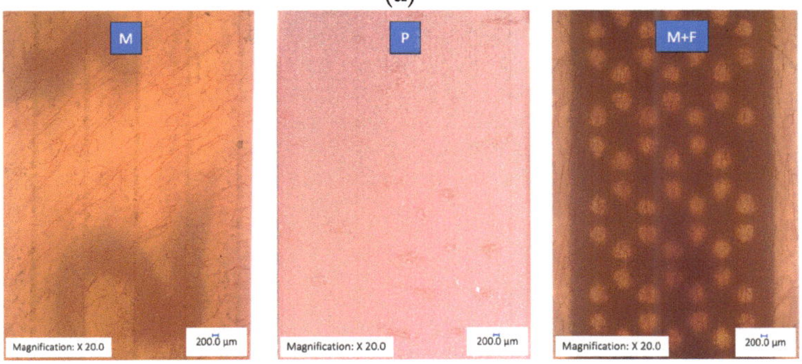

(**b**)

Figure 2. (**a**) Image showing samples of one bar from each experimental group: Milled (M), Printed (P), and Milled + Framework (M + F). The white arrows point to the location of the framework toward the left side of the M + F bar. The photos were obtained using the digital microscope Keyence VHX-6000, Keyence, Itasca, MN, USA. The magnification is 20×. (**b**) Image composition showing one sample of each experimental group observed under a transmitted light microscope, Milled (M), Printed (P), and Milled + Framework (M + F). The milled samples under transmitted light look orange and possess more characterization, including the simulated blood vessels, and the M + F shows the framework. The printed sample is pick and plain without color characterization. The photos were obtained using the digital microscope Keyence VHX-6000, Keyence, Itasca, MN, USA. Magnification 20×.

2.3. Deformation (Displacement) and Stress/Strain Curves within the Elastic Area

To evaluate the amount of deformation (displacement) under a standardized compressive force, a three point-bending test was completed using a Dynamic Mechanical Analysis (DMA-850) from TA Instruments (New Castle, DE, USA). A strain ramp from 0.1% as a constant rate was applied until the axial force reached the instrument limits of 18 N. The test was completed at a temperature of 37 °C.

The deformation (displacement under the compressive force) was measured in microns. In addition, the stress/strain curves for the elastic portion were recorded for all the groups. Each group consisted of 11 samples, for a total of 33 samples. One sample from each group was used for calibration purposes. The calibration sample was set in the testing area, and a repeated axial force of 18 N was applied to verify the reliability of the 0.1% strain ramp.

2.4. Impact Fracture Test

After the bending tests and stress/strain curves were completed, the remaining 33 samples were tested using a Tinius Olsen IT-503/504 impact tester machine (Tinius Olsen Testing Machine Co., Horsham, PA, USA) equipped with a 5.5 J pendulum. Un-notched impact tests were performed on all the samples to evaluate the energy required to fracture them. Initially, calibration was necessary to ensure the stabilization of the samples in the sample holder and to verify that the centers of the samples were aligned with the tip of the pendulum. The center was identified using a digital caliper, and the location was marked with a pen. Calibration confirmed that 30 mm was the center of the samples, coinciding with the pendulum tip. Finally, impact tests were conducted using samples with the following dimensions: width 10 mm, thickness 4 mm, and length 60 mm. The impact fracture values were recorded in kJ/m^2 (kilojoules per square meter of cross-section)

2.5. Fracture Analysis

To evaluate qualitatively the fracture characteristics of the samples, a digital microscope (Keyence VHX-6000, Keyence, Itasca, MN, USA) and a 3D-laser confocal microscope (Keyence VK-250, Keyence, Itasca, MN, USA) were used. A fractography analysis was completed, including the analysis of the impact zone, middle zone, and the side opposite to the impact, which were evaluated with the digital microscope at different magnifications. Three different types of fractures (clean, shattered, bent) were observed. A clean fracture resulted in two fragments that could be matched. A shattered fracture resulted in multiple fragments that were impossible to match. A bent fracture resulted in two fragments still connected by the framework.

3. Results

3.1. Stress/Strain Curves within the Elastic Area

A maximum standardized force of 18 N was applied to all the samples. The three-point bending test showed higher elastic deformation for the printed group as compared to the other groups (milled and milled with titanium reinforcement). The lowest deformation occurred in the metal-reinforced milled samples.

Figure 3a shows ten stress/strain curves obtained with the three-point bending test for the printed (P) group. Initially, the stress increased linearly with the strain, indicating elastic behavior where the material returns to its original shape when the stress is removed. The slopes of the curves appear smaller than the milled and milled and reinforced samples. Some samples of the printed group showed outlier behavior.

Figure 2b shows ten stress/strain curves obtained with the three-point bending test for the milled (M) group. The curves were linear, demonstrating an elastic behavior, and the slopes were higher than the printed group.

Figure 3c shows ten stress/strain curves obtained with the three-point bending test for the milled and reinforced group (M + F). The lowest strain was observed in this group, in addition, indicating a stiffer group.

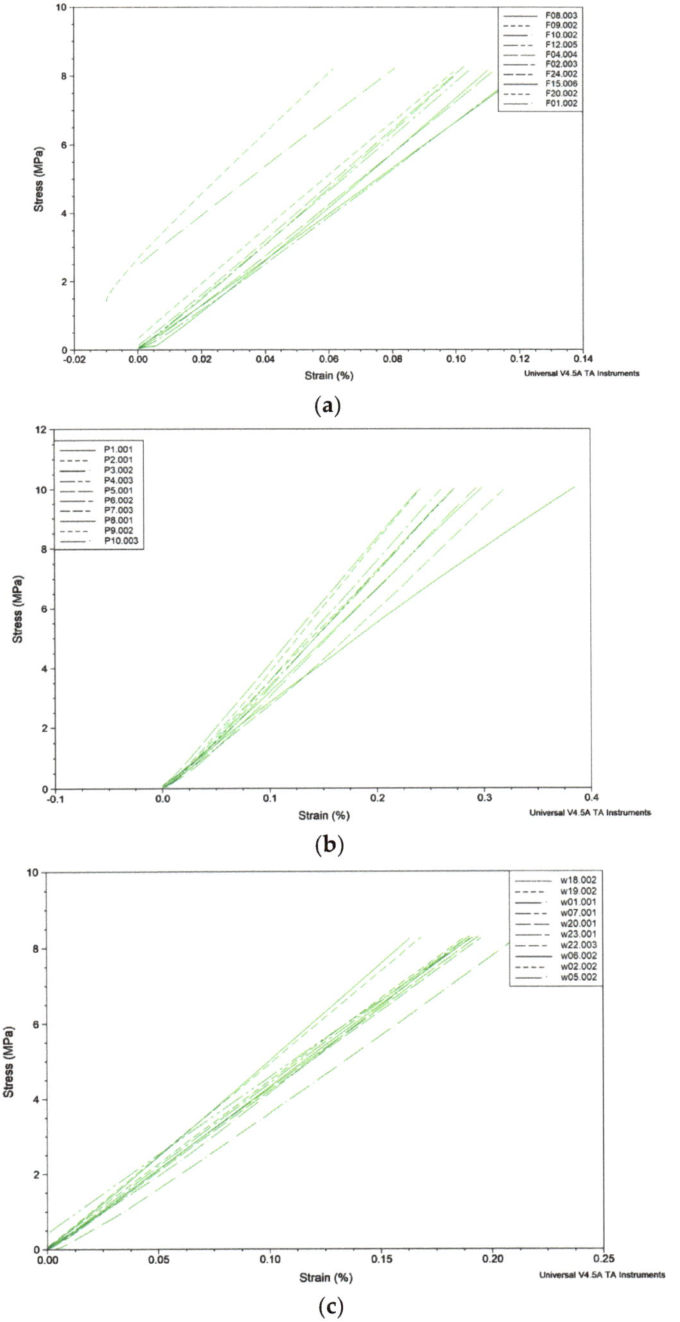

Figure 3. (**a**) Stress/strain curves for 10 printed samples. (**b**) Stress/strain curves for 10 milled samples. (**c**) Stress/strain cures for 10 milled and reinforced samples.

3.2. *Deformation*

All the samples experienced some elastic deformation under vertical load (displacement). The printed group suffered more elastic deformation than the other groups (milled and milled with reinforcement). Typically, the printed samples suffered elastic deformations in the range of 160 µm to 300 µm. The milled samples showed elastic deformations in the range of 75 µm to 140 µm. The smallest deformation was observed in the milled samples with reinforcement, with a range of 40 µm to 90 µm. Furthermore, the values were highly variable for the printed group and homogeneous for the milled groups (Figure 4 and Table 1).

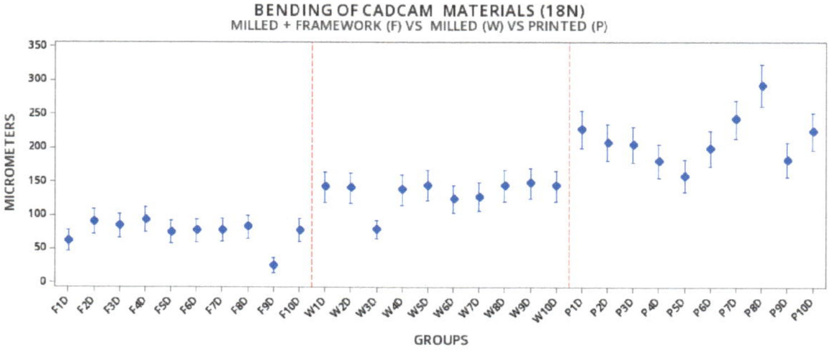

Figure 4. Deformation during the three-point bending test was present in all the groups. The displacement is measured in micrometers, and ten samples are measured per group: P (Printed), W (milled without reinforcement), F (milled with framework), and P (212 µm) groups. The vertical dotted lines separate each group (F, W, and P).

Table 1. Elastic deformation for the three groups, including mean, standard deviation, and median distributions. Additional details and information are included in Supplementary Materials.

Factor	N	Mean	StDev
Printed	10	212.0	129.03
Milled	10	133.01	85.12
Milled + reinforcement	10	74.48	47.57

Statistical Comparisons of the Elastic Deformation

The statistical comparisons showed that milled-with-reinforcement materials were superior as compared to the milled and printed materials and confirmed that the milled material is superior to the printed material (Table 2).

Table 2. Statistical comparisons. Differences between means and *p* values.

Comparisons between Groups	Difference of Means	95% CI	Adjusted *p*-Value
Milled Vs Printed	−79.0	(−103.9, −54.0)	0.001
Milled + reinforced Vs Printed	−137.5	(−162.5, −112.6)	0.001
Milled + reinforced Vs Milled	−58.5	(−83.5, −33.6)	0.002

3.3. *Fracture Toughness Analysis*

The impact fracture test showed higher fracture toughness for the milled samples reinforced with metallic frameworks, followed by the milled samples. The lowest values were observed in the printed group (Table 3).

Table 3. Descriptive statistics of the fracture toughness of CAD/CAM denture base materials: kJ/m^2 (Kilojoules/sectional area).

Groups	N	Mean	StDev
Milled kJ/m^2	10	8.634	1.225
Milled + reinforcement kJ/m^2	10	15.203	2.244
Printed kJ/m^2	10	6.304	2.600

Statistical Comparisons of the Fracture Toughness of CAD/CAM Materials

Statistical analysis showed higher fracture toughness in the milled and reinforced group as compared to the other groups (milled and printed) (Table 4).

Table 4. Multiple group comparisons: Tukey post-test.

Difference of Levels	Difference of Means	95% CI	Adjusted *p*-Value
Milled + reinforcement Vs Milled	6.569	(4.968, 8.170)	0.001
Printed Vs Milled	−2.330	(−3.931, −0.729)	0.003
Printed Vs Milled + reinforcement	−8.899	(−10.499, −7.298)	0.001

3.4. *Fractographic Analysis*

The type of fracture, the fracture propagation characteristics, and the fracture lines were different in the milled and in the printed groups. Figure 5 shows samples of each group immediately after the impact test.

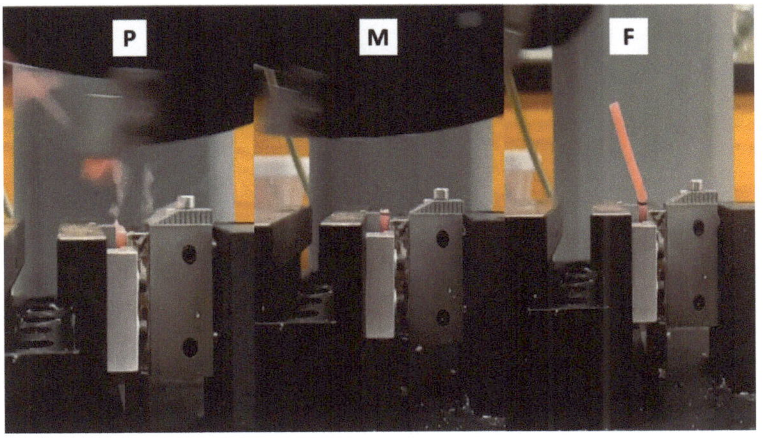

Figure 5. Representative photos of printed (P), milled (M), and reinforced samples (F). Different fractures occurred; the P group showed mainly shattered fractures, the M group showed clean fractures, and the F group showed a bent fracture (the framework keeps the segments united).

3.4.1. Fracture Analysis Printed Group

The samples exposed to the impact test fractured in multiple pieces, and the fragments showed multiple fracture lines extending from the impact areas in random directions along the samples. In addition, multiple sharp edges with different heights were appreciated. At the area of impact, a dark zone indicated the compression produced by the impact, and multiple microfractures extended toward the middle zone. The middle zone was less rough; different fracture faces could be observed that resulted in multiple chipped parts. The zone opposite to the impact also showed multiple facets in multiple directions; small and parallel microfracture lines were observed near the surface (Figure 6).

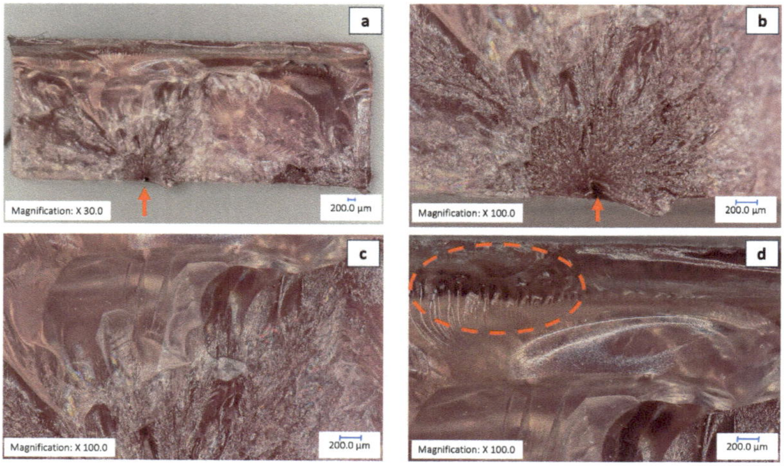

Figure 6. Printed samples: analysis of the fragments after the impact fracture test. (**a**) This image illustrates an overview of one fragment. Multiple shattered zones can be observed. The red arrow points to the area where the impact occurred. Magnification is 30×. (**b**) This image shows the impact zone (darker area) in a close view. Magnification is 100×. (**c**) This image shows the middle portion of the sample. Magnification is 100×. (**d**) This image demonstrates the opposite side to the impact where the microfractures extend. The red circle illustrates an area with microfractures. Magnification is 100×.

3.4.2. Fracture Analysis Milled Group

The milled group showed a different fracture pattern. First, the fractures followed the direction of the impact. Horizontal compression radial bands appeared, extending from the impact point toward the opposite side. The surfaces of the samples were less irregular than in the printed group. The samples showed fewer fracture facets. There was a color change (darkening) observed at the side opposite to the fracture. A closer view of compression bands showed increased diameters as they progressed to the opposite side. Some microfractures were observed running perpendicular to the compression bands. There was not a clear transition between the impact and the middle zone. The compression bands tended to disappear near the end of the middle zone. In some instances, microfractures were observed near the facets. Horizontal facets can be observed near the opposite side. There was a band of microfractures perpendicular to the facets. The surface was less irregular than at the middle and impact zones (Figure 7).

3.4.3. Fracture Analysis of the Milled-with-Framework Group

The milled group with reinforcement showed a pattern comparable to the milled group. However, not all samples showed fragment separation after the impact test. Smaller compression bands were observed, extending from the impact side along the sample. The impact occurred at the bottom of the samples, and the metal reinforcement could be seen at the side opposite to the impact (grey circles). In addition, an area compatible with an opaque or a coating was observed around the metal reinforcement. A close view of the impact zone showed the smaller size of compression bands as compared to the milled samples. The red fibers that simulate blood vessels were also observed. A close view of the impact zone showed the smaller size of the compression bands and their changing directions. The surface was slightly irregular. There was not a clear fracture orientation. Some fractures ran perpendicular to the metal reinforcement, then, when the fracture reached the reinforcement, stopped or disappeared. The layer covering the framework presented some microfractures in different directions (Figure 8).

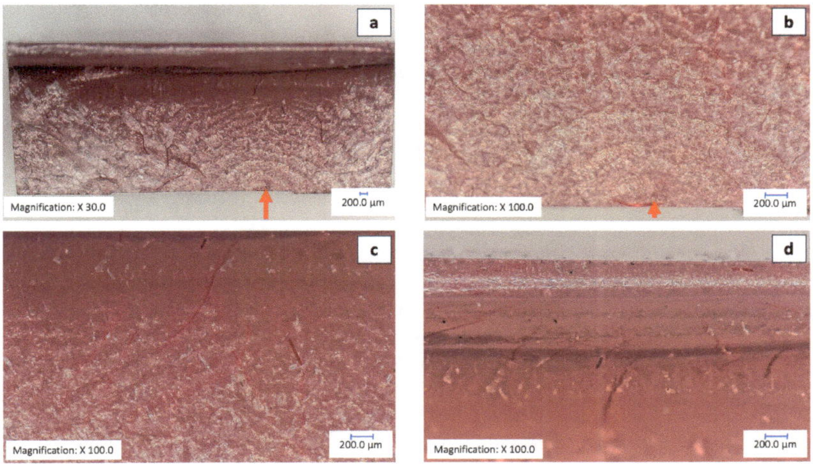

Figure 7. Milled samples without reinforcement. Analysis of the fragments after the impact fracture test. In general, the fracture is cleaner and not shattered. (**a**) Overview of one fragment. The red arrow points to the area where the impact occurred. Compression bands irradiate from the impact point toward the opposite side. Magnification is 30×. (**b**) This image shows a close view of the impact zone. Alternant clear and dark bands can be observed. Magnification is 100×. (**c**) This image shows the middle portion of the sample. Also, red filaments (simulating blood vessels) can be seen embedded in the sample. Magnification is 100×. (**d**) This image demonstrates the opposite side to the impact where some microfractures can also be seen. Magnification is 100×.

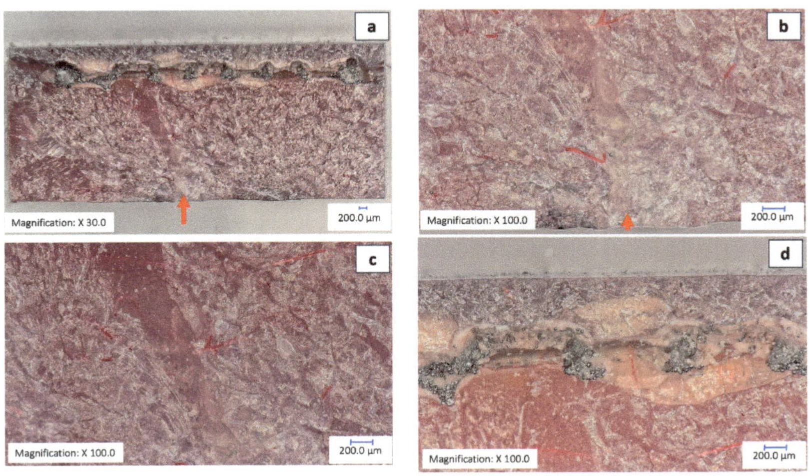

Figure 8. Milled and reinforced samples. Analysis of the fragments after the impact fracture test. (**a**), Overview. The red arrow at the base of the sample shows the impact zone. At the top of the sample the reinforcement can be observed. Some compression bands can be observed. Magnification is 30×. (**b**) The red arrow illustrates the impact zone in a close view. Magnification is 100× (**c**) This image shows the middle portion of the sample. Magnification is 100×. (**d**) This image demonstrates the side opposite to the impact and the framework section. Magnification is 100×.

4. Discussion

This study aimed to evaluate the elastic deformation, fracture toughness, and fracture characteristics following the impact test of three CAD/CAM denture base materials, including 3D-printed, milled, and milled with an embedded 3D-printed titanium framework. Our results showed that the milled samples suffered less deformation and possessed higher fracture toughness as compared to the 3D-printed samples. This is in agreement with the studies by Fouda et al. [26], who evaluated the flexural strength and hardness of conventional heat-polymerized acrylic resins, milled resins, and 3D-printed resins used for the fabrication of denture bases. Their results showed that milled resins possess higher flexural strength, elastic modulus, and hardness as compared to conventional resins and 3D-printed resins. Valenti et al. [27] evaluated the mechanical properties of 3D-printed prosthetic materials as compared to milled and conventional materials in in vitro studies. The materials included ceramics, polymers, and metals. Seventy-six studies were included, and their analysis concluded that 3D-printed polymeric materials possessed inferior flexural strength, fracture load, and hardness. Thus, their rigidity and fracture resistance does not support mastication forces for extended periods. Finally, Prpić et al. [28] found the lowest flexural strength in 3D-printed denture base samples in comparison to conventionally manufactured and milled CAD/CAM denture base materials.

If milled CAD/CAM denture base material is mechanically superior to conventional and 3D-printed denture base materials, why do we need to use reinforcements? Takahashi et al. [29] completed a comprehensive review of reinforcement in removable prosthodontics and its impact on the fracture and deformation of the prostheses and the quality of life of the patients who use them. Specifically, fracture and flexural strength and elastic modulus were compared in prostheses with and without reinforcement. Their results showed that metal reinforcements placed in thin and deformable areas effectively improved the mechanical properties of the prostheses and, indirectly, the patient's quality of life by reducing the maintenance and repair of the prostheses. In addition, any material (conventional, milled, and 3D-printed) exposed to the oral environment for enough time will experience a decrease in the original values of fracture strength, modulus, and hardness. Therefore, including a reinforcement will counterbalance for these changes [24,25].

Denture bases with implant attachments, including balls or locators, have been associated with increased deformation and higher stress around the attachments, indicating the need for reinforcement in the denture bases [30]. Finally, it seems that the incorporation of a reinforcement reduces and redistribute the strains on the supporting structures, reducing the incidence of fractures in implant overdentures [31].

The thickness of our samples, 4 mm, was selected for standardization based on different in vitro studies. These studies determined that a thickness of 4 mm exhibited higher fracture toughness as compared to thicknesses of 3 mm and 2 mm [32]. Furthermore, CAD/CAM-manufactured samples of different thicknesses were tested to determine the minimal thickness that can satisfactorily withstand mechanical loads. This study concluded that CAD/CAM denture base resins with a thickness of 2 mm do not exhibit better mechanical properties as compared to conventional resins. Therefore, reinforcement was recommended for both types of resins when thicknesses are lower than 4 mm [33]. Given that in the clinical settings patients often prefer a minimal thickness for the denture base, the benefits of embedding 3D-printed titanium reinforcement could be applied.

The elastic deformations experienced by the 3D-printed samples resulted in higher standard deviations. This outcome is attributed more to the intrinsic nature of the 3D-printed material than to experimental inconsistencies. Printed materials fabricated from liquid resins possess inherent defects created during the printing process, such as porosity, layer separation, bubbling, and gaps, all of which adversely affect mechanical strength. In contrast, the pucks used for milling denture base materials are fabricated under standardized conditions of pressure and temperature, resulting in a denser structure with minimal porosity and improved mechanical properties [34–36].

A striking finding of the present study was that the fractures differed between samples. For example, the 3D-printed samples suffered from shattered fractures and were more brittle. Meanwhile, the milled samples showed mostly clean fractures, resulting in two or three pieces, and the milled and reinforced samples showed fractures but not the separation of the segments.

The limitations of this study include the absence of a control group fabricated by conventional methods and the omission of evaluation of other thicknesses of denture base materials. However, we used the printed group as a control to the milled and the milled-with-reinforced samples. To increase the validity and reproducibility, we followed ISO standards for the samples used in the mechanical tests. This allowed the comparisons to be centered on the fabrication method.

4.1. Practical Implications

The results of the present experimental study demonstrate that 3D-printed samples experience higher elastic deformation. Thus, 3D-printed complete dentures will experience deformation (flexing) and unstable occlusion under increased axial and non-axial loads (like these produced through clenching and bruxism). Furthermore, the lack of rigidity of 3D-printed denture bases can result in inefficient or reduced bite force, thus reducing the chewing efficiency.

In contrast, milled samples with reinforcement, as well as milled samples, exhibit lower elastic deformation, potentially resulting in more stable occlusion and higher masticatory efficiency. The resistance to impact fracture is superior in milled samples with a titanium framework as compared to milled samples without titanium frameworks and printed samples.

Therefore, based on these results, it is recommended to add a reinforcement method to any CAD/CAM denture base materials. This is particularly important for 3D-printed denture base materials. The fracture type of the printed samples was characterized by shattering, with fractures occurring in multiple directions, multiple fracture facets, material chipping, and material loss. Clinically, this implies that in the event of an impact, a 3D-printed denture may break into multiple pieces, which could preclude repair.

4.2. Opportunities for Research

Several aspects require further exploration, including the effect of CAD/CAM material thickness on reinforced versus non-reinforced CAD/CAM denture bases and the mechanical strength of milled denture bases incorporating 3D-printed frameworks with different thicknesses. It is also necessary to include a control group with conventionally fabricated denture bases for comparison with both the reinforced and non-reinforced CAD/CAM denture bases. Additionally, the effect of aging (thermocycling) on the mechanical properties of reinforced versus non-reinforced CAD/CAM denture base materials should be evaluated. Finally, it is important to investigate whether the reinforcement material leaches into the oral environment

5. Conclusions

Within the limitations of this experimental study, the null hypothesis can be rejected. This study found that milled denture base material, printed denture base material, and milled denture base material with an embedded metallic framework exhibit different flexural strength and impact fracture toughness.

Thus, the following can be concluded:

First, milled samples with an embedded 3D-printed titanium framework demonstrate higher resistance to impact as compared to milled samples without a framework and printed samples.

Second, milled samples with an embedded 3D-printed titanium framework show increased flexural strength and lower elastic deformation as compared to milled samples without a framework and printed samples.

Third, printed denture base material exhibits the lowest resistance to impact and the lowest flexural strength as compared to milled denture base materials with and without a framework.

Supplementary Materials: The following supporting information can be downloaded at: https://www.mdpi.com/article/10.3390/prosthesis6040053/s1.

Author Contributions: Conceptualization, R.D.-R.; data curation, I.B., A.A. and Y.F.; formal analysis, I.B. and A.A.; investigation, R.D.-R., I.B., A.A. and Y.F.; methodology, R.D.-R., G.R. and M.R.; resources, R.D.-R., G.R. and M.R.; software, R.D.-R. and Y.F.; supervision, R.D.-R.; validation, R.D.-R., I.B. and A.A.; visualization, R.D.-R., I.B. and A.A.; writing—original draft, I.B. and A.A.; writing—review and editing, R.D.-R., I.B., A.A., Y.F., G.R. and M.R. All authors have read and agreed to the published version of the manuscript.

Funding: This research received no external funding.

Institutional Review Board Statement: Not applicable.

Informed Consent Statement: Not applicable.

Data Availability Statement: Raw data are available in Supplementary Materials and further data will be made available upon request to the authors.

Acknowledgments: The authors acknowledgment the technical support of Axel Calderon from the Center for Implant Digital Technology (CIDT) of the School of Dental Medicine at Stony Brook University. The authors also acknowledge the access to the ThINC facilities at the Advanced Energy Center, Stony Brook University, supported by Myriam Rafailovich.

Conflicts of Interest: The authors declare no conflicts of interest.

References

1. Xie, Q.; Ding, T.; Yang, G. Rehabilitation of Oral Function with Removable Dentures--Still an Option? *J. Oral. Rehabil.* **2015**, *42*, 234–242. [CrossRef] [PubMed]
2. Salloum, A.M. Effect of Adding Basalt Powder on Flexural Strength of a Denture Base Acrylic Resin. *BJSTR* **2019**, *18*, 13205–13210. [CrossRef]
3. Rickman, L.J.; Padipatvuthikul, P.; Satterthwaite, J.D. Contemporary Denture Base Resins: Part 1. *Dent. Update* **2012**, *39*, 25–30. [CrossRef] [PubMed]
4. Zafar, M.S. Prosthodontic Applications of Polymethyl Methacrylate (PMMA): An Update. *Polymers* **2020**, *12*, 2299. [CrossRef] [PubMed]
5. Beyli, M.S.; von Fraunhofer, J.A. An Analysis of Causes of Fracture of Acrylic Resin Dentures. *J. Prosthet. Dent.* **1981**, *46*, 238–241. [CrossRef] [PubMed]
6. Nandal, S.; Ghalaut, P.; Shekhawat, H.; Gulati, M. New Era in Denture Base Resins: A Review. *Dent. J. Adv. Stud.* **2013**, *01*, 136–143. [CrossRef]
7. Prombonas, A.E.; Vlissidis, D.S. Comparison of the Midline Stress Fields in Maxillary and Mandibular Complete Dentures: A Pilot Study. *J. Prosthet. Dent.* **2006**, *95*, 63–70. [CrossRef] [PubMed]
8. Cilingir, A.; Bilhan, H.; Baysal, G.; Sunbuloglu, E.; Bozdag, E. The Impact of Frenulum Height on Strains in Maxillary Denture Bases. *J. Adv. Prosthodont.* **2013**, *5*, 409–415. [CrossRef] [PubMed]
9. Polyzois, G.L.; Tarantili, P.A.; Frangou, M.J.; Andreopoulos, A.G. Fracture Force, Deflection at Fracture, and Toughness of Repaired Denture Resin Subjected to Microwave Polymerization or Reinforced with Wire or Glass Fiber. *J. Prosthet. Dent.* **2001**, *86*, 613–619. [CrossRef]
10. Wiskott, H.W.; Nicholls, J.I.; Belser, U.C. Stress Fatigue: Basic Principles and Prosthodontic Implications. *Int. J. Prosthodont.* **1995**, *8*, 105–116.
11. Steinmassl, P.-A.; Wiedemair, V.; Huck, C.; Klaunzer, F.; Steinmassl, O.; Grunert, I.; Dumfahrt, H. Do CAD/CAM Dentures Really Release Less Monomer than Conventional Dentures? *Clin. Oral. Investig.* **2017**, *21*, 1697–1705. [CrossRef] [PubMed]
12. Aldegheishem, A.; AlDeeb, M.; Al-Ahdal, K.; Helmi, M.; Alsagob, E.I. Influence of Reinforcing Agents on the Mechanical Properties of Denture Base Resin: A Systematic Review. *Polymers* **2021**, *13*, 3083. [CrossRef] [PubMed]
13. Gad, M.M.; Abualsaud, R.; Al-Thobity, A.M.; Baba, N.Z.; Al-Harbi, F.A. Influence of Addition of Different Nanoparticles on the Surface Properties of Poly(Methylmethacrylate) Denture Base Material. *J. Prosthodont.* **2020**, *29*, 422–428. [CrossRef] [PubMed]
14. Abdulrazzaq Naji, S.; Behroozibakhsh, M.; Jafarzadeh Kashi, T.S.; Eslami, H.; Masaeli, R.; Mahgoli, H.; Tahriri, M.; Ghavvami Lahiji, M.; Rakhshan, V. Effects of Incorporation of 2.5 and 5 Wt% TiO_2 Nanotubes on Fracture Toughness, Flexural Strength, and Microhardness of Denture Base Poly Methyl Methacrylate (PMMA). *J. Adv. Prosthodont.* **2018**, *10*, 113. [CrossRef] [PubMed]

15. Khan, A.A.; Fareed, M.A.; Alshehri, A.H.; Aldegheishem, A.; Alharthi, R.; Saadaldin, S.A.; Zafar, M.S. Mechanical Properties of the Modified Denture Base Materials and Polymerization Methods: A Systematic Review. *Int. J. Mol. Sci.* **2022**, *23*, 5737. [CrossRef] [PubMed]
16. Bangera, M.K.; Kotian, R.; Ravishankar, N. Effect of Titanium Dioxide Nanoparticle Reinforcement on Flexural Strength of Denture Base Resin: A Systematic Review and Meta-Analysis. *Jpn. Dent. Sci. Rev.* **2020**, *56*, 68–76. [CrossRef] [PubMed]
17. Alla, R.K.; Sajjan, S.; Alluri, V.R.; Ginjupalli, K.; Upadhya, N. Influence of Fiber Reinforcement on the Properties of Denture Base Resins. *JBNB* **2013**, *4*, 91–97. [CrossRef]
18. Oluwajana, F.; Walmsley, A.D. Titanium Alloy Removable Partial Denture Framework in a Patient with a Metal Allergy: A Case Study. *Br. Dent. J.* **2012**, *213*, 123–124. [CrossRef] [PubMed]
19. Mourouzis, P.; Pandoleon, P.; Tortopidis, D.; Tolidis, K. Clinical Evaluation of Removable Partial Dentures with Digitally Fabricated Metal Framework after 4 Years of Clinical Service. *J. Prosthodont.* **2024**, *33*, 5–11. [CrossRef]
20. Park, K.; Kang, N.-G.; Lee, J.-H.; Srinivasan, M. Removable Complete Denture with a Metal Base: Integration of Digital Design and Conventional Fabrication Techniques. *J. Esthet. Restor. Dent.* **2023**, *36*, 255–262. [CrossRef]
21. Peterson, L.J.; Ellis, E.; Hupp, J.R.; Tucker, M.R. *Contemporary Oral and Maxillofacial Surgery*; Elsevier Health Sciences: Philadelphia, PA, USA, 2018.
22. de Oliveira Limírio, J.P.J.; de Luna Gomes, J.M.; Alves Rezende, M.C.R.; Lemos, C.A.A.; Rosa, C.D.D.R.D.; Pellizzer, E.P. Mechanical Properties of Polymethyl Methacrylate as a Denture Base: Conventional versus CAD-CAM Resin—A Systematic Review and Meta-Analysis of in Vitro Studies. *J. Prosthet. Dent.* **2022**, *128*, 1221–1229. [CrossRef]
23. Abdul-Monem, M.M.; Hanno, K.I. Effect of Thermocycling on Surface Topography and Fracture Toughness of Milled and Additively Manufactured Denture Base Materials: An in-Vitro Study. *BMC Oral Health* **2024**, *24*, 267. [CrossRef]
24. Hanno, K.I.; Abdul-Monem, M.M. Effect of Denture Cleansers on the Physical and Mechanical Properties of CAD-CAM Milled and 3D Printed Denture Base Materials: An in Vitro Study. *J. Prosthet. Dent.* **2023**, *130*, 798.e1–798.e8. [CrossRef]
25. Amaral, C.F.; Gomes, R.S.; Rodrigues Garcia, R.C.M.; Del Bel Cury, A.A. Stress Distribution of Single-Implant-Retained Overdenture Reinforced with a Framework: A Finite Element Analysis Study. *J. Prosthet. Dent.* **2018**, *119*, 791–796. [CrossRef]
26. Fouda, S.M.; Gad, M.M.; Abualsaud, R.; Ellakany, P.; AlRumaih, H.S.; Khan, S.Q.; Akhtar, S.; Al-Qarni, F.D.; Al-Harbi, F.A. Flexural Properties and Hardness of CAD-CAM Denture Base Materials. *J. Prosthodont.* **2023**, *32*, 318–324. [CrossRef]
27. Valenti, C.; Isabella Federici, M.; Masciotti, F.; Marinucci, L.; Xhimitiku, I.; Cianetti, S.; Pagano, S. Mechanical Properties of 3D-Printed Prosthetic Materials Compared with Milled and Conventional Processing: A Systematic Review and Meta-Analysis of in Vitro Studies. *J. Prosthet. Dent.* **2022**. *Online ahead of print*. [CrossRef]
28. Prpić, V.; Schauperl, Z.; Ćatić, A.; Dulčić, N.; Čimić, S. Comparison of Mechanical Properties of 3D-Printed, CAD/CAM, and Conventional Denture Base Materials. *J. Prosthodont.* **2020**, *29*, 524–528. [CrossRef]
29. Takahashi, T.; Gonda, T.; Mizuno, Y.; Fujinami, Y.; Maeda, Y. Reinforcement in Removable Prosthodontics: A Literature Review. *J. Oral Rehabil.* **2017**, *44*, 133–143. [CrossRef]
30. ELsyad, M.A.; Errabti, H.M.; Mustafa, A.Z. Mandibular Denture Base Deformation with Locator and Ball Attachments of Implant-Retained Overdentures. *J. Prosthodont.* **2016**, *25*, 656–664. [CrossRef]
31. Gibreel, M.F.; Khalifa, A.; Said, M.M.; Mahanna, F.; El-Amier, N.; Närhi, T.O.; Perea-Lowery, L.; Vallittu, P.K. Biomechanical Aspects of Reinforced Implant Overdentures: A Systematic Review. *J. Mech. Behav. Biomed. Mater.* **2019**, *91*, 202–211. [CrossRef]
32. Tokgoz, S.; Ozdiler, A.; Gencel, B.; Bozdag, E.; Isık-Ozkol, G. Effects of Denture Base Thicknesses and Reinforcement on Fracture Strength in Mandibular Implant Overdenture with Bar Attachment: Under Various Acrylic Resin Types. *Eur. J. Dent.* **2019**, *13*, 64–68. [CrossRef] [PubMed]
33. Steinmassl, O.; Offermanns, V.; Stöckl, W.; Dumfahrt, H.; Grunert, I.; Steinmassl, P.-A. In Vitro Analysis of the Fracture Resistance of CAD/CAM Denture Base Resins. *Materials* **2018**, *11*, 401. [CrossRef] [PubMed]
34. Galati, M.; Minetola, P. On the measure of the aesthetic quality of 3D printed plastic parts. *Int. J. Interact. Des. Manuf.* **2020**, *14*, 381–392. [CrossRef]
35. Erokhin, K.; Naumov, S.; Ananikov, V. Defects in 3D Printing and Strategies to Enhance Quality of FFF Additive Manufacturing. A Review. 2023. Available online: https://chemrxiv.org/engage/chemrxiv/article-details/6516b0aba69febde9eea81b1 (accessed on 1 June 2024).
36. Holzmond, O.; Li, X. In situ real time defect detection of 3D printed parts. *Addit. Manuf.* **2017**, *17*, 135–142. [CrossRef]

prosthesis MDPI

Article

Citric Acid-Based Solutions as Decontaminant Mouthwash in Titanium and Dental Prostheses Materials in Implantoplasty Processes

Pilar Fernández-Garrido [1], Pedro Fernández-Dominguez [2], Laura Fernández De La Fuente [1], Barbara Manso De Gustin [1], José Felipe Varona [2], Begoña M. Bosch [3], Javier Gil [3,*] and Manuel Fernández-Domínguez [2]

1 Department of Translational Medicine, CEU San Pablo University, Urbanización Montepríncipe, 28925 Madrid, Spain; pilarfergarrido@gmail.com (P.F.-G.); laurafernandezdelafuente@gmail.com (L.F.D.L.F.); barbaramdeg@gmail.com (B.M.D.G.)
2 Facultad de Odontología, Universidad Camilo José Cela, C/Castillo de Alarcón, 49. Urb. Villafranca del Castillo, 28691 Villanueva de la Cañada, Spain; pfernandezdominguez@gmail.com (P.F.-D.); jfvarona@hmhospitales.com (J.F.V.); clinferfun@yahoo.es (M.F.-D.)
3 Bioengineering Institute of Technology, Facultad de Medicina y Ciencias de la Salud, Universitat Internacional de Catalunya, Josep Trueta S/n, 08195 Sant Cugat del Vallés, Spain; bbosch@uic.es
* Correspondence: xavier.gil@uic.es

Abstract: The machining of implants and parts for dental prostheses to eliminate biofilm in the implantoplasty process causes a loss of mechanical properties and also characteristics of the surfaces, making tissue regeneration difficult. In the present work, treatments consisting of elements that can reduce infection, such as citric acid and magnesium, together with elements that can improve cell adhesion and proliferation, such as collagen, are proposed for implant–crown assembly. Titanium, zirconia, composite (PMMA + feldspar) and cobalt–chromium discs were immersed in four different solutions: 25% citric acid, 25% citric acid with the addition of collagen 0.25 g/L, 25% citric acid with the addition of 0.50 g/L and the latter with the addition of 1% Mg $(NO_3)_2$. After immersion was applied for 2 and 10 min, the roughness was determined by interferometric microscopy and the contact angle (CA) was evaluated. Human fibroblastic and osteoblastic line cells (HFFs and SaOS-2) were used to determine cell viability and proliferation capacity. Cell binding and cytotoxicity were determined by resazurin sodium salt assay (Alamar Blue) and cell morphology by confocal assay (immunofluorescence F-actin (phalloidin)) after 3 days of incubation. For the evaluation of bacterial activity, the bacterial strains *Sptreptococcus gordonii* (Gram+) and *Pseudomonas aeruginosa* (Gram−) were used. The antibacterial properties of the proposed treatments were determined by means of the resazurin sodium salt (Alamar Blue) assay after 1 day of incubation. The treatments considerably decreased the contact angle of the treated samples with respect to the control samples. The treatments endowed the surfaces of the samples with a hydrophilic/super-hydrophilic character. The combination of elements proposed for this study provided cell viability greater than 70%; considering the absence of cytotoxicity, it therefore promotes the adhesion and proliferation of fibroblasts and osteoblasts. In addition, it also endows the surface with antibacterial characteristics against from Gram+ and Gram− bacteria without damaging the cells. These results show that this mouthwash can be useful in oral applications to produce a new passivation layer that favors the hydrophilicity of the surface and promotes cellular activity for the formation of fibroblasts and osteoblasts, as well as showing bactericidal activity.

Keywords: citric acid; fibroblasts; osteoblasts; wettability; bactericide effect; mouthwash

1. Introduction

Dental implants have emerged as the preferred solution for restoring both aesthetics and function lost due to missing teeth, given their high success rates [1–3]. However, alongside the rising popularity of dental implants, there has been a corresponding increase in the occurrence of biological complications, in particular, peri-implantitis, a destructive biofilm-mediated inflammatory condition characterized by inflammation in the peri-implant connective tissue and progressive loss of supporting bone [4]. Currently, the main challenge in oral implantology is the bacterial infection of dental implants causing diseases such as periodontitis and peri-implantitis. This issue results in 24% of dental implants requiring revision within 10 years of implantation [4,5].

Given the infectious nature of this condition, the primary therapeutic goal is to modify the environment to promote an aerobic ecosystem, fostering health and stability. In order to accomplish this, it is vital to disrupt biofilm formation on the surface of the affected implant and to address any local factors that may have contributed to the onset and progression of the disease [5]. For this purpose, different surgical and non-surgical measures have been proposed.

One of the solutions is to replace the infected dental implant with a new one. However, in some cases, a calcium phosphate filling should be produced for bone regeneration, and once sufficient bone formation is achieved, the new implant should be placed [6,7]. In other situations, the removal of the infected implant does not allow for the placement of a new dental implant because there is not enough space [8,9]. This fact means that the clinician must extract neighboring teeth to achieve the placement of a new implant. In some cases, narrow dental implants can be placed to avoid the removal of a healthy tooth [9]. As can be seen, the techniques are complicated, expensive for the patient, and have long treatment times [10–13].

This fact makes implantoplasty, which consists of the mechanization of the dental implant and part of its connection with the prosthesis to eliminate biofilm, a viable treatment option. This approach avoids clinical complications but causes a loss in the mechanical properties of the implant and the prosthesis. In the case of metal components, it reduces corrosion resistance and increases the release of ions into the physiological environment [14–19]. Figure 1 shows the surfaces of dental implants and prostheses machined in order to remove biofilm.

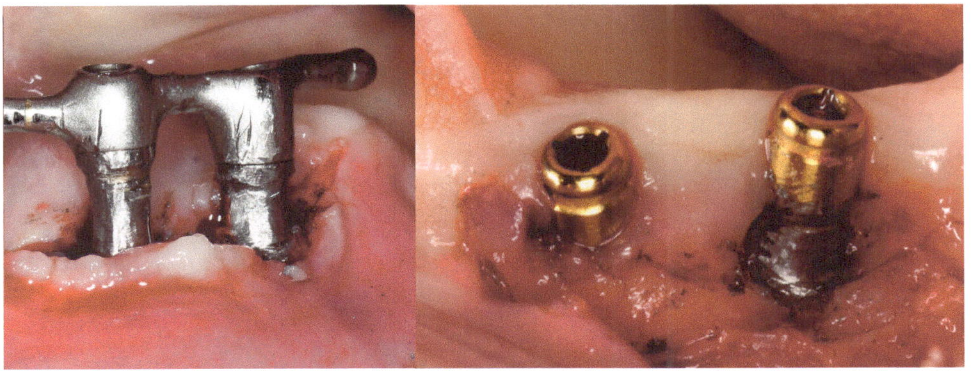

Figure 1. Dental implants and abutments treated by implantoplasty. Metallic particles in the tissue and grinding marks on the titanium surface produced by machining of the samples can be observed. These marks produce an increase in roughness.

Kotsakis et al. [20] demonstrated that the machining of titanium causes a reduction in oxygen concentration on the machined surface produced by the inflammation process, which degrades the protective titanium oxide layer (TiO_2) that acts as a passivation layer. Due to the lack of oxygen, pure titanium is formed and oxidized in a mixed way without reaching the stoichiometry of TiO_2. These so-called mixed oxides have lower corrosion resistance and exhibit toxicity that hinders cellular activities, both in fibroblastic and osteoblastic functions [21,22].

Citric acid treatments, being weak acid treatments, do not chemically attack the titanium as more aggressive agents such as hypochlorous acid do. It has been observed that treatments with hypochlorous acid and hydrogen peroxide eliminate bacteria very effectively but cause a very significant increase in roughness that favors the rapid recolonization of bacteria. Other treatments, such as ozone gas, have been used but large doses cannot be given due to the risk of soft tissue necrosis. Similarly, strong acids are not suitable for oral treatment, as this acidic composition burns soft and hard tissues. For example, treatments with hydrofluoric acid react with apatites, generating fluorapatite crystals, which can cause severe pain to the patient [23,24]. These limitations have led to the consideration of citric acid as the best candidate to obtain mouthwashes for clinical application. In this research, citric acid solutions with weak pH are investigated in order to obtain a mouthwash solution. Beyond its bactericidal effect due to citric acid, it generates a titanium oxide layer approximately of 7 nanometers thick, which protects the dental implant from chemical degradation and electrochemical corrosion. This layer is homogeneous, compact and stoichiometric, leading to an increase in corrosion resistance and reduced ion release, as demonstrated by different authors [25,26]. Citric acid treatment favors the bactericidal character for different types of bacteria, both Gram-positive and Gram-negative. In order to improve cellular activity, different concentrations of collagen and divalent magnesium salts were added. These additions have been proven to increase cell adhesion, proliferation and differentiation [27–29]. The main goal of this research is to develop a solution that can be applied to patients undergoing implantoplasty, in order to mitigate or reduce the associated problems of this technique.

The hypothesis of this research is that citric acid solutions with collagen and magnesium cations will promote osteoblastic and fibroblastic cell behavior, presenting a bactericidal character in Gram-positive and Gram-negative bacteria.

2. Materials and Methods

2.1. Materials

For this study, 320 discs of 4 different materials were used in a dental implant system. The experimental design allowed for determining the number of samples, indicating 15 samples for each biological and microbiological test. The roughness and wettability studies did not affect the samples. The total was 60 but we incorporated 5 more per experiment for possible unforeseen events. This meant that there were 80 discs for each material:

- Ti: commercially pure Titanium (Ti), grade 3.
- Zr: zirconia (ZrO_2 with 2.5% in weight of yttria) (Y_2O_3).
- Composite formed by polymethyl methacrylate (PMMA) with feldspar $CaAl_2Si_2O_8$ with 38% in volume.
- CrCo: the chromium content was 30 wt%, the Mo content was 7 wt% and the W content was 0.1 wt%; cobalt was the balance.

Eighty discs for each material were supplied by the company SOADCO S.L. (SOADCO, Escaldes Engordany, Andorra).

Implantoplasty was carried out by the same researcher (JG) using the drilling protocol. To achieve this, a GENTLEsilence LUX 8000B turbine (KaVo Dental GmbH, Biberach an der Riß, Germany) under constant irrigation was used; the surface was sequentially modified with a fine-grained tungsten carbide bur (ref. number H379.314. 014 KOMET; GmbH & Co. KG, Lemgo, Germany). Tungsten carbide burs are the main tool for the initial shaping of implant prostheses, with the bur size adjusted to the specific area of treatment. Generally, larger burs are used on the vestibular and palatal sides, while smaller-diameter burs are employed in limited-access areas or interproximal spaces. These burs effectively eliminate implant threads, providing a smooth surface texture. Moreover, to create a refined finish, a series of polishing drills (from coarse- to fine-grained) are used. The references of the silicon carbide polishers are (order no. 9608.314.030 KOMET; GmbH & Co. KG, Lemgo,

Germany) for the coarse-grained and (order no. 9618.314.030 KOMET; GmbH & Co. KG, Lemgo, Germany) for the fine-grained [30–32]. The disks were sterilized at a temperature of 121 °C for 30 min.

The immersions were performed in four different dissolutions of chemical compositions, as shown in Table 1. The concentration proposed in this study were chosen because of different previous studies [25,26] that refer to passivation processes with citric acid and that do not cause irritation to soft tissues when applied. Also, the references on the effect of collagen and divalent cations, such as magnesium or calcium, made us introduce these elements in the formulation [23–26,28]. The different materials are immersed in the different solutions for 2 and 10 min. These times have been suggested by clinicians estimating between 2 and 10 min. These values are the most common in treatments with antibiotic agents and ozone treatments, among others, that are performed on the patient. Ten minutes is considered the maximum recommended value for the well-being of the patient.

Table 1. Citric-based dissolutions.

Dissolution	Chemical Composition
25% Citric acid (AC)	Citric acid 25% in volume (v)
25% Citric acid + collagen 250 (AC 250)	Citric acid 25% (v) with 0.25 g collagen/L
25% Citric acid + collagen 500 (AC 500)	Citric acid 25% (v) with 0.50 g collagen/L
25% Citric acid + collagen 500 + 1% Mg (AC 500/Mg)	Citric acid 25% (v) with 0.50 g collagen/L and 10% $Mg(NO_3)_2 \cdot 6H_2O$

2.2. Roughness Analysis

The smooth and micro-roughened surfaces were analyzed using a white light interferometer microscope (Wyko NT9300 Optical Profiler, Veeco Instruments, New York, NY, USA) in vertical scanning interferometry mode. A minimum of three measurements were taken from three different samples of each series. Approximately 230 imaging frames were used, enabling rapid and highly accurate measurements of the grooves. Surface analysis covered areas of 127.7 × 95.8 μm for groove imaging and 63.1 × 47.3 μm for plain regions within the grooves. Data filtering and analysis were conducted with Wyko Vision 4.10 software (Veeco Instruments), with a Gaussian filter applied to remove curvature and tilt from every surface analysis. Sa (average roughness) was measured, which represents the arithmetic average of the absolute values of the surface deviations from the mean plane [33–35].

2.3. Wettability

The contact angle (CA) was determined to evaluate the surface wettability of the titanium with treatments except the control for 2 and 10 min of immersion, using 5 samples per material. The wettability measurements of the samples were measured using the contact angle system "OCA 15 plus" (Dataphysics Instrument Company, Filderstadt, Germany) and the results were analyzed with "SCA20" software 123.45 (Dataphysics Instrument Company, Filderstadt, Germany).

For droplet deposition, a 1 mL "Braun" syringe was employed in a droplet generation system with micrometer displacement control, allowing for a precise dosing volume of 2 μL at a rate of 1 μL/s. The liquid droplets were backlit with LEDs through ground glass and the contact angle was measured 5 s after placing the droplets on the surface. MiliQ water was used for contact angle measurement, which was conducted on both untreated and treated samples with a "Citric Acid 25% + Collagen 500" solution after 2 and 10 min of immersion [36].

2.4. Fibroblast Culture

The objective of fibroblast cultures is to indicate the degree of cytocompatibility and the ease of regenerating soft tissue at the bone–soft tissue interface. This fact is of great importance for the formation of a biological seal to prevent bacterial leakage. Human foreskin fibroblast (Millipore, Billerica, MA, USA) primary cells (HFFs) were cultured in phenol red-free Dulbecco's Minimum Essential Medium (DMEM; Invitrogen, Carlsbad, CA, USA) supplemented with 10% fetal bovine serum (FBS), L-glutamine (2 mM) and penicillin/streptomycin (50 U/mL and 50 mg/mL, respectively) at 37 °C in a humidified incubator at 5% CO_2, with media changed every 2 days. Cells between the sixth and tenth passages were used in all the experiments. Subconfluent cells were trypsinized, centrifuged and seeded at a density of 6×10^3 cells/disc with serum-free DMEM without phenol red onto different micro-grooved titanium discs in a 48-well microplate with an agarose layer (in order to prevent cell attachment to the dish). Tissue culture polystyrene (TCPS) and polished c.p. titanium served as reference substrates. Cellular analyses were performed at 4 h, 24 h and 72 h after seeding.

HFFs were cultured on the different surfaces. Then, cell adhesion and proliferation were analyzed using Cell Proliferation Reagent WST-1 (Roche Applied Science, Penzberg, Germany). This colorimetric protocol measures the creation of the formazan dye by cellular activity. The tetrazolium salts incorporated to the medium are cleaved by mitochondrial dehydrogenases of living cells, and the resulting soluble formazan dye can be analyzed spectrophotometrically. There is a direct correlation between the absorbance of the dye solution and the cell number. Viability was evaluated at the specified culture times by incubating for 2 h with 1:10 WST-1 in serum-free DMEM without phenol red. The optical density (OD) at 440 nm of cell supernatant was evaluated with an EL × 800 Universal Microplate Reader (Bio-Tek Instruments, Inc., Winooski, VT, USA). Three different samples for every surface and two different experiments were measured in parallel. A standard curve was performed using cell numbers ranging from 3×10^3 to 50×10^3.

Non-viable cells were quantified by means of measurement of released lactate dehydrogenase (LDH) enzyme at the specified culture times. For that purpose, the cell-free culture supernatant was collected, centrifuged at 250× g for 5 min and then analyzed with Cytotoxicity Detection Kit LDH (Roche Applied Science, Basel, Switzerland) as per the manufacturer's instructions. The reduction of tetrazolium salts into the formazan dye by LDH activity was measured spectrophotometrically at 490 nm. TCPS was used as a low control sample and lysed cells were utilized as a high control sample (maximum releasable LDH activity). Three different samples of each series in two experiments were analyzed.

2.5. Osteoblasts Culture

The objective of osteoblast cultures is to determine the degree of osteoblastic cytocompatibility and the ease of hard tissue regeneration to achieve bone regeneration and increase the mechanical fixation of the implant–abutment system to the bone with osseointegration. For the cell adhesion assay, osteoblastic SaOS-2 cells, a cell line with epithelial morphology derived from bone, were used. Six to seven cell passages were performed before seeding the cells onto the study samples. During cell passages, a control of the growth and cell viability was tested. Cells were initially thawed by gently shaking the cryovial in a 37 °C water bath for 1–2 min. From this point onward, everything was performed under sterile conditions. Once thawed, the content was transferred to a falcon with 9 mL of culture medium and centrifuged at 300 G for 3 min. Then, the supernatant was aspirated, and the pellet was resuspended with 1 mL of cell culture medium. Cells were seeded in flasks F175 and were kept at 37 °C with 5% CO_2, with the cell culture medium being changed 2–3 times per week.

The cell culture medium for this cell line consists of McCoy's 5a Medium Modified with L-glutamine 1.5 mM and 2200 mg/L sodium bicarbonate. This medium was supplemented with 15% fetal bovine serum (FBS), 1% penicillin/streptomycin (P/S) and 2% sodium pyruvate solution (NaPyr).

Cell passage was carried out at 90% confluence. Therefore, the cells were detached from the flasks by removing the medium, washing twice with 5 mL of PBS (37 °C) to remove dead cells, and then adding 5 mL of 0.05% trypsin. The flasks were left in the incubator at 37 °C with 5% CO_2 for 2–3 min. Afterward, trypsin was neutralized with

7 mL of cell culture medium (37 °C) and the content was transferred to a Falcon tube and centrifuged for 5 min at 300 G. The supernatant was then aspirated, and the pellet was resuspended with cell culture medium. Cell counting with Tripan Blue was then performed. For this, 10 µL of previously resuspended cells was prepared in an Eppendorf tube and mixed with Tripan Blue. A volume of 10 µL of the total was transferred to a Neubauer chamber for counting using phase contrast microscopy. The corresponding calculations were then carried out [37–39].

A resazurin salt assay (Alamar Blue) was used to assess cell proliferation and cell viability. The protocol was as follows: 5 mg resazurin salt (Sigma-Aldrich, St. Louis, MO, USA) was added to 1 mL of PBS, obtaining a stock solution of 5 mg/mL. Then, 100 µL was transferred to a Falcon tube with 50 mL of cell culture medium, and the solution was filtered to ensure sterile conditions. The final solution had a concentration of 10 µg/mL. This solution was protected from light.

After three days of cell culture, the culture medium was removed, and each well was washed with 500 µL of pre-warmed PBS. Then, 300 µL of 10 µg/mL resazurin solution was added in each well and incubated at 37 °C and 5% CO_2 for 3 h. Afterward, 200 µL from each well was transferred to a 96-well plate, transparent, and finally, the absorbance was analyzed. The wavelength was of 570 nm and 600 nm, and an Infinite® 200 PRO Multimode Absorbance Multimode Microplate Reader (TECAN, Männedorf, Switzerland) was used.

2.6. *Immunofluorescence*

For the immunofluorescence assay, a working solution of actin 488-stained phalloidin (100 nM) was prepared by diluting 58.8 µL of the 14 µM stock in 8.4 mL of PBS. Additionally, a DAPI solution was prepared by diluting 10 µL in 10 mL of PBS. Both solutions were kept at room temperature, without light exposure. A 0.1% Triton-X solution was made by diluting 1 mL of Triton-X in 9 mL of PBS. The samples were analyzed using the STELLARIS 5 Cryo Confocal Light Microscope.

After three days of cell culture, the culture medium was removed, and cells were washed with 500 µL of pre-warmed PBS. Then, 350 µL of 4% PFA/PBS was added in the wells at room temperature, in order to fix the cells. Following fixation, cells were washed with 500 µL of PBS and then permeabilization was performed using 350 µL of 0.1% Triton-X/PBS. After 10 min, the cells were washed with 500 µL of PBS and after, 350 µL of actin 488-stained phalloidin was incorporated.

Cells were then incubated for 30 min at room temperature without light exposure. After actin staining, cells were washed three times with 500 µL of PBS and cells were incubated with 350 µL of DAPI solution at room temperature in the dark for 2–3 min. Finally, they were washed again with 500 µL of PBS and cells were kept with 500 µL of PBS at 4 °C, without light exposure.

2.7. *Bacterial Culture*

Bacterial cultures were performed to determine the bactericidal capacity of the different solutions studied as a way to prevent bacterial recolonization. Gram+ and Gram− bacteria were used to see their behavior. The results give us important information for determining the best solution for oral application. We must take into account a limitation of this study, which is to determine the behavior with biofilm and not with isolated strains. In any case, the results allowed for us to characterize the bactericidal capacity of the solutions. Bacterial assays were performed using two oral pathogens, representing a Gram-negative and a Gram-positive bacterial strain. *Pseudomonas aeruginosa*, a Gram-negative bacterial strain, was sourced from Colección española de cultivos tipo (CECT 110, Valencia, Spain). For the Gram-positive strain, *Streptococcus gordonii* were used, and were obtained from Colección española de cultivos tipo (CECT 804, Valencia, Spain).

A total of six samples (n = 6) were used for the bacterial adhesion test, with three samples from each study group dedicated to the Gram-positive and three to the Gram-negative bacteria. Prior to the test, the culture media and material (PBS) were sterilized by autoclaving at 121 °C for 30 min using autoclave oven SELECTA model Sterilmax (SELECTA, Abrera, Spain). As previously described, samples were also sterilized by incubating in alcohol three times for 5 min in sterile culture plates. Afterward, the samples were exposed to ultraviolet light for another 30 min [40–42].

Agar plates were incubated at 37 °C for 24 h. The bacterial inoculum was prepared by suspending the bacteria in 5 mL of Brain Heart Infusion Broth (BHI) (Sigma Aldrich, St. Louis, MO, USA) followed by an incubation for 24 h at 37 °C. The medium was then adjusted to an optical density of 0.1 at a wavelength of 600 nm (OD600 = 0.1). For the bacterial adhesion test, 500 μL of the suspension (OD600 = 0.1) was added to each well of the culture plate and incubated at 37 °C for 1 h, using an incubator oven MEMMERT BE500 (MEMMERT Gmbh, Scheabach, Germany). All tests were carried out under static conditions without external stirring.

Then, the samples were rinsed twice with PBS for 5 min each and fixed with a 2.5% glutaraldehyde solution in PBS for 30 min at 4 °C. Following fixation, the glutaraldehyde solution was removed, and the samples were rinsed three times with PBS for 5 min each.

For viability analysis, a confocal microscope and the LIVE/DEAD Backlight bacterial viability kit (Thermo Fisher, Barcelona, Spain) were used [35]. A solution was prepared by mixing 1.5 μL of propidium with 1 mL of PBS. Using a micropipette, a drop of this solution (approximately 50 μL/sample) was applied to the surface. After incubating at room temperature without light exposure for 15 min, the samples were rinsed three times with PBS for 5 min.

The surfaces were then examined by laser scanning microscopy (CLSM). Three images per sample were captured at 630× magnification. Live and dead bacteria were detected using a wavelength of 488 nm and 561 nm. This analysis enabled both the assessment of bacterial viability on each surface and an initial comparison of the of bacterial count present in the different group of samples.

2.8. Statistical Analysis

Statistical analysis was carried out using the comparative T.TEST (with the Excel software version 16.0.18025.20104). This was performed between the different groups at 95% of confidence. Therefore, statistically significant differences are with values of ($p < 0.05$).

3. Results

The roughness measurements (Sa) in Table 2, reveal that the different immersion treatments carried out on the discs increases slightly the roughness, as no statistically significant differences ($p < 0.05$) were observed with respect to the control group in Ti and Zr. However, regarding the roughness for Comp and CrCo, the implantoplasty produces higher roughness in relation to the control. In these cases, the differences are statistically significant $p < 0.05$.

Table 2. Roughness parameters values (Sa) in micrometers of the samples studied. The asterisks (*) mean differences statistically significant at $p < 0.05$ for each material and the different treatments. The double asterisks (**) mean differences statistically significant at $p < 0.05$ in relation to the values with single and without asterisks.

Treatment	Ti	Zr	Comp	CrCo
As-received	0.15 ± 0.09 *	0.10 ± 0.07	0.12 ± 0.09 *	0.18 ± 0.09 *
Implantoplasty (Ctrl)	0.25 ± 0.15 *	0.15 ± 0.05	0.28 ± 0.07 **	0.37 ± 0.10 **
AC	0.33 ± 0.13 **	0.17 ± 0.08	0.29 ± 0.09 **	0.40 ± 0.12 **
AC 250	0.30 ± 0.10 **	0.14 ± 0.09	0.27 ± 0.07 **	0.43 ± 0.13 **
AC 500	0.25 ± 0.11 **	0.17 ± 0.05	0.25 ± 0.08 **	0.44 ± 0.14 **
AC 500/Mg	0.27 ± 0.10 **	0.18 ± 0.04	0.26 ± 0.09 **	0.42 ± 0.15 **

From the roughness results, it can be observed that the implantoplasty generates a higher roughness than the control samples due to the machining processes (Table 1). Statistically significant differences can be seen in the surfaces with Ti implantoplasty when treated with the different citric acid solutions, since they present a slight acid attack that slightly increases the roughness values. For the other three materials used, the effect of immersion in the citric acid solution does not cause any statistically significant difference with the surface of the material that has undergone implantoplasty [32].

Figure 2 shows the results obtained from the surface contact angle measurements for each material treated with the different solutions for 2 and 10 min. First, the results obtained from the "Control" samples were only with implantoplasty treatment, but were only sterilized in an autoclave at 121 °C for 30 min. These values show values from 65° for composite to 99° for CrCo. These values show a considerable hydrophobic character. Of the four materials studied, the composite (PMMA) is the most hydrophilic material. On the other hand, the CrCo alloy has a hydrophobic surface because the contact angle is greater than 90°.

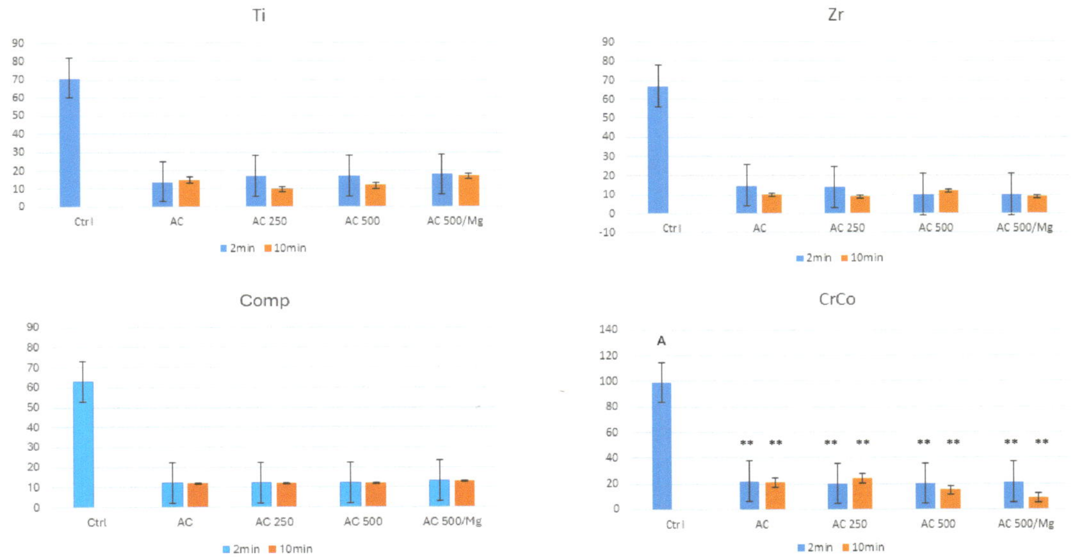

Figure 2. Contact angles of the different surfaces without treatment (Ctrl) and with immersion in different solutions based in citric acid studied. The samples were immersed for 2 and 10 min. The capital letter shows the statistical difference significance between CrCo, and the others surfaces with $p < 0.05$, and the asterisks the statistical difference significance between the immersion treatments of the CrCo and the other surfaces with $p < 0.05$. No statistical difference significance was found between the different times of treatment in any surface.

Secondly, the results obtained from the treated samples, i.e., immersed for 2 and 10 min in the citric acid solutions, show in all cases a very significant decrease in the contact angle, almost in all cases not exceeding 10° of contact angle, which makes the surfaces super-hydrophilic. The differences between the immersion times of 2 and 10 min do not show statistically significant differences ($p < 0.005$) in any of the treated materials. Moreover, no statistically significant differences are observed in the treated samples in general except in the case of CrCo. Specifically, CrCo values are around 20° while the other materials (Ti, Comp and Zr) present values around 10°.

Based on the results between 2 and 10 min, where no significant changes in behavior were observed, we focused this study on the two-minute immersion treatments. This treatment time was selected because this solution is intended to function as a mouthwash, so the minimum time enhances patient comfort.

The results of the cell cytotoxicity test are presented in Figure 3 for the surface of the four proposed materials with the five conditions.

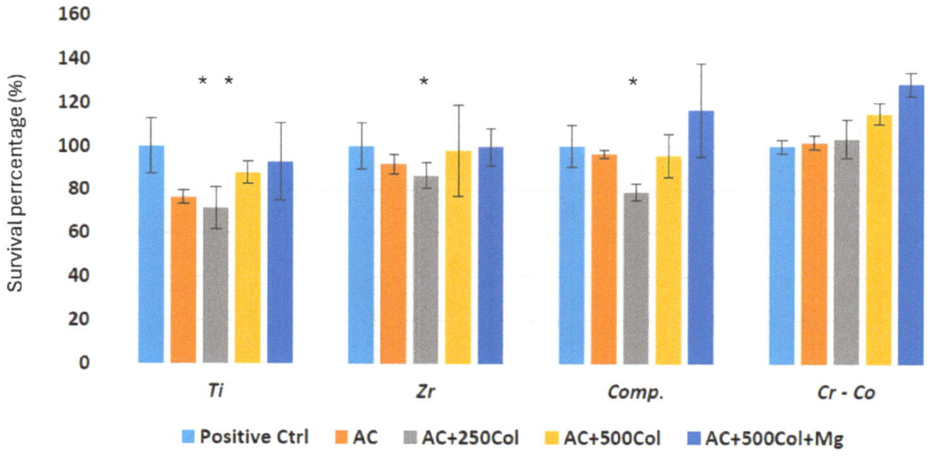

Figure 3. Survival percentage of fibroblasts for different treatments. Asterisks mean statistical differences significance $p < 0.05$.

Figure 3 shows that all citric acid treatments are cytocompatible, since the survival of the cultured fibroblast cells exceeds the 70% survival criterion. It can be seen that the AC + 500 Col and AC + 500 Col + Mg solutions show excellent behavior with fibroblasts.

Figure 4 demonstrates that collagen enhances the adhesion of fibroblasts, as more fibroblasts are present in this surface. The increase in collagen concentration does not offer statistically significant differences compared to the CrCo surface. However, it can be observed that there is a significant difference in the increase in the number of fibroblasts for the concentration of 500 in comparison to the lower concentration (250). No differences were observed between the other conditions studied.

An F-actin (phalloidin/DAPI) immunofluorescence assay was performed using a confocal light microscope in order to determine the presence and distribution of osteoblastic cells. This allowed for determining whether the material surfaces are favorable to the adhesion of this cells. Figure 5 shows that Ti and Zr surfaces show good cell viability for each treatment applied. Regarding the composite surfaces, treatments with 25% citric acid + 250 collagen and 25% citric acid + 500 collagen show a reduced number of cells, while the "Control" sample is the surface with the highest cell density. Regarding the CrCo surfaces, treatments with the 25% citric acid + 500 collagen and 25% citric acid + 500 collagen + 1% Mg present a decreased cell adhesion in comparison to the remaining surfaces.

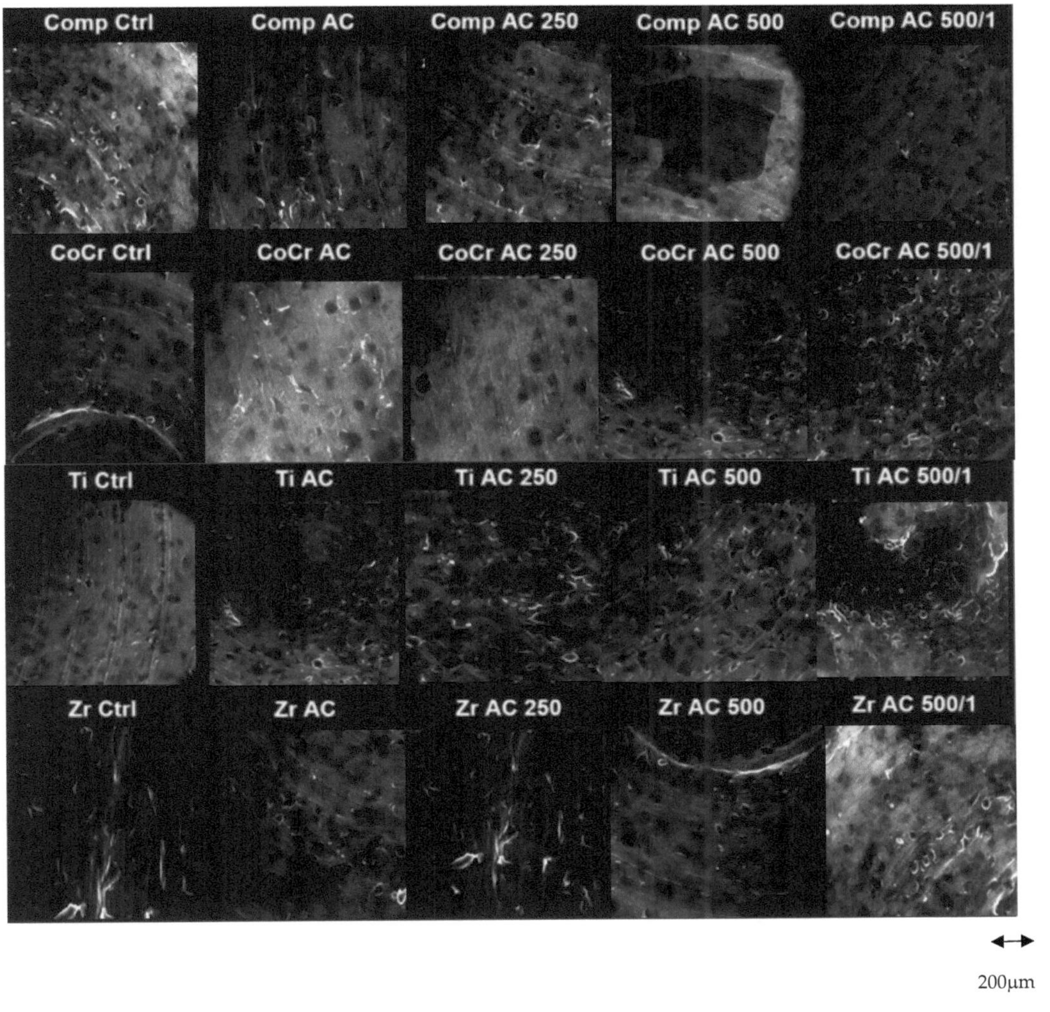

Figure 4. Fibroblasts cultured on different surfaces and with different dissolutions observed by scanning electron microscopy.

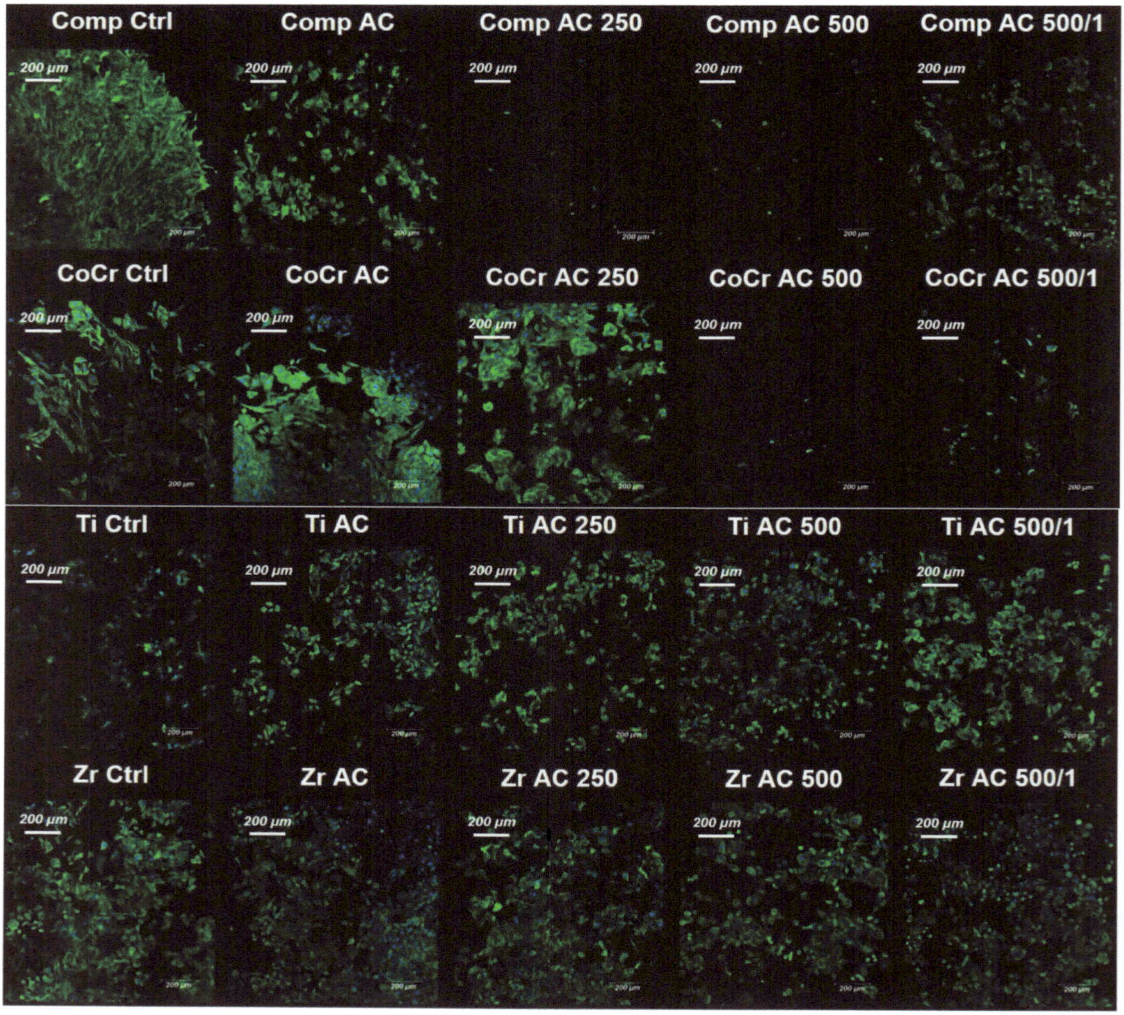

Figure 5. Immunofluorescence tests results of Ti, Zr, composite and CrCo surfaces with the applied treatments, showing the presence of osteoblastic cells.

Bacterial Culture

Figures 6 and 7 show the results obtained after the Alamar Blue test to determine the adhesion and bacterial growth on the surfaces proposed for the study for *Streptococcus gordonii* (Gram-positive) and *Pseudomonas aeruginosa* (Gram-negative). Significant bacterial colonization on all the surfaces in the control treatment can be observed, especially in the CrCo alloy. In all of them, both Gram-positive and Gram-negative bacteria show a notable reduction in bacterial activity when using citric acid treatments. It can be seen that the collagen and magnesium contents do not have a statistically significant effect on the reduction of bacterial colonization. It can also be seen that the action of citric acid with collagen causes a CFU reduction around of 75% for *Streptococcus gordonii* and 80% for *Pseudomonas aeruginosa*.

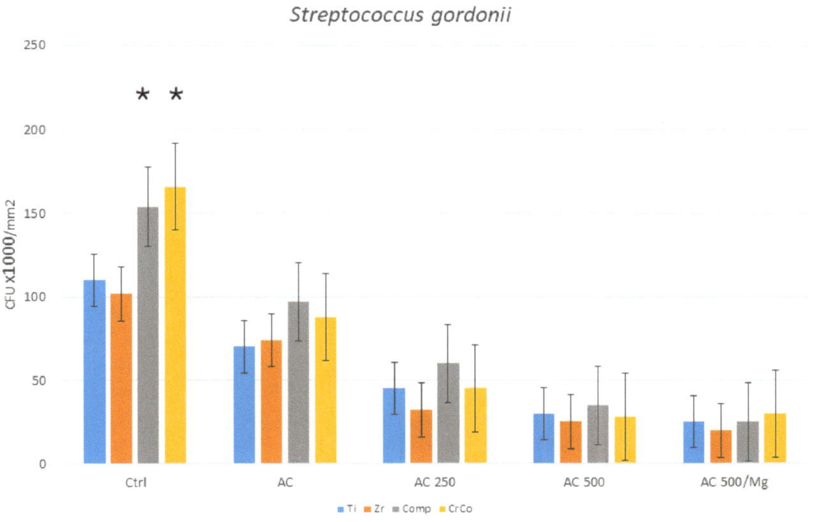

Figure 6. Quantitative analysis of *Streptococcus gordonii* (Gram+) for the different treatments and surfaces. Asterisks mean a statistically significant difference at $p < 0.05$. There are significant differences between the control and the four solutions studied for the materials studied. The different materials do not offer statistically significant differences with the treatments with the citric acid-based solutions and all of them offer important reductions in bacterial colonies. For the control samples, composite and CrCo present the worst bacterial behavior.

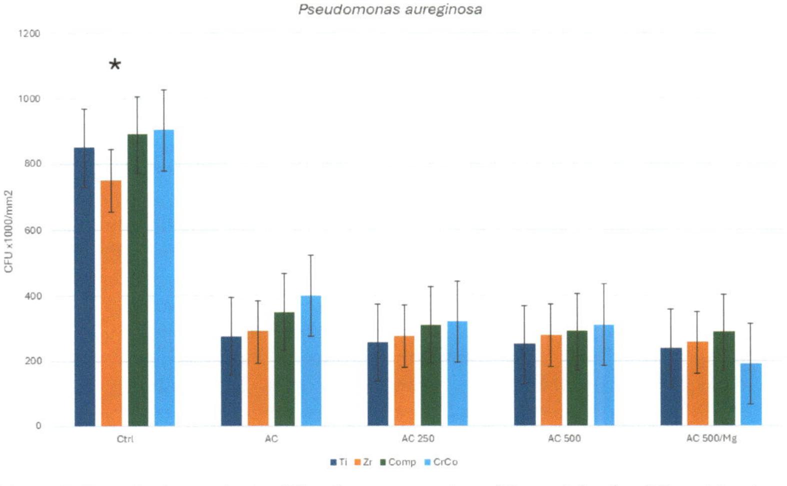

Figure 7. Quantitative analysis of *Pseudomonas aeruginosa* (Gram−) for the different treatments and surfaces. Asterisk means statistical differences significance $p < 0.05$. There are significant differences between the control and the four solutions studied for the materials studied in the colonies of Gram− bacteria. For the control samples, zirconia presents statistically significant differences with respect to the rest of the materials in relation to bacterial colonization. The different materials do not offer statistically significant differences among them with the treatments with the citric acid-based solutions and all of them offer significant reductions in bacterial colonies, but with a lower efficacy than for the Gram+ bacteria studied.

4. Discussion

It is well known that rough surfaces promote better cell adhesion, proliferation and differentiation, especially of osteoblastic cells [36,37]. Surface roughness or treated surfaces can interfere with cell morphology and orientation [38]. In this project, it is shown how the two metallic surfaces present greater roughness than the zirconia or the composite. It is also known that the increase in hydrophilicity favors wettability and, in consequence, this results in greater protein adsorption, which favors cell activity [43–46]. As has been observed, Ti presents higher wettability than CrCo, and therefore titanium has favorable factors for osteoblastic activity. Although CrCo surface exhibits good roughness, it has lower osteoblastic activity compared to Ti due to its low hydrophilic nature, which does not favor osteoblastic activity.

Based on the results obtained in the cell cytotoxicity test, it was determined that none of the treatments proposed for this study presented cytotoxicity for the SaOS-2 osteoblastic cells in any of the materials studied. Analyzing the values obtained, all of them are in the "non-cytotoxic" range, as determined by ISO 10993-5 [47]. The aforementioned ISO establishes that "the reduction of cell viability by more than 30% is considered a cytotoxic effect" [47]. Our results indicate a reduced adhesion of osteoblastic cells in the presence of collagen for the composite sample. This fact seems to be due to the lack of anchorage of the collagen molecules on the surface of the composite. This has been previously described by other authors, who concluded that the polymeric compound does not favor the adhesion of the collagen molecule and therefore inhibits the accelerating effect on both fibroblastic and osteoblastic cell adhesion [48,49].

Collagen is a well-known biomaterial commonly used in films, composites and three-dimensional matrixes, as it can enhance the recomposition and granulation of tissues, and also has the ability to act as protection of wounds and tissues against infection. Therefore, collagen is used as a support material in healing processes, and it is widely used in dental therapy. Moreover, it is a non-toxic, biodegradable and bioabsorbable material [32]. Also, collagen is the main component of ECM; type I is the most abundant, about 85%, together with proteins such as laminins, fibronectin and vitronectin. This fact favors cellular activity and should be a key element in triggering soft tissue formation to achieve a biological seal at the implant–abutment connection. In an in vivo study, Maria Sartori et al. investigated the effects of dental implants coated with type I collagen on bone regeneration and osseointegration in osteopenic rats, in which they found greater mechanical stability and a higher rate of osseointegration [50]. Other studies have affirmed the ability of type I collagen to promote osseointegration by stimulating bone formation at the cellular and molecular level [51]. In all cases, the collagen was dissolved in an acidic solution since it dissolves in an acid medium, which favors homogeneity and application.

The incorporation of magnesium is due to the fact that it has properties capable of improving bone bonding if implemented in implant surface modifications. It is currently used for titanium implant surfaces, providing improved properties. Jiang et al. were able to decrease Young's modulus, increase strength and provide improved biocompatibility in titanium implants. Veronese et al. combined titanium dioxide (TiO_2) with magnesium and obtained anti-inflammatory properties [52]. The inflammatory response plays an important role in the implantation of dental implants and may help to moderate osteogenesis [53,54].

The application of a citric acid treatment generates a small increase in surface roughness, which leads to an increase in bacterial adherence [55,56], but the acidic character provided to the sample surfaces prevents or decreases microbiological colonization [57]. The concentration of citric acid is related to the antibacterial action it provides, and thus causes a reduction in the pH of the extracellular matrices [23–26,28]. It is hypothesized that the presence of citric acid modifies the permeability of the bacterial membrane, varying the hydrogen gradient between intracellular and extracellular sites [52]. In addition, it has an antioxidant capacity to prevent or delay some type of cell damage and also has a negative effect on mycobacteria [52].

Focusing on the results of antibacterial activity obtained in Figures 5 and 6, it has been demonstrated that all treatments show antibacterial activity. In the case of pure titanium, in addition to the treatment provided, it is capable of forming a biocompatible titanium oxide layer, providing high resistance to corrosion with an oxidizing character, thus reducing bacterial activity [21,22,58].

A study has determined the existence of a relationship between wettability and bacterial colonization, based on the hydrophobicity of the surface [59,60]. For example, metallic surfaces that exhibit a hydrophobic nature result in a higher adhesion of hydrophobic bacteria. In consequence, lower bacterial adhesion could correlate with an increase in surface hydrophobicity [36,59]. Magnesium has also been shown to have an antibacterial effect, which could explain the lower levels of bacterial survival in Mg-containing samples, although the differences were not statistically significant [52,61].

Finally, of the four treatments studied, the one that stands out most for the osteointegrative and antibacterial properties provided is the Citric Acid 25% + Collagen 500 + Magnesium 1%. This has been validated by other studies that show how citric acid provides good antibacterial properties without damaging osteoblastic cells. Moreover, this is aligned with collagen and magnesium biological function, as collagen is the main component of the extracellular matrix capable of improving tissue recomposition, and magnesium improve bone union and presents antibacterial properties.

The action of citric acid in increasing surface hydrophilicity, producing a stable titanium oxide layer and bactericidal character improves systems based on strong acids or bases or ozone flux treatments. Firstly, because it does not improve the surface properties of titanium and in many cases, it affects its roughness or causes the incorporation of hydrogen into the titanium, which can cause the so-called hydrogen embrittlement. Moreover, in the case of treatment with citric acid-based solutions, it does not affect the health of the soft tissues. Finally, collagen can be dissolved in an acidic medium and can be incorporated into the solution in order to increase cell adhesion as well as to have a synergistic effect with divalent magnesium cations.

This work is preliminary but served in preparing cytocompatible citric acid-based solutions that favor both fibroblastic and osteoblastic cell activity and are clearly bactericidal against both Gram-positive and Gram-negative bacteria. However, concentrations should be optimized to improve cellular and bactericidal response. This study should be carried out in dental biofilm to complete the studies of two strains that are common in the mouth but do not respond reliably to what occurs in the mouth [62–65]. This study should be completed with in vivo studies with infection and to evaluate disinfection, as well as possible tissue regeneration in dental implants that have undergone implantoplasty. It is important to obtain a product that helps disinfection and favors bone growth for the new osseointegration of the implant, as well as the formation of a biological seal produced by the regeneration of the soft tissue [66,67]. Implantoplasty also faces other challenges such as the effect of the small particles, with different sizes and materials, that are present in the biological bed and present a toxic nature. Further studies should address this issue.

5. Conclusions

This contribution studies one of the most common materials in dental implant with prosthesis. The results verified the significant decrease in the contact angle for titanium, zirconia and PMMA composite with feldspar. Specifically, the values decreased from 70 to 15 and for CrCo from 100 to 25. All treatments showed that an increase in wettability causes higher cellular activity. Moreover, all treatments demonstrated cytocompatibility and good osteoblastic behavior. This suggests a promising solution for the regeneration of soft and hard tissues around the dental implant and the biological seal during prosthesis. It has been demonstrated that the different solutions have a strong bactericidal effect on both *Streptococcus gordonii* (Gram+) and *Pseudomonas aeruginosa* (Gram−) strains, reducing colonies around 72% and 64%, respectively. From the results obtained, the mouthwash with the best cellular activity and bactericidal capacity is the citric acid solution with

500 collagen and 1% magnesium. Although the in vitro results of this preliminary study are encouraging, further in vivo tests are needed before clinical application.

Author Contributions: Conceptualization, J.G., B.M.B. and M.F.-D.; methodology, P.F.-D., L.F.D.L.F., B.M.D.G. and J.F.V.; software, P.F.-G.; validation, J.G., P.F.-D., L.F.D.L.F., B.M.D.G. and J.F.V.; formal analysis, M.F.-D.; investigation, P.F.-G.; resources, M.F.-D.; data curation, P.F.-D., L.F.D.L.F., B.M.D.G. and J.F.V.; writing—original draft preparation, J.G. and B.M.B.; writing—review and editing, J.G.; visualization, M.F.-D.; supervision, J.G.; project administration, M.F.-D.; funding acquisition, M.F.-D. All authors have read and agreed to the published version of the manuscript.

Funding: This research received no external funding.

Institutional Review Board Statement: Not applicable.

Informed Consent Statement: Not applicable.

Data Availability Statement: The original contributions presented in this study are included in the article, further inquiries can be directed to the corresponding author.

Acknowledgments: The authors are grateful to Klockner Dental Implants. This work was supported by the Spanish Government and the Ministry of Science and Innovation of Spain through research projects CONCEPTO PDC2022-133628-C22 (co-funded by the European Regional Development Fund (ERDF), a way to build Europe) and research project MINECO (PID2022-137496OB-I00).

Conflicts of Interest: The authors declare no conflicts of interest.

References

1. Sailer, I.; Makarov, N.A.; Thoma, D.S.; Zwahlen, M.; Pjetursson, B.E. All-ceramic or metal-ceramic tooth-supported fixed dental prostheses (FDPs)? A systematic review of the survival and complication rates, Part I: Single crowns (SCs). *Dent. Mater.* **2015**, *31*, 603–623. [CrossRef]
2. Roos-Jansåker, A.M.; Lindahl, C.; Renvert, H.; Renvert, S. Nine- to fourteen-year follow-up of implant treatment, Part I: Implant loss and associations to various factors. *J. Clin. Periodontol.* **2006**, *33*, 283–289. [CrossRef]
3. Pjetursson, B.E.; Valente, N.A.; Strasding, M.; Zwahlen, M.; Liu, S.; Sailer, I. A systematic review of the survival and complication rates of zirconia-ceramic and metal-ceramic single crowns. *Clin. Oral Implant. Res.* **2018**, *29* (Suppl. S16), 199–214. [CrossRef]
4. Schwarz, F.; Derks, J.; Monje, A.; Wang, H.L. Peri-implantitis. *J. Periodont.* **2018**, *89* (Suppl. S1), S267–S290. [CrossRef]
5. Daubert, D.M.; Weinstein, B.F. Biofilm as a risk factor in implant treatment. *Periodontology 2000* **2019**, *81*, 29–40. [CrossRef]
6. Englezos, E.; Cosyn, J.; Koole, S.; Jacquet, W.; De Bruyn, H. Resective Treatment of Peri-implantitis: Clinical and Radiographic Outcomes After 2 Years. *Int. J. Periodontics Restor. Dent.* **2018**, *38*, 729–735. [CrossRef]
7. Romeo, E.; Lops, D.; Chiapasco, M.; Ghisolfi, M.; Vogel, G. Therapy of peri-implantitis with resective surgery. A 3-year clinical trial on rough screw-shaped oral implants. Part II: Radiographic outcome. *Clin. Oral Implant. Res.* **2007**, *18*, 179–187. [CrossRef]
8. Monje, A.; Schwarz, F. Principles of Combined Surgical Therapy for the Management of Peri-Implantitis. *Clin. Adv. Periodontics* **2022**, *12*, 57–63. [CrossRef]
9. Bertl, K.; Isidor, F.; von Steyern, P.V.; Stavropoulos, A. Does implantoplasty affect the failure strength of narrow and regular diameter implants? A laboratory study. *Clin. Oral Investig.* **2021**, *25*, 2203–2211. [CrossRef]
10. Leitão-Almeida, B.; Camps-Font, O.; Correia, A.; Mir-Mari, J.; Figueiredo, R.; Valmaseda-Castellón, E. Effect of bone loss on the fracture resistance of narrow dental implants after implantoplasty. An in vitro study. *Med. Oral Patol. Oral Cir. Bucal* **2021**, *26*, e611–e618. [CrossRef] [PubMed]
11. Hentenaar DF, M.; De Waal YC, M.; Stewart, R.E.; Van Winkelhoff, A.J.; Meijer HJ, A.; Raghoebar, G.M. Erythritol airpolishing in the non-surgical treatment of peri-implantitis: A randomized controlled trial. *Clin. Oral Implant. Res.* **2021**, *32*, 840–852. [CrossRef]
12. Tomasi, C.; Regidor, E.; Ortiz-Vigón, A.; Derks, J. Efficacy of reconstructive surgical therapy at peri-implantitis-related bone defects. A systematic review and meta-analysis. *J. Clin. Periodontol.* **2019**, *46* (Suppl. S21), 340–356. [CrossRef]
13. Schwarz, F.; Schmucker, A.; Becker, J. Efficacy of alternative or adjunctive measures to conventional treatment of peri-implant mucositis and peri-implantitis: A systematic review and meta-analysis. *Int. J. Implant. Dent.* **2015**, *1*, 22. [CrossRef]
14. Renvert, S.; Roos-Jansåker, A.M.; Claffey, N. Non-surgical treatment of peri-implant mucositis and peri-implantitis: A literature review. *J. Clin. Periodontol.* **2022**, *35* (Suppl. S8), 305–315. [CrossRef]
15. Monje, A.; Pons, R.; Amerio, E.; Wang, H.L.; Nart, J. Resolution of peri-implantitis by means of implantoplasty as adjunct to surgical therapy: A retrospective study. *J. Periodontol.* **2022**, *93*, 110–122. [CrossRef]
16. Barrak, F.N.; Li, S.; Muntane, A.M.; Jones, J.R. Particle release from implantoplasty of dental implants and impact on cells. *Int. J. Implant. Dent.* **2020**, *6*, 50. [CrossRef]
17. Toledano-Serrabona, J.; Sánchez-Garcés, M.A.; Gay-Escoda, C.; Valmaseda-Castellon, E.; Camps-Font, O.; Verdeguer, P.; Molmeneu, M.; Gil, F.J. Mechanical properties and corrosión behavior of Ti6Al4V particles obtained by Implatoplasty. An in vivo study. Part. II. *Materials* **2021**, *14*, 6519. [CrossRef]

18. Toledano-Serrabona, J.; Gil, F.J.; Camps-Font, O.; Valmaseda-Castellón, E.; Gay-Escoda, C.; Sánchez-Garcés, M.Á. Physicochemical and Biological Characterization of Ti6Al4V Particles Obtained by Implantoplasty: An In Vitro Study. Part I. *Materials* **2021**, *14*, 6507. [CrossRef]
19. Rupp, F.; Liang, L.; Geis-Gerstorfer, J.; Scheideler, L.; Hüttig, F. Surface characteristics of dental implants: A review. *Dent. Mater.* **2018**, *34*, 40–57. [CrossRef]
20. Kotsakis, G.A.; Lan, C.; Barbosa, J.; Lill, K.; Chen, R.; Rudney, J.; Aparicio, C. Antimicrobial Agents Used in the Treatment of Peri-Implantitis Alter the Physicochemistry and Cytocompatibility of Titanium Surfaces. *J. Periodontol.* **2016**, *87*, 809–819. [CrossRef]
21. Gil, F.J.; Planell, J.; Proubasta, I.; Vazquez, J. Fundamentos de biomecánica y biomateriales. In *Fundamentos de Biomecánica y Biomateriales*; Ergon: Barcelona, Spain, 1997; pp. 125–132.
22. Gil, F.J.; Planell, J.A. Aplicaciones biomédicas del titanio v sus aleaciones. *Biomecánica* **1993**, *1*, 34–43. [CrossRef]
23. Souza, J.G.S.; Cordeiro, J.M.; Lima, C.V.; Barão, V.A.R. Citric acid reduces oral biofilm and influences the electrochemical behavior of titanium: An in situ and in vitro study. *J. Periodontol.* **2019**, *90*, 149–158. [CrossRef] [PubMed]
24. Książek, E. Citric Acid: Properties, Microbial Production, and Applications in Industries. *Molecules* **2024**, *29*, 22. [CrossRef]
25. Cordeiro, J.M.; Pires, J.M.; Souza, J.G.S.; Lima, C.V.; Bertolini, M.M.; Rangel, E.C.; Barão, V.A.R. Optimizing citric acid protocol to control implant-related infections: An in vitro and in situ study. *J. Periodontal Res.* **2021**, *56*, 558–568. [CrossRef] [PubMed]
26. Fonseca, V.C.P.D.; Abreu, L.G.; Andrade, E.J.; Asquino, N.; Esteves Lima, R.P. Effectiveness of antimicrobial photodynamic therapy in the treatment of peri-implantitis: Systematic review and meta-analysis. *Lasers Med. Sci.* **2024**, *39*, 186. [CrossRef]
27. Wang, S.; Xu, C.; Yu, S.; Wu, X.; Jie, Z.; Dai, H. Citric acid enhances the physical properties, cytocompatibility and osteogenesis of magnesium calcium phosphate cement. *J. Mech. Behav. Biomed. Mater.* **2019**, *94*, 42–50. [CrossRef]
28. Brynhildsen, L.; Rosswall, T. Effects of cadmium, copper, magnesium, and zinc on the decomposition of citrate by a *Klebsiella* sp. *Appl. Environ. Microbiol.* **1989**, *55*, 1375–1379. [CrossRef]
29. Gil, F.J.; Solano, E.; Pena, J.; Engel, E.; Mendoza, A.; Planell, J.A. Microstructural, mechanical and citotoxicity evaluation of different NiTi and NiTiCu shape memory alloys. *J. Mater. Sci. Mater. Med.* **2004**, *15*, 1181–1185. [CrossRef]
30. Janson, O.; Gururaj, S.; Pujari-Palmer, S.; Karlsson Ott, M.; Strømme, M.; Engqvist, H. Titanium surface modification to enhance antibacterial and bioactive properties while retaining biocompatibility. *Mater. Sci. Eng. C* **2019**, *96*, 272–279. [CrossRef]
31. Costa-Berenguer, X.; Garcia-Garcia, M.; Sanchez-Torres, A.; Sanz-Alonso, M.; Figueiredo, R.; Valmaseda-Castellon, E. Effect of implantoplasty on fracture resistance and surface roughness of standard diameter dental implants. *Clin. Oral Implant. Res.* **2018**, *29*, 46–54. [CrossRef]
32. Francis, S.; Iaculli, F.; Perrotti, V.; Piattelli, A.; Quaranta, A. Titanium Surface Decontamination: A Systematic Review of In Vitro Comparative Studies. *Int. J. Oral Maxillofac. Implant.* **2022**, *37*, 76–84. [CrossRef] [PubMed]
33. Martin, J.Y.; Schwartz, Z.; Hummert, T.W.; Costochondral Schraub, D.M.; Simpson, J.; Lankford, J., Jr.; Dean, D.D.; Cochran, D.L.; Boyan, B.D. Effect of titanium surface roughness on proliferation, differentiation, and protein synthesis of human osteoblast-like cells (MG63). *J. Biomed. Mater. Res.* **1995**, *29*, 389–401. [CrossRef] [PubMed]
34. Boyan, B.D.; Lincks, J.; Lohmann, C.H.; Sylvia, V.L.; Cochran, D.L.; Blanchard, C.R.; Dean, D.D.; Schwartz, Z. Effect of surface roughness and composition on chondrocytes is dependent on cell maturation state. *J. Orthop. Res.* **1999**, *17*, 446–457. [CrossRef]
35. Godoy-Gallardo, M.; Guillem-Marti, J.; Sevilla, P.; Manero, J.M.; Gil, F.J.; Rodriguez, D. Anhydride-functional silane immobilized onto titanium surfaces induces osteoblast cell differentiation and reduces bacterial adhesion and biofilm formation. *Mater. Sci. Eng. C Mater. Biol. Appl.* **2016**, *59*, 524–532. [CrossRef]
36. Rabel, K.; Kohal, R.J.; Steinberg, T.; Tomakidi, P.; Rolauffs, B.; Adolfsson, E.; Palmero, P.; Fürderer, T.; Altmann, B. Controlling osteoblast morphology and proliferation via surface micro-topographies of implant biomaterials. *Sci. Rep.* **2020**, *10*, 12810. [CrossRef]
37. Esposito, M. Titanium for Dental Implants (I). In *Titanium in Medicine: Material Science, Surface Science, Engineering, Biological Responses and Medical Applications*; Brunette, D.M., Tengvall, P., Textor, M., Thomsen, P., Eds.; Springer: Berlin/Heidelberg, Germany, 2001.
38. Charest, J.L.; Eliason, M.T.; García, A.J.; King, W.P. Combined microscale mechanical topography and chemical patterns on polymer cell culture substrates. *Biomaterials* **2006**, *27*, 2487–2494. [CrossRef]
39. Mustafa, K.; Wennerberg, A.; Wroblewski, J.; Hultenby, K.; Lopez, B.S.; Arvidson, K. Determining optimal surface roughness of TiO(2) blasted titanium implant material for attachment, proliferation and differentiation of cells derived from human mandibular alveolar bone. *Clin. Oral Implant. Res.* **2001**, *12*, 515–525. [CrossRef]
40. Fielding, G.A.; Roy, M.; Bandyopadhyay, A.; Bose, S. Antibacterial and biological characteristics of silver containing, and strontium doped plasma sprayed hydroxyapatite coatings. *Acta Biomater.* **2012**, *8*, 3144–3152. [CrossRef] [PubMed]
41. Yamaguchi, S.; Nath, S.; Sugawara, Y.; Divakarla, K.; Das, T.; Manos, J.; Chrzanowski, W.; Matsushita, T.; Kokubo, T. Two-in-one biointerfaces—Antimicrobial and bioactive nanoporous gallium titanate layers for titanium implants. *Nanomaterials* **2017**, *7*, 229. [CrossRef]
42. Godoy-Gallardo, M.; Manzanares-Céspedes, M.C.; Sevilla, P.; Nart, J.; Manzanares, N.; Manero, J.M.; Gil, F.J.; Boyd, S.K.; Rodríguez, D. Evaluation of bone loss in antibacterial coated dental implants: An experimental study in dogs. *Mater. Sci. Eng. C Mater. Biol. Appl.* **2016**, *69*, 538–545. [CrossRef]
43. Medvedev, A.E.; Neumann, A.; Ng, H.P.; Lapovok, R.; Kasper, C.; Lowe, T.C. Combined effect of grain refinement and surface modification of pure titanium on the attachment of mesenchymal stem cells and osteoblast-like SaOS-2 cells. *Mater. Sci. Eng. C* **2017**, *71*, 483–497. [CrossRef] [PubMed]

44. Tao, Z.S.; Zhou, W.S.; He, X.W.; Liu, W.; Bai, B.L.; Zhou, Q. A comparative study of zinc, magnesium, strontium-incorporated hydroxyapatite-coated titanium implants for osseointegration of osteopenic rats. *Mater. Sci. Eng. C* **2016**, *62*, 226–232. [CrossRef]
45. Mitchel, J.A.; Hoffman-Kim, D. Cellular scale anisotropic topography guides Schwann cell motility. *PLoS ONE* **2011**, *6*, e24316. [CrossRef]
46. Huang, Q.; Elkhooly, T.A.; Liu, X.; Zhang, R.; Yang, X.; Shen, Z.; Feng, Q. Effects of hierarchical micro/nano-topographies on the morphology, proliferation and differentiation of osteoblast-like cells. *Colloids Surf. B Biointerfaces* **2016**, *145*, 37–45. [CrossRef]
47. *ISO 10993—5 (55)*; Biological Evaluation of Medical Devices—Part 5, Tests for In Vitro Cytotoxicity. ISO: Geneva, Switzerland, 2009.
48. Bax, D.V.; Smalley, H.E.; Farndale, R.W.; Best, S.M.; Cameron, R.E. Cellular response to collagen-elastin composite materials. *Acta Biomater.* **2019**, *86*, 158–170. [CrossRef]
49. Yoon, B.H.; Kim, H.W.; Lee, S.H.; Bae, C.J.; Koh, Y.H.; Kong, Y.M.; Kim, H.E. Stability and cellular responses to fluorapatite-collagen composites. *Biomaterials* **2005**, *26*, 2957–2963. [CrossRef]
50. Sartori, M.; Giavaresi, G.; Parrilli, A.; Ferrari, A.; Aldini, N.N.; Morra, M.; Cassinelli, C.; Bollati, D.; Fini, M. Collagen type I coating stimulates bone regeneration and osteointegration of titanium implants in the osteopenic rat. *Int. Orthop.* **2015**, *39*, 2041–2052. [CrossRef]
51. Kämmerer, P.W.; Scholz, M.; Baudisch, M.; Liese, J.; Wegner, K.; Frerich, B.; Lang, H. Guided Bone Regeneration Using Collagen Scaffolds, Growth Factors, and Periodontal Ligament Stem Cells for Treatment of Peri-Implant Bone Defects In Vivo. *Stem Cells Int.* **2017**, *2017*, 3548435. [CrossRef]
52. Veronese, N.; Pizzol, D.; Smith, L.; Dominguez, L.J.; Barbagallo, M. Effect of Magnesium Supplementation on Inflammatory Parameters: A Meta-Analysis of Randomized Controlled Trials. *Nutrients* **2022**, *14*, 679. [CrossRef]
53. Buser, D.; Schenk, R.K.; Steinemann, S.; Fiorellini, J.P.; Fox, C.H.; Stich, H. Influence of surface characteristics on bone integration of titanium implants. A Histomorphometric Study Miniat. *Pigs. J. Biomed. Mater. Res.* **1991**, *25*, 889–902. [CrossRef]
54. Schwarz, F.; Sahm, N.; Becker, J. Combined surgical therapy of advanced peri-implantitis lesions with concomitant soft tissue volume augmentation. A case series. *Clin. Oral Implant. Res.* **2014**, *25*, 132–136. [CrossRef] [PubMed]
55. Cochis, A.; Azzimonti, B.; Della Valle, C.; De Giglio, E.; Bloise, N.; Visai, L.; Cometa, S.; Rimondini, L.; Chiesa, R. The effect of silver or gallium doped titanium against the multidrug resistant Acinetobacter baumannii. *Biomaterials* **2016**, *80*, 80–95. [CrossRef]
56. Avila, G.; Misch, K.; Galindo-Moreno, P.; Wang, H.L. Implant surface treatment using biomimetic agents. *Implant. Dent.* **2009**, *18*, 17–26. [CrossRef]
57. Buxadera-Palomero, J.; Calvo, C.; Torrent-Camarero, S.; Gil, F.J.; Mas-Moruno, C.; Canal, C.; Rodríguez, D. Biofunctional polyethylene glycol coatings on titanium: An in vitro-based comparison of functionalization methods. *Colloids Surf. B Biointerfaces* **2017**, *152*, 367–375. [CrossRef]
58. Gosau, M.; Hahnel, S.; Schwarz, F.; Gerlach, T.; Reichert, T.E.; Bürgers, R. Effect of six different peri-implantitis disinfection methods on in vivo human oral biofilm. *Clin. Oral Implant. Res.* **2010**, *21*, 866–872. [CrossRef]
59. Zhang, X.; Bai, R.; Sun, Q.; Zhuang, Z.; Zhang, Y.; Chen, S.; Han, B. Bio-inspired special wettability in oral antibacterial applications. *Front. Bioeng. Biotechnol.* **2022**, *10*, 1001616. [CrossRef]
60. Hemmatian, T.; Lee, H.; Kim, J. Bacteria Adhesion of Textiles Influenced by Wettability and Pore Characteristics of Fibrous Substrates. *Polymers* **2021**, *13*, 223. [CrossRef]
61. Vallet-Regí, M.; Román, J.; Padilla, S.; Doadrio, J.C.; Gil, F.J. Bioactivity and mechanical properties of SiO2–CaO–P2O5 glassceramics. *J. Mater. Chem.* **2005**, *15*, 1353–1359. [CrossRef]
62. Xia, Z.; Yu, X.; Wei, M. Biomimetic collagen/apatite coating formation on Ti6Al4V substrates. *J. Biomed. Mater. Res.—Part B Appl. Biomater.* **2012**, *100*, 871–881. [CrossRef]
63. Belloni, A.; Argentieri, G.; Orilisi, G.; Notarstefano, V.; Giorgini, E.; D'Addazio, G.; Orsini, G.; Caputi, S.; Sinjari, B. New insights on collagen structural organization and spatial distribution around dental implants: A comparison between machined and laser-treated surfaces. *J. Transl. Med.* **2024**, *22*, 120. [CrossRef]
64. Zhao, S.F.; Jiang, Q.H.; Peel, S.; Wang, X.X.; He, F.M. Effects of magnesium-substituted nanohydroxyapatite coating on implant osseointegration. *Clin. Oral Implant. Res.* **2013**, *24* (Suppl. A100), 34–41. [CrossRef]
65. Louropoulou, A.; Slot, D.E.; Van der Weijden, F. The effects of mechanical instruments on contaminated titanium dental implant surfaces: A systematic review. *Clin. Oral Implant. Res.* **2014**, *25*, 1149–1160. [CrossRef]
66. de Tapia, B.; Valles, C.; Ribeiro-Amaral, T.; Mor, C.; Herrera, D.; Sanz, M.; Nart, J. The adjunctive effect of a titanium brush in implant surface decontamination at peri-implantitis surgical regenerative interventions: A randomized controlled clinical trial. *J. Clin. Periodontol.* **2019**, *46*, 586–596. [CrossRef]
67. Dostie, S.; Alkadi, L.T.; Owen, G.; Bi, J.; Shen, Y.; Haapasalo, M.; Larjava, H.S. Chemotherapeutic decontamination of dental implants colonized by mature multispecies oral biofilm. *J. Clin. Periodontol.* **2017**, *44*, 403–409. [CrossRef]

prosthesis

MDPI

Article

Impact of Surgeons' Experience on Implant Placement Accuracy Using a Dynamic Navigation System: A Cadaver Pilot Study

Francesco Pera [1,*,†], Camillo Vocaturo [1,*,†], Armando Crupi [1], Beatrice Longhi [1], Alessandro Campagna [1], Antonino Fiorino [2], Umberto Gibello [1,3,‡] and Andrea Roccuzzo [4,5,6,‡]

1 C.I.R. Dental School, Department of Surgical Sciences, University of Turin, 10126 Turin, Italy; armando.crupi@unito.it (A.C.); beatrice.longhi@unito.it (B.L.); alessandro.campag181@edu.unito.it (A.C.); umberto.gibello@polito.it (U.G.)

2 Department of Neuroscience, Reproductive and Odontostomatological Sciences, University of Naples Federico II, 80131 Naples, Italy; fiorinodr.antonino@gmail.com

3 DIMEAS, Politecnico di Torino, 10129 Turin, Italy

4 Shanghai Perio-Implant Innovation Center, Institute for Integrated Oral, Craniofacial and Sensory Research & Department of Oral and Maxillofacial Implantology Shanghai Ninth People Hospital, Shanghai Jiao Tong University School of Medicine, Shanghai 200115, China; andrea.roccuzzo@unibe.ch

5 Department of Periodontology, School of Dental Medicine, University of Bern, 3010 Bern, Switzerland

6 Unit for Practice-Based Research, School of Dental Medicine, University of Bern, 3010 Bern, Switzerland

* Correspondence: francesco.pera@unito.it (F.P.); camillovocaturo@gmail.com (C.V.)

† These authors contributed equally to this work.

‡ Co-shared senior author position.

Abstract: Objectives: The study's objective was to evaluate the accuracy of dynamic computer-assisted surgical implant placement systems during practical training on fresh defrozen cephali. **Methods:** Three defrozen cephali with terminal dentition received a total of 26 implants (15 4.3 × 13 mm and 11 4.3 × 13 mm, Nobel Biocare Service AG (Zrich-Flughafen Switzerland)) following a standardized protocol: a digital scanning and planning protocol followed by dynamic navigation surgery (X-Guide, X-Nav Technologies, LLC, Lansdale, PA, USA). All surgical interventions were performed by two surgeons: a senior oral surgeon (OE) with more than 5 years of implant dentistry experience and a non-experienced surgeon (NE). **Results:** Different linear and angular measurements (i.e., deviation shoulder point; deviation tip point; depth deviation shoulder point; depth deviation tip point; B/L and M/D angular deviations) were calculated in duplicate to estimate the discrepancy of the virtual digital planning with respect to the real clinical scenario. The differences between the two operators were also explored. The results of the bivariate analysis detected clinical negligible differences between the operators, without any statistically significant differences for all investigated parameters ($p > 0.05$). **Conclusions:** The preliminary positive findings of this pilot study suggest that the investigated dynamic navigation system could be a viable and safe technique for implant surgery and may offer additional safety benefits to non-experienced operators, despite the required learning.

Keywords: dental implants; dynamic computer-assisted surgery; computer-guided implantology; navigation systems

1. Introduction

It was widely reported that, despite the decrease over the last two decades of the prevalence of tooth loss, edentulism is still a global prevalent condition with differences among countries, age groups, and socioeconomic status [1]. Consequently, to overcome this problem, during the last three decades, the use of implant-supported oral rehabilitations has become the standard of care dramatically improving individuals' chewing function, esthetics, and patient-reported quality of life (PRQoL) [2,3].

From a diagnostic point of view, the implementation of cone beam computed tomography (CBCT) imaging prior to dental implant has increased significantly in recent years with documented advantages such as a decrease in radiation exposure compared with conventional computed tomography [4,5]. However, implant placement might present some clinical challenges, such as the presence of narrow ridges and slim margins and the need for preserving critical anatomical structures (i.e., mandibular nerve) [6]. In this respect, the use of pre-operative planning including CBCTs do represent a crucial step to minimize complications and maximize proper 3-D implant placement [7]. From a technical aspect, recent developments have allowed the use of voxel sizes down to tenths of a millimeter and the ability to visualize and measure anatomic structures in all spatial dimensions [5,8]. From a descriptive point of view, computer-assisted surgical (CAS) implant placement systems were the first to be introduced. Such systems can be categorized as either static or dynamic [5,9]. In the first clinical scenario, CAS systems use a drill within templates with embedded sleeves, which could be placed on the neighboring side of the surgical site to facilitate the transferring of the position of implant planning [10,11]. These templates could be supported by teeth, mucosa, or bone, helping conduct implants in an optimal position [10,12]. The reliability of this procedure has been clinically proven [10,13–15]. However, several studies have already shown that different factors could influence the accuracy of static CAS, e.g., the fabrication technology of surgical guides, the axial accuracy of the sleeves' housings, the lack of direct visual contact with the surgical site, intraoral positioning, and template fixation [10,16–19].

Static CAS systems use guides fabricated with a computer-aided design (CAD/CAM) based on 3D scans of the patient [5,9,19]. It is essential to highlight the differences in terms of digital impression accuracy on full-arch implant rehabilitations depending on the timing; in fact, immediate post-surgical intraoral digital scans may show a higher risk of imprecision than those obtained after tissue healing [20].

In contrast, dynamic CAS systems track the patient and surgical instruments and present real-time positional and guidance feedback on a computer display [5,16].

To improve these two technologies, the use of dynamic navigation systems was developed based on "motion-tracking technology", which tracks the position of the surgical site and drill real-time behaviors combined to the patient's presurgical CBCT [10,21–23]. In addition, during implant placement, tracking cameras are in use to continuously monitor the attached marker on the patient's jaw and surgical handpiece. This procedure is displayed in real-time on a screen superimposed on implant planning. Therefore, from a clinical perspective, any potential deviation of the implant and drill axis could be controlled and corrected [10,24]. Furthermore, the real-time monitoring of the procedure allows the surgeon to modify virtual planning during surgery [10,15].

Nevertheless, it is clinical experience that dynamic CAS requires higher training and experience than static CAS [25]. In this respect, considering the increasing use of dCAS among oral surgeons, and although previous investigations evaluated several factors in the accuracy of implant placement, there is a lack of evidence on the role of the surgeon's experience in the use of such technology and surgery accuracy [25,26].

To the best of the authors' knowledge, this study is the first cadaver study evaluating the influence of operator's experience on the accuracy of implant placement using dCAS. The primary aim was to investigate differences in terms of implant placement accuracy between expert and non-expert operators by evaluating discrepancies superimposing pre-operative and post-operative CBCT, measuring linear (mm) and angular (degrees) deviations.

2. Materials and Methods

2.1. Study Design

This pilot study evaluated the accuracy of implant placement using dCAS during practical training on fresh defrozen cephali. The samples were donated by individuals for scientific purposes and official laboratory permission to work on the cadavers was obtained from the Italian competent authority. This study was conducted according to the revised guidelines of the Declaration of Helsinki and did not require any ethical approval.

For this study, surgical sessions were performed on adult fresh defrozen cephalus, fixed with 10% formalin. Three partial edentulous cadavers with terminal dentition were selected to be treated with conventional implants.

The operators had different degrees of experience: one (NE) was a senior oral surgery resident at University of Turin, with implant dentistry experience of more than 5 years, while the second (OE) was a non-experienced oral surgeon (less than 5 years of clinical experience in implant dentistry).

2.2. Inclusion Criteria

The following inclusion criteria were used: (1) the absence of macroscopic pathology in the bony regions of the maxilla and mandible; (2) at least 3 residual teeth, required for the positioning and stabilization of thermoplastic devices (clip) with 3 radiopaque fiducials (X-Clip, X-Nav Technologies, Lansdale, PA, USA) during CBCT scan and surgical procedures.

2.3. Scanning and Planning Protocol

In this case, considering that fresh defrozen cephali had terminal dentition with >3 stable adjacent teeth and located in an area of the dental arch that did not violate the implant milling, a fiducial-based registration protocol was performed.

This protocol involves, before the acquisition of the pre-operative CBCT scan, placing a thermoplastic clip with three radiopaque fiducials (X-Clip, X-Nav Technologies) on the remaining teeth of the dental arch involved in the implant surgery (Figure 1).

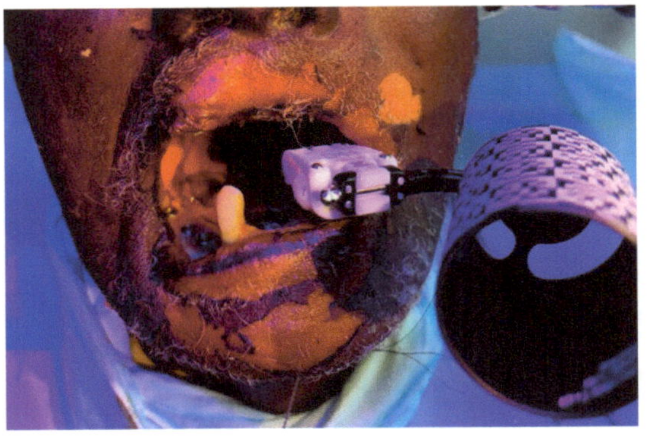

Figure 1. X-Clip placement on the fresh defrozen cephalus.

Then, a CBCT scans (NNT—Medical Suite®; NewTom, Imola, Italy) was performed with the following set up: 110 Kv, 1.94 mA, 3.6 s, 685.41 DAP (mGy×m^2), 100 × 140 FOV (mm) (Figure 2).

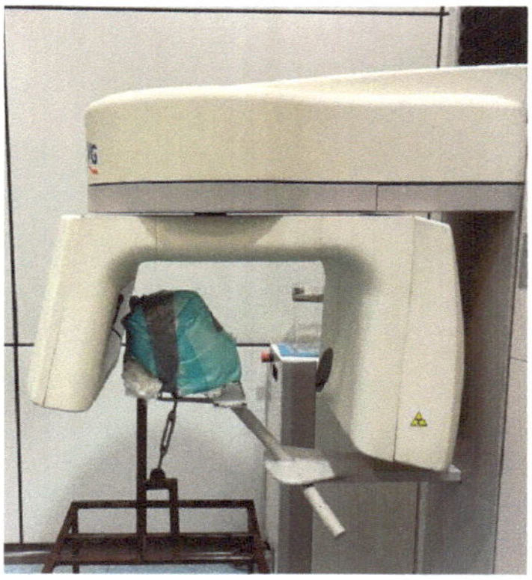

Figure 2. Representation of CBCT scan processing.

The digital information (.dicom data set) was uploaded to the dynamic navigation planning system (DTX StudioTM Implant 3.4.3.3, Nobel Biocare AB, 402 26 Vasta Hamngatan 1, 411 17 Göteborg, Sweden).

This software allowed us to define the arch, nerve mapping, and implant planning through MPR (multiplanar reformation). We used it to plan the ideal implant placement, which was identical to the implant placed in the cadavers (diameter, apical diameter, length, shape), thanks to an implant library contained on DTX.

Files from intraoral scanners were superimposed on a DICOM data set and the combined images allowed us to plan, with the osseous, dental, and soft tissue structures visible along with the patient's occlusion, prosthetic-guided implant placement (Figure 3).

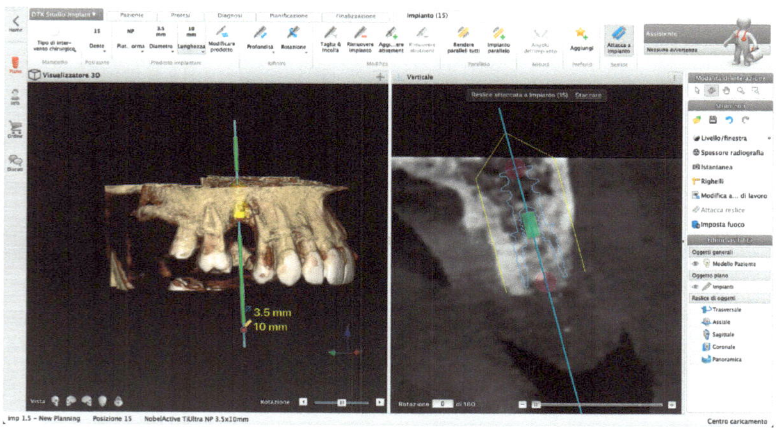

Figure 3. Example of implant placement planning.

When starting the planning, a panoramic curve for the arch is drawn on the axial plane of the patient's scan, and the inferior alveolar nerve also can be identified and referenced on the mandible.

Once the surgical plan was defined, the data set was exported by DTX (DTX StudioTM Implant 3.4.3.3, Nobel Biocare AB, 402 26 Vasta Hamngatan 1, 411 17 Göteborg, Sweden) and imported as an .stl file into the DN software.

A total of 26 implants were successfully placed by the two operators. At the end of the surgical interventions, a post-operative CBCT scan was taken to perform the accuracy evaluation.

2.4. Calibration Sessions and Surgical Procedures

Each surgeon performed a surgical session after five days of training, which included training on manikins, mentoring with over-the-shoulder observation and hands-on mentoring. No surgery was performed prior to the assurance of a high level of agreement between the two operators.

Before the initiation of this study, all practitioners received standard hands-on training for virtual planning with implant treatment planning software (DTX StudioTM Implant 3.4.3.3, Nobel Biocare AB, 402 26 Vasta Hamngatan 1, 411 17 Göteborg, Sweden) and surgical procedure simulation with the navigation system to achieve minimal proficiency.

All implants were positioned using a dynamic navigation surgery system (X-Guide, X-Nav Technologies). Depending on the implant site characteristics, conventional (with flap) or flapless surgical procedure was performed. Standardized implants were placed in all cases (15 NobelReplace Conical Connection 4.3 × 13 mm in mandible and 11 NobelActive TiUltra 4.3 × 13 mm in maxilla, Nobel Biocare, Zürich-Flughafen, Switzerland).

Prior to each surgical procedure, the clip was mounted on the teeth in the same position as CBCT scanning and attached fiducial markers and the cylinder of the attached patient tracking matrix, extraorally oriented. Likewise, the handpiece, patient tracking array, and drills were calibrated. All these instruments must be within the line of sight of the overhead stereo cameras to be tracked on the monitor.

Calibration of the surgical handpiece was performed before the surgical acts. Handpiece calibration relates the geometry of the handpiece tracking array to the drill axis and CBCT fiducials, hence providing a link between the pre-operative planning coordinate system and a trackable coordinate system. After calibration, the operators performed the surgery. Real-time checks were performed through patient's CBCT anatomy, and the implant coordinates were pre-planned to guarantee the accuracy of the tracking, all using the navigation screen on the monitor (Figure 4).

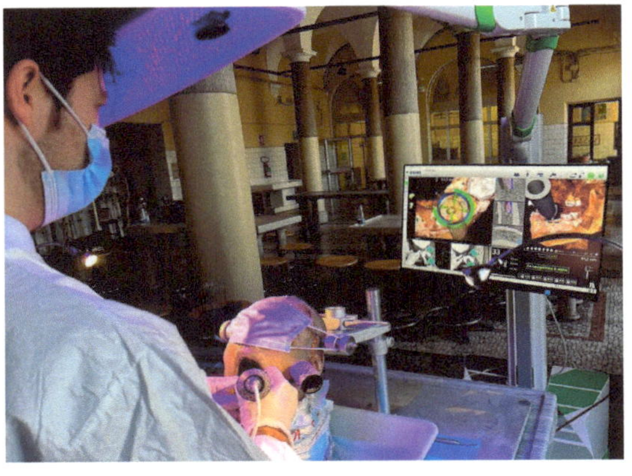

Figure 4. Real-time monitoring of implant placement.

If necessary, changes in the plan were made during surgical acts, including implant size, length, width, shape, and positioning, to achieve an accurate implant position.

Dynamic reference frame calibration relates the geometry of the patient-tracking array to the CT fiducials (i.e., three radiopaque fiducials of thermoplastic clip placed during

CBCT). This provides a link between the pre-operative planning coordinate system and a trackable coordinate system. The stereo-tracking system simultaneously triangulated each tracking array to determine their precise position and orientation in a common coordinate frame. In combination with the calibrations, this real-time link allowed the drill's body and tip to be related.

The patient dynamic reference frames included the clip with the connected patient-tracking cylinder. It was placed onto the teeth in the same location for CBCT acquisition. The tracking software algorithm triangulated the 2 arrays continuously. Two live video windows allowed the surgical team to obtain virtual feedback from the navigation system to visualize site preparation and monitor the quality of tracking in the surgical field volume.

2.5. Accuracy Analysis

The evaluated outcome variables were previously described [5]. After surgery, post-operative CBCT was performed with the same FOV and resolution of the pre-operative CBCT (110 Kv, 1.94 mA, 3.6 s, 685.41 DAP (mGy$\times$m^2), 100 $\times$ 140 FOV (mm)), to compare deviations between the planned and placed implants.

The accuracy of the implant placement of the two operators was assessed by superimposing the pre-operative virtual surgical plan and the post-operative CBCT scan and quantifying deviations of the delivered implant from the planned position and orientation. The same methodology proposed and validated by Block MS. et al. and Jorba-García A. et al. was implemented [25,27].

The DICOM images of the post-operative CT were uploaded in a dedicated software.

To obtain very precise results, the implants were planned and superimposed on placed implants to perfectly replicate the morphology of the implants (centroid apex and shoulder, and their spatial coordinates).

The accuracy was assessed overlapping the post-operative CT scan (with placed implants) with the pre-operative one (with planned implants). The accuracy evaluation involved angular and linear (coronal, apical, and depth) deviations. The DICOM images of the post-operative CT were uploaded in a dedicated software (NobelGuide validation study tool in DTX Studio Implant 3.6). A segmentation based on tissue density was carried out to separate implants from the surrounding bone.

The STL files of the maxillary and mandible bone with the planned implants obtained from the pre-operative CBCT were uploaded into the software. The superimposition of the pre-operative and post-operative CT images was achieved by using the best-fit alignment tool. The planned and inserted implants were considered as cones with a base and a centroid apex and shoulder and their spatial coordinates (the center of the base and the apex) were registered by using DTX and were exported in an Excel sheet to calculate coronal, apical, depth, and angular deviations.

A mathematical algorithm was used on the presurgical case with the plan, the post-surgical case with the virtual implant overlaid on the actual implant, and the meshed CBCT scans to calculate angular and positional deviations between the planned and actual implant positions in 3 dimensions.

The following deviations (mean $\pm$ standard deviation) from the virtual plan were calculated and listed in Figure 5:

- Mesiodistal (M/D) Angular Deviation: The mesiodistal angle between the vertical axes of the planned and placed implants.
- Buccolingual (B/L) Angular Deviation: The buccolingual angle between the vertical axes of the planned and placed implants.
- Deviation Shoulder Point (mm): A 2-dimensional distance between the shoulder centroids of the planned and placed implants.

- Deviation Tip Point (mm): A 2-dimensional distance between the apex centroids of the planned and placed implants.
- Depth Deviation Shoulder Point (mm): Depth distance between the shoulder centroids of the planned and placed implants on the *z*-axis.
- Depth Deviation Tip Point (mm): Depth distance between the apex centroids of the planned and placed implants on the *z*-axis.

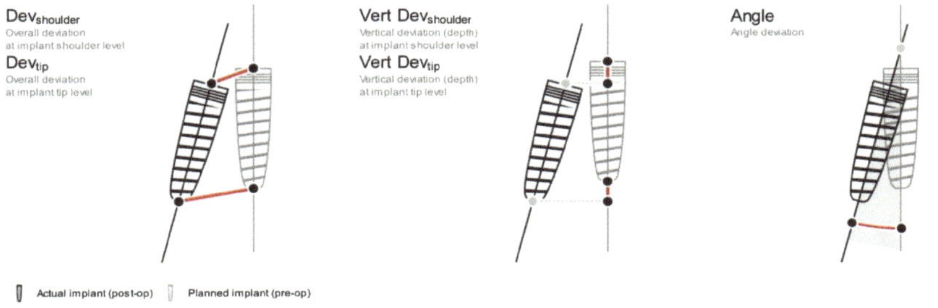

Figure 5. Representation of positional and angular deviations: first two images illustrate positional deviations; third image illustrate exemplification of angular deviations (M/D and B/L).

For each parameter, the mean of the two measurements performed by the two blinded and previously calibrated operators (Cohen's kappa coefficient = 0.72) was calculated and used for statistical analysis.

2.6. Statistical Analysis

To ensure objectivity and avoid bias, the operator variable was anonymized and a blinded external statistician, independent of the research team, conducted the data analysis. A database was created using Microsoft Excel (Microsoft, Redmond, WA, USA) for data recording and management.

Statistical analysis was performed using Stata (StataCorp LLC, College Station, TX, USA). The distribution of all continuous numerical variables was assessed using the Skewness/Kurtosis test for normality (sktest in Stata). Descriptive statistics were reported for each variable as mean, standard deviation (SD), median, and interquantile range as none of the variables satisfied the assumption of normality. Data were stratified according to operator experience level (i.e., experienced or non-experienced) as the primary outcome, and the jaw (maxilla or mandible) position (anterior or posterior). Statistical differences between the groups were evaluated using the Wilcoxon–Mann–Whitney test for non-parametric comparisons. A *p*-value < 0.05 was considered the threshold for statistical significance.

3. Results

A total of twenty-six implants were placed and analyzed in three fresh defrozen cephali, without deviations from the original digital surgical planning. All specimens were available for analysis since no intra-surgical complications were recorded.

Table 1 shows the deviation from the virtual plan of the main outcomes for each surgeon, in terms of mean, standard deviation (SD), median, and interquantile range of all placed implants and apex linear (shoulder, apex and depth) and angular deviations (M/D, B/L) stratified according to the analyzed variables.

Table 1. Results of main outcome variables for expert surgeon (OE), non-experienced surgeon (NE), and overall.

Group	Stats	N	Deviation Shoulder Point (mm)	Deviation Tip Point (mm)	Depth Deviation Shoulder Point (mm)	Depth Deviation Tip Point (mm)	B/L Angular Deviation (Grades)	M/D Angular Deviation (Grades)
OE	Mean	13	3.08	2.38	1.02	1.03	3.34	3.75
	SD		1.95	1.85	1.05	1.04	2.23	5.97
	p50		2.41	1.61	0.49	0.53	3.6	1.8
	IQR		2.43	1.96	1.22	1.22	3.5	2.8
NE	Mean	13	2.91	2.53	1.02	1.03	3.34	3.75
	SD		3.08	3.36	1.05	1.04	2.23	5.97
	p50		1.48	1.46	0.7	0.68	2.4	2.3
	IQR		1.57	1.38	0.55	0.51	4.2	3.8
p-value			0.23	0.52	0.98	0.88	0.78	0.54
TOTAL	Mean	13	3	2.45	0.88	0.86	3.76	3.38
	SD		2.53	2.66	0.81	0.80	2.97	4.44
	p50		1.94	1.60	0.62	0.66	3	2.15
	IQR		2.82	1.58	0.65	0.63	3.6	3

Data (mean values) with respect to the stratification according to jaw (i.e., maxilla vs. mandible) and position (i.e., anterior vs. posterior) are reported in Tables 2 and 3.

Table 2. Results of secondary outcome variables for site and overall.

Site	Stats	N	Deviation Shoulder Point (mm)	Deviation Tip Point (mm)	Depth Deviation Shoulder Point (mm)	Depth Deviation Tip Point (mm)	B/L Angular Deviation (Grades)	M/D Angular Deviation (Grades)
Mand	Mean	15	2.80	2.53	0.64	0.65	2.79	1.79
	SD		3.05	3.40	0.63	0.63	1.87	1.20
	p50		1.73	1.29	0.42	0.44	2.3	1.8
	IQR		1.25	0.7	0.65	0.59	3.1	1.5
Max	Mean	11	3.27	2.36	1.20	1.15	5.08	5.53
	SD		1.67	1.22	0.94	0.92	3.71	6.29
	p50		3.48	1.85	0.92	0.83	4.1	5.1
	IQR		3.42	1.94	1.37	0.93	6.3	3.9
p-value			0.17	0.15	0.14	0.10	0.12	0.018 *

* Indixates Statistically significant difference.

Analyzing the results regarding the role of the surgeon's experience, expert and non-expert surgeons showed an analogous accuracy during implant placement for each variable studied, so deviations were negligible ($p > 0.05$) in all outcomes. Details of all recorded data are listed in Tables 1–3.

Table 3. Results of secondary outcome variables for position and overall.

Position	Stats	N	Deviation Shoulder Point (mm)	Deviation Tip Point (mm)	Depth Deviation Shoulder Point (mm)	Depth Deviation Tip Point (mm)	B/L Angular Deviation (Grades)	M/D Angular Deviation (Grades)
Ant	Mean	13	2.92	1.83	0.78	0.78	4.59	2.93
	SD		1.51	0.99	0.65	0.59	3.39	1.81
	p50		2.62	1.61	0.49	0.64	4.4	2.6
	IQR		2.26	1.58	0.55	0.59	3.6	2.6
Post	Mean	13	3.07	3.08	0.97	0.94	2.93	3.82
	SD		3.32	3.59	0.96	0.97	2.31	2.31
	p50		1.68	1.6	0.7	0.68	2.4	2
	IQR		1.4	1.21	0.65	0.57	2.3	2.3
p-value			0.23	0.63	0.74	0.94	0.18	0.50

4. Discussion

To the best of the authors' knowledge, this is the first cadaver pilot study investigating the influence of surgeons' experience on the accuracy of implant placement using dCAS. The goal of the present study was to assess the potential differences between the operators' experiences and the definite outcomes: implant placement accuracy.

Since the introduction of three-dimensional imaging and visualization software before implant placement in the 1990s and the further introduction of dynamic navigation in the 2000s, the majority of recent research has been conducted on computer-assisted guidance to create the most accurate device for prosthetic-guided implant placement [28].

In this context, the goal of the present study was to investigate the association between surgeons' experiences and implant placement accuracy. The findings revealed encouraging results concerning the role of surgeons' experiences. Specifically, unexperienced surgeons showed higher performance than those with more experience in terms of linear and angular deviations during navigated implant placement, but no significant differences were found for any variables evaluated between the two operators ($p > 0.05$).

These results may seem counterintuitive, as one would typically expect higher accuracy from experienced surgeons because, with this surgical system, they might further improve their implant placement accuracy [29]. However, a possible reason could be that unexperienced surgeons adhered more strictly to the dynamic navigation system's guidance because of their short surgical experience, resulting in similar precision. In contrast, experienced surgeons may have relied more on their prior experience, which could have led to slight deviations from the system's guidance [30].

In light of the obtained descriptive results of this study, the outcomes aimed to evaluate the depth accuracy, requiring a more detailed analysis, as shown by the overall mean deviation at tip depth (0.86 mm—SD 0.79) and shoulder depth (0.88 mm—SD 0.81). It is a large discrepancy from the planning that is unacceptable in critical anatomical areas where the inferior alveolar nerve is at risk [31].

Consequently, a risk of implant deviation still exists with the navigation system due to the errors that might be generated during the workflow steps of image acquisition, tracking clip stability, registration and calibration, and errors when overlaying the two CBCT scans, as reported by prior studies [9,25,32–34].

Thus, in the authors' opinions, for other surgical approaches, a 2 mm safety margin should be applied to all important anatomical structures in presurgical planning [25]. From a clinical perspective, it should be recalled that the application of such a margin might raise some questions on the overall reliability of navigation systems. In other words, the clinical question, "Does this tool help implant placement in complex cases?", remains open.

When comparing the presented results with those available in the literature, it should be underlined that an in vitro study by Jorba-Garcià et al. revealed negligible differences between two operators ($p > 0.05$) using a dynamic navigation system. Consequently, the outcomes evaluated in the present cadaver study are in accordance with this paper [25]. Similarly, Pellegrino et al. showed an analogous implant placement accuracy between four operators with different grades of implant surgery experience, resulting in not statistically significant differences in the majority of the evaluated outcomes in terms of two/three-dimensional deviations [35].

Furthermore, Wang et al. investigated three different approaches (i.e., free hand, sCAS, and dCAS) and showed that experienced vs. non-experienced operators had an analogous accuracy, and differences between operators were not statistically significant ($p > 0.05$) [34,36].

Finally, Sun et al. and Wu et al. showed that the surgeons' experience levels did not influence implant placement accuracy with dCAS [37,38].

All cited studies revealed partially lower linear and angular deviations than the present study, but such results should be interpreted with more caution because the data concerning the accuracy of dCAS are obtained during in vitro trainings (i.e., using artificial models), which can lead to higher accuracy in comparison to real clinical scenarios, such as fresh defrozen cephalus [9,25,39,40].

In addition, a direct comparison with the pre-clinical scenario with respect to the type of implant placement (i.e., flapless vs. open surgery) might be an important confounding factor requiring further investigation.

Although the dynamic navigation system offers excellent accuracy, its application is limited in clinical practice mainly because of the required learning curve, the risk of inaccurate implant placement, as mentioned above, due to system error, and the high cost of the device. It represents a large economic investment for oral surgeons, which includes the cost per single case of fiducial clips, markers, and plates [25,36,41,42].

Despite the limitations, the present results can be considered promising positive preliminary results acting as a starting point for future clinical research with a larger sample size. Moreover, an important aspect that should be investigated is a bivariate analysis considering the implant planning (i.e., the gold-standard) and each operator's effective implant placement.

5. Conclusions

Within the limitations of this study in terms of sample size and clinical parameters evaluated (such as bone density), the present findings suggest that dynamic computer-assisted surgical implant placement systems could be a viable and safe technique for implant surgery by any operator, independently of surgical experience.

Based on the obtained results, this system might offer additional clinical benefits to an unexperienced operator, despite the required learning curve and the cost originated by the initial investment.

Author Contributions: Conceptualization, F.P. and C.V.; methodology, F.P., C.V. and U.G.; formal analysis, C.V., A.R. and A.F.; investigation, F.P., C.V. and U.G.; resources, F.P.; data curation, C.V.; writing—original draft preparation, F.P., C.V., U.G. and A.R.; writing—review and editing, C.V., A.C. (Armando Crupo), A.C. (Alessandro Campagna), U.G. and A.R.; visualization, F.P., A.C. (Alessandro Campagna), U.G. and B.L.; supervision, A.R.; project administration, F.P and A.R. All authors have read and agreed to the published version of the manuscript.

Funding: This research received no external funding.

Institutional Review Board Statement: Due to the nature of the study not involving human beings, this study did not require ethical approval. This study was conducted according to the revised guidelines of the Declaration of Helsinki.

Informed Consent Statement: Not applicable.

Data Availability Statement: Data are available after the authors' consultations.

Acknowledgments: The authors would like to thank the company Nobel Biocare for providing the materials.

Conflicts of Interest: The authors declare no conflicts of interest.

References

1. Borg-Bartolo, R.; Roccuzzo, A.; Molinero-Mourelle, P.; Schimmel, M.; Gambetta-Tessini, K.; Chaurasia, A.; Koca-Ünsal, R.B.; Tennert, C.; Giacaman, R.; Campus, G. Global prevalence of edentulism and dental caries in middle-aged and elderly persons: A systematic review and meta-analysis. *J. Dent.* **2022**, *127*, 104335. [CrossRef] [PubMed]
2. Duong, H.-Y.; Roccuzzo, A.; Stähli, A.; Salvi, G.E.; Lang, N.P.; Sculean, A. Oral health-related quality of life of patients rehabilitated with fixed and removable implant-supported dental prostheses. *Periodontol. 2000* **2022**, *88*, 201–237. [CrossRef] [PubMed]
3. Schimmel, M.; Araujo, M.; Abou-Ayash, S.; Buser, R.; Ebenezer, S.; Fonseca, M.; Heitz-Mayfield, L.J.; Holtzman, L.P.; Kamnoedboon, P.; Levine, R.; et al. Group 4 ITI Consensus Report: Patient benefits following implant treatment in partially and fully edentulous patients. *Clin. Oral Implant. Res.* **2023**, *34* (Suppl. S26), 257–265. [CrossRef] [PubMed]
4. Harris, D.; Horner, K.; Gröndahl, K.; Jacobs, R.; Helmrot, E.; Benic, G.I.; Bornstein, M.M.; Dawood, A.; Quirynen, M.E.A.O. guidelines for the use of diagnostic imaging in implant dentistry 2011. A consensus workshop organized by the European Association for Osseointegration at the Medical University of Warsaw. *Clin. Oral Implant. Res.* **2012**, *23*, 1243–1253. [CrossRef] [PubMed]
5. Emery, R.W.; Merritt, S.A.; Lank, K.; Gibbs, J.D. Accuracy of Dynamic Navigation for Dental Implant Placement-Model-Based Evaluation. *J. Oral Implantol.* **2016**, *42*, 399–405. [CrossRef] [PubMed]
6. Pjetursson, B.E.; Brägger, U.; Lang, N.P.; Zwahlen, M.E. Brägger, U.; Lang, N.P.; Zwahlen, M. Comparison of survival and complication rates of tooth-supported fixed dental prostheses (FDPs) and implant-supported FDPs and single crowns (SCs). *Clin. Oral Implant. Res.* **2007**, *18* (Suppl. S3), 97–113. [CrossRef] [PubMed]
7. Saini, R.S.; Bavabeedu, S.S.; Quadri, S.A.; Gurumurthy, V.; Kanji, M.A.; Kuruniyan, M.S.; Binduhayyim, R.I.H.; Avetisyan, A.; Heboyan, A. Impact of 3D imaging techniques and virtual patients on the accuracy of planning and surgical placement of dental implants: A systematic review. *Digit. Health* **2024**, *10*, 20552076241253550. [CrossRef]
8. Widmann, G.; Stoffner, R.; Schullian, P.; Widmann, R.; Keiler, M.; Zangerl, A.; Puelacher, W.; Bale, R.J. Comparison of the accuracy of invasive and noninvasive registration methods for image-guided oral implant surgery. *Int. J. Oral Maxillofac. Implant.* **2010**, *25*, 491–498.
9. Jung, R.E.; Schneider, D.; Ganeles, J.; Wismeijer, D.; Zwahlen, M.; Hämmerle, C.H.; Tahmaseb, A. Computer technology applications in surgical implant dentistry: A systematic review. *Int. J. Oral Maxillofac. Implant.* **2009**, *24*, 92–109.
10. Taheri Otaghsara, S.S.; Joda, T.; Thieringer, F.M. Accuracy of dental implant placement using static versus dynamic computer-assisted implant surgery: An in vitro study. *J. Dent.* **2023**, *132*, 104487. [CrossRef]
11. Ruppin, J.; Popovic, A.; Strauss, M.; Spüntrup, E.; Steiner, A.; Stoll, C. Evaluation of the accuracy of three different computer-aided surgery systems in dental implantology: Optical tracking vs. stereolithographic splint systems. *Clin. Oral Implant. Res.* **2008**, *19*, 709–716. [CrossRef]
12. Zhou, M.; Zhou, H.; Li, S.Y.; Zhu, Y.B.; Geng, Y.M. Comparison of the accuracy of dental implant placement using static and dynamic computer-assisted systems: An in vitro study. *J. Stomatol. Oral Maxillofac. Surg.* **2021**, *122*, 343–348. [CrossRef]
13. Tahmaseb, A.; Wu, V.; Wismeijer, D.; Coucke, W.; Evans, C. The accuracy of static computer-aided implant surgery: A systematic review and meta-analysis. *Clin. Oral Implant. Res.* **2018**, *29* (Suppl. S16), 416–435. [CrossRef] [PubMed]
14. Al Yafi, F.; Camenisch, B.; Al-Sabbagh, M. Is Digital Guided Implant Surgery Accurate and Reliable? *Dent. Clin. N. Am.* **2019**, *63*, 381–397. [CrossRef] [PubMed]
15. Gargallo-Albiol, J.; Barootchi, S.; Salomó-Coll, O.; Wang, H.L. Advantages and disadvantages of implant navigation surgery. A systematic review. *Ann. Anat. Anat. Anz. Off. Organ Anat. Ges.* **2019**, *225*, 1–10. [CrossRef]
16. Brief, J.; Edinger, D.; Hassfeld, S.; Eggers, G. Accuracy of image-guided implantology. *Clin. Oral Implant. Res.* **2005**, *16*, 495–501. [CrossRef] [PubMed]

17. Lo Russo, L.; Guida, L.; Mariani, P.; Ronsivalle, V.; Gallo, C.; Cicciù, M.; Laino, L. Effect of Fabrication Technology on the Accuracy of Surgical Guides for Dental-Implant Surgery. *Bioengineering* **2023**, *10*, 875. [CrossRef] [PubMed]
18. Ronsivalle, V.; Venezia, P.; Bennici, O.; D'Antò, V.; Leonardi, R.; Giudice, A.L. Accuracy of digital workflow for placing orthodontic miniscrews using generic and licensed open systems. A 3d imaging analysis of non-native .stl files for guided protocols. *BMC Oral Health* **2023**, *23*, 494. [CrossRef] [PubMed]
19. Tahmaseb, A.; Wismeijer, D.; Coucke, W.; Derksen, W. Computer technology applications in surgical implant dentistry: A systematic review. *Int. J. Oral Maxillofac. Implant.* **2014**, *29*, 25–42. [CrossRef] [PubMed]
20. Bagnasco, F.; Menini, M.; Pesce, P.; Crupi, A.; Gibello, U.; Delucchi, F.; Carossa, M.; Pera, F. Comparison of Full-Arch Intraoral Scans Immediately After Implant Insertion Versus Healed Tissue: A Multicentric Clinical Study. *Prosthesis* **2024**, *6*, 1359–1371. [CrossRef]
21. Vercruyssen, M.; Fortin, T.; Widmann, G.; Jacobs, R.; Quirynen, M. Different techniques of static/dynamic guided implant surgery: Modalities and indications. *Periodontol. 2000* **2014**, *66*, 214–227. [CrossRef] [PubMed]
22. Mediavilla Guzmán, A.; Riad Deglow, E.; Zubizarreta-Macho, Á.; Agustín-Panadero, R.; Hernández Montero, S. Accuracy of Computer-Aided Dynamic Navigation Compared to Computer-Aided Static Navigation for Dental Implant Placement: An In Vitro Study. *J. Clin. Med.* **2019**, *8*, 2123. [CrossRef] [PubMed]
23. Du, Y.; Wangrao, K.; Liu, L.; Liu, L.; Yao, Y. Quantification of image artifacts from navigation markers in dynamic guided implant surgery and the effect on registration performance in different clinical scenarios. *Int. J. Oral Maxillofac. Implant.* **2019**, *34*, 726–736. [CrossRef] [PubMed]
24. Kaewsiri, D.; Panmekiate, S.; Subbalekha, K.; Mattheos, N.; Pimkhaokham, A. The accuracy of static vs. dynamic computer-assisted implant surgery in single tooth space: A randomized controlled trial. *Clin. Oral Implant. Res.* **2019**, *30*, 505–514. [CrossRef]
25. Jorba-García, A.; Figueiredo, R.; González-Barnadas, A.; Camps-Font, O.; Valmaseda-Castellón, E. Accuracy and the role of experience in dynamic computer guided dental implant surgery: An in-vitro study. *Med. Oral Patol. Oral Cir. Bucal* **2019**, *24*, e76–e83. [CrossRef]
26. Pozzi, A.; Pozzi, A.; Carosi, P.; Laureti, A.; Mattheos, N.; Pimkhaokham, A.; Chow, J.; Arcuri, L. Accuracy of navigation guided implant surgery for immediate loading complete arch restorations: Prospective clinical trial. *Clin. Implant Dent. Relat. Res.* **2024**, *26*, 954–971. [CrossRef] [PubMed]
27. Block, M.S.; Emery, R.W. Static or Dynamic Navigation for Implant Placement-Choosing the Method of Guidance. *J. Oral Maxillofac. Surg. Off. J. Am. Assoc. Oral Maxillofac. Surg.* **2016**, *74*, 269–277. [CrossRef]
28. Miranda, N.O.; Dos Anjos, L.M.; Rocha, A.O.; Dallepiane FG da Cruz, A.C.C.; Cardoso, M.; Henriques, B.; Benfatti, C.A.M.; Magini, R.S. Global research interest and publication trends on guided surgery in implant dentistry: A metrics-based analysis. *J. Prosthet. Dent.* **2025**, in press. [CrossRef] [PubMed]
29. Einsiedel, D.; Giacaman, S.K.; Seidel, A.; Berger, L.; Buchbender, M.; Wichmann, M.; Matta, R.E. Accuracy of full-guided versus half-guided implant procedures carried out with digital implant planning software by students as part of a university curriculum. *BMC Med. Educ.* **2024**, *24*, 1316. [CrossRef]
30. Seth, C.; Bawa, A.; Gotfredsen, K. Digital versus conventional prosthetic workflow for dental students providing implant-supported single crowns: A randomized crossover study. *J. Prosthet. Dent.* **2024**, *131*, 450–456. [CrossRef] [PubMed]
31. Sammartino, G.; Marenzi, G.; Citarella, R.; Ciccarelli, R.; Wang, H.L. Analysis of the occlusal stress transmitted to the inferior alveolar nerve by an osseointegrated threaded fixture. *J. Periodontol.* **2008**, *79*, 1735–1744. [CrossRef] [PubMed]
32. Elian, N.; Jalbout, Z.N.; Classi, A.J.; Wexler, A.; Sarment, D.; Tarnow, D.P. Precision of flapless implant placement using real-time surgical navigation: A case series. *Int. J. Oral Maxillofac. Implant.* **2008**, *23*, 1123–1127.
33. Gaggl, A.; Schultes, G. Assessment of accuracy of navigated implant placement in the maxilla. *Int. J. Oral Maxillofac. Implant.* **2002**, *17*, 263–270.
34. Wang, X.; Shaheen, E.; Shujaat, S.; Meeus, J.; Legrand, P.; Lahoud, P.; do Nascimento Gerhardt, M.; Politis, C.; Jacobs, R. Influence of experience on dental implant placement: An in vitro comparison of freehand, static guided and dynamic navigation approaches. *Int. J. Implant Dent.* **2022**, *8*, 42. [CrossRef]
35. Pellegrino, G.; Bellini, P.; Cavallini, P.F.; Ferri, A.; Zacchino, A.; Taraschi, V.; Marchetti, C.; Consolo, U. Dynamic Navigation in Dental Implantology: The Influence of Surgical Experience on Implant Placement Accuracy and Operating Time. An in Vitro Study. *Int. J. Environ. Res. Public Health* **2020**, *17*, 2153. [CrossRef]
36. Wang, X.; Shujaat, S.; Meeus, J.; Shaheen, E.; Legrand, P.; Lahoud, P.; Gerhardt, M.D.N.; Jacobs, R. Performance of novice versus experienced surgeons for dental implant placement with freehand, static guided and dynamic navigation approaches. *Sci. Rep.* **2023**, *13*, 2598. [CrossRef]
37. Sun, T.-M.; Lee, H.E.; Lan, T.H. The influence of dental experience on a dental implant navigation system. *BMC Oral Health* **2019**, *19*, 222. [CrossRef] [PubMed]

38. Wu, D.; Zhou, L.; Yang, J.; Zhang, B.; Lin, Y.; Chen, J.; Huang, W.; Chen, Y. Accuracy of dynamic navigation compared to static surgical guide for dental implant placement. *Int. J. Implant Dent.* **2020**, *6*, 78. [CrossRef]
39. Jorba-García, A.; González-Barnadas, A.; Camps-Font, O.; Figueiredo, R.; Valmaseda-Castellón, E. Accuracy assessment of dynamic computer-aided implant placement: A systematic review and meta-analysis. *Clin. Oral Investig.* **2021**, *25*, 2479–2494. [CrossRef] [PubMed]
40. Wei, S.-M.; Zhu, Y.; Wei, J.X.; Zhang, C.N.; Shi, J.Y.; Lai, H.C. Accuracy of dynamic navigation in implant surgery: A systematic review and meta-analysis. *Clin. Oral Implant. Res.* **2021**, *32*, 383–393. [CrossRef] [PubMed]
41. Spille, J.; Jin, F.; Behrens, E.; Açil, Y.; Lichtenstein, J.; Naujokat, H.; Gülses, A.; Flörke, C.; Wiltfang, J. Comparison of implant placement accuracy in two different preoperative digital workflows: Navigated vs. pilot-drill-guided surgery. *Int. J. Implant Dent.* **2021**, *7*, 45. [CrossRef] [PubMed]
42. Panchal, N.; Mahmood, L.; Retana, A.; Emery, R., 3rd. Dynamic Navigation for Dental Implant Surgery. *Oral Maxillofac. Surg. Clin. N. Am.* **2019**, *31*, 539–547. [CrossRef] [PubMed]

prosthesis

Article

Adaptation of 3D-Printed and Milled Titanium Custom Post and Core

Abdulaziz A. Alzaid [1,*], Sarah Bukhari [2], Mathew T. Kattadiyil [3], Hatem Alqarni [1], Abdulaziz A. AlHelal [4], Khalid K. Alanazi [5], Montry S. Suprono [6], Rami Jekki [3] and Erik F. Sahl [7]

1 Department of Restorative and Prosthetic Dental Sciences, College of Dentistry, King Saud bin Abdulaziz University for Health Sciences, King Abdullah International Medical Research Center, Riyadh 14611, Saudi Arabia; qarnih@ksau-hs.edu.sa
2 Division of Prosthodontics, Department of Oral and Maxillofacial Rehabilitation, Ibn Sina National College for Medical Studies, Jeddah 22421, Saudi Arabia; sbukhari@ibnsina.edu.sa
3 Advanced Specialty Education Program in Prosthodontics, School of Dentistry, Loma Linda University, Loma Linda, CA 92354, USA; mkattadiyil@llu.edu (M.T.K.); rjekki@llu.edu (R.J.)
4 Department of Prosthetic Dental Sciences, College of Dentistry, King Saud University, Riyadh 11451, Saudi Arabia; abalhelal@ksu.edu.sa
5 Conservative Dental Science Department, College of Dentistry, Prince Sattam bin Abdulaziz University, Al-Kharj 16278, Saudi Arabia; kk.alanazi@psau.edu.sa
6 Center for Dental Research, School of Dentistry, Loma Linda University, Loma Linda, CA 92354, USA; msuprono@llu.edu
7 Advanced Specialty Education Program in Periodontics, School of Dentistry, Loma Linda University, Loma Linda, CA 92354, USA; esahl06d@llu.edu
* Correspondence: zaidab@ksau-hs.edu.sa; Tel.: +966-114299999 (ext. 95916)

Abstract: Background/Objectives: The purpose of this in vitro study was to evaluate and compare the internal adaptation and cement film thickness of cast-gold custom post and core (CPC), three-dimensionally (3D)-printed titanium (Ti) CPC, and milled Ti CPC. Methods: Forty-eight 3D printed resin models, simulating a tooth prepared to receive a CPC, were fabricated. Models were randomly assigned to one of three groups ($n = 16$ per group): (A) cast-gold CPC (control group), (B) 3D-printed Ti CPC, and (C) milled Ti CPC. Following the manufacturing of CPCs, each CPC was cemented using dual-cure polymerizing resin cement. Then, each model/post-and-core assembly was sectioned at the coronal, middle, and apical thirds of the post at a specific point. Each section was photographed using a microscope in a standardized setting (25×). The pixel count for cement surface area was calculated for each image using Adobe Photoshop software. Descriptive statistics of the mean and standard deviation of the cement film thickness around posts were calculated. Kruskal–Wallis and Dwass–Steel–Critchlow–Fligner tests were used for statistical analysis, with a significance level of $\alpha = 0.05$. Results: Pairwise comparisons in the coronal section revealed a statistically significant difference ($p < 0.05$) between groups A and B and groups B and C. In the middle section, there was a statistically significant difference ($p < 0.05$) between groups A and B only. In the apical section, there was a statistically significant difference ($p < 0.05$) between all groups. Conclusions: Within the limitation of the present study, neither 3D printed nor milled Ti CPC could achieve comparable cement film thickness to cast-gold CPC in all three sections. Cast-gold CPC cement film thickness was found to be more reduced and consistent, thus having superior internal adaptation to 3D-printed and milled Ti CPCs.

Keywords: post and core; adaptation; CAD/CAM; 3D printing; milling

1. Introduction

Multiple factors have been associated with the success of post and core when restoring endodontically treated teeth. These factors include the amount of remaining tooth structure (structural integrity), the composition of the post material, the modulus of elasticity of the post alloy, post diameter, cement layer thickness, and the length of the post [1–4]. Cast-gold custom post and core (CPC) has been considered the "gold standard" in CPC restorations due to its superior long-term success rate [5–7]. However, due to higher fabrication costs, to reduce chair time, and to simplify the restorative procedure, alternative treatment modalities to CPC have been developed. This has resulted in the use of less expensive metal alloys, prefabricated posts, and core buildups with either amalgam or composite resin materials [3–6,8].

With the advancement of computer-aided design and computer-aided manufacturing (CAD/CAM) technology, potential inaccuracies in the dental casting technique have been eliminated with the introduction of milling and 3-dimensional (3D) printing [9–14]. The application of titanium (Ti) alloy has been successful in multiple aspects of dentistry, with very promising clinical outcomes. This has been made possible with the advancements in CAD/CAM dental technology. The favorable mechanical and physical properties of Ti allow for both milling and printing methods of fabrication. When comparing CAD/CAM technologies to conventional methods, CAD/CAM technologies have been reported to reduce manufacturing time and inter-operator errors, and improve the overall efficiency of dental treatment. Another advantage of 3D printing is that material waste can potentially be kept to a minimum [13,14].

The clinical success of custom posts could be significantly impacted by adaptation and cement film thickness [15]. Cement film thickness uniformity is an essential factor when considering stress distribution. A less than ideal adaptation of the post can lead to an excessively thick cement layer, which is a negative factor for the long-term success of post-and-core treatment and correlates with higher frequencies of post debonding [16,17]. A minimum and uniform cement layer indicates that the post is well adapted to the canal space [18], thereby enhancing tooth fracture resistance [19] and reducing the risk for post debonding [20]. A poorly adapted post could increase the risk of tooth fracture [21] and microleakage [20], which can progress to cause marginal discoloration [22], secondary caries [23,24], and even compromise the apical seal [25].

The objective of this study is to evaluate and compare the internal adaptation of cast-gold CPC, 3D-printed Ti alloy CPC, and milled Ti alloy CPC. The null hypothesis was that no difference between cast-gold CPC, 3D-printed Ti alloy CPC, and milled Ti alloy CPC in regard to internal adaptation and cement film thickness will be found.

2. Materials and Methods

For the purpose of standardization in this study, a digital light processing (DLP) 3D printer (NextDent 5100; NextDent, Soesterberg, The Netherlands) was used to print 48 resin models. These models were made to simulate a tooth that was prepared to receive a CPC. Each tooth model was 36 mm in height, with post space occupying the coronal 8 mm, with a taper of 6 degrees, while maintaining a ferrule of 2 mm in height and 1 mm in width circumferentially (Figure 1). An impression was made for each model using a light body polyvinyl siloxane (PVS) impression material (Examix, GC America Inc., Alsip, IL, USA). PVS impression material was mixed and injected into the post space. To ensure an accurate impression of the post space, a plastic Para-post system pattern (ParaPost XP; Coltene/Whaledent Inc., Cuyahoga Falls, OH, USA) was inserted, followed by a sectional impression tray that was loaded with the heavy-body PVS material (Examix, GC America Inc., Alsip, IL, USA). The sectional tray was painted with tray adhesive material (VPS Tray Adhesive; 3M ESPE, St. Paul, MN, USA) prior to injecting the heavy-body PVS, and was left to dry for 7 min. Models were randomly assigned to one of three groups: (A) cast-gold CPC, which served as the control group, (B) 3D-printed titanium CPC, and (C) milled titanium CPC.

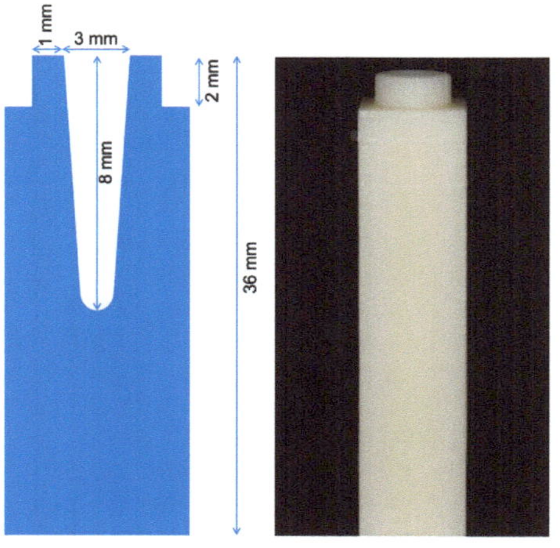

Figure 1. The standardized resin model design and dimensions, which simulate a tooth prepared to receive a CPC.

For group A, the cast-gold CPC group, impressions were poured in type V dental stone (Die-Keen; Whip Mix Corp, KY, USA). Cast-gold CPCs were then fabricated using pattern resin (GC Pattern Resin; GC America Inc., Alsip, IL, USA). Following post space lubrication using petroleum jelly (Vaseline, Uniliver, NJ, USA), the plastic Para-post system pattern (ParaPost XP; Coltene/Whaledent Inc., Cuyahoga Falls, OH, USA) was layered with pattern resin, and an impression of the post space was captured. The core was built with the same material. The prepared patterns for the cast-gold CPCs were then invested in suitable investment material (Beauty-Cast; Whip Mix, Louisville, KY, USA) without a ring liner and cast in Type-III gold alloy (Jensen Dental, North Haven, CT, USA).

For group B, the 3D-printed titanium alloy CPC group, impressions were scanned (3Shape D900L; 3Shape, Copenhagen, Denmark) and the obtained standard tessellation language (STL) files were used for designing and generating CAM files (Dental System; 3Shape, Copenhagen, Denmark) for printing the titanium (Ti-6Al-4V alloy) CPCs through direct metal laser sintering (DMLS) technology (Renovis Surgical Technologies, Redlands, CA, USA). The same technique for fabricating group B was used to fabricate group C, the milled titanium alloy (Ti-6Al-4V alloy) CPCs, except that the generated STL files were sent out for milling the titanium CPCs (Core3dcentres, Las Vegas, NV, USA) (Figures 2 and 3).

Figure 2. Schematic diagram of study material and methods.

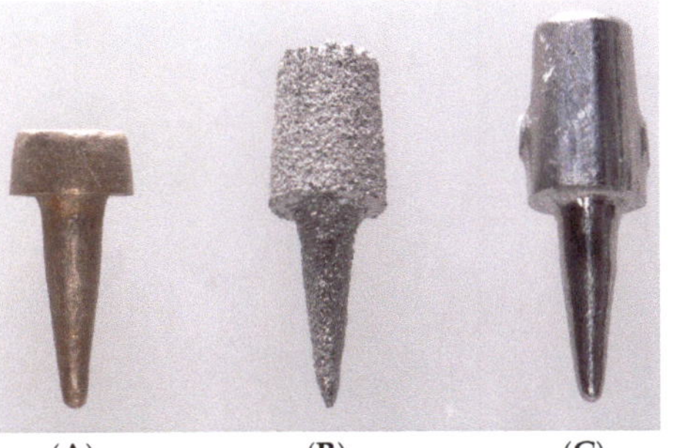

(A) (B) (C)

Figure 3. A representative sample of group (**A**): cast-gold CPC; group (**B**): 3D-printed Ti alloy CPC; and group (**C**): milled Ti alloy CPC, respectively.

All CPCs were evaluated visually and by using a dental explorer along the margin for full seating. Aerosol indicator spray (Occlude; Pascal Company Inc., Bellevue, WA, USA) was used to check for premature contacts that prevented the complete seating of the post, and adjustments were made using a fine diamond bur under copious water irrigation. Each adjustment was made as a single uniform stroke over the high spot, and it was repeated until the indicator spray mark appeared homogenous with the other parts of the post. The number of adjustments needed to achieve complete seating was recorded for each group.

Before cementation, airborne particle abrasion was performed using 250 μm Al_2O_3 (Renfert, St. Charles, IL, USA) particles under a pressure of 0.4 MPa, which was followed by cleaning using 70% ethanol. Cementation was completed using dual-cure polymerizing resin cement (Relyx Unicem; 3M ESPE, St. Paul, MN, USA). Cement mixing was achieved following manufacturer instructions (3M ESPE, St. Paul, MN, USA). All posts were coated with the cement. Cement was also extruded into the canal space by using a syringe with a 0.36 mm capillary tip (Ultradent Products Inc., South Jordan, UT, USA). Posts were then introduced gently into the canals with a gentle rocking motion to decrease hydrostatic pressure and to ensure complete seating. Once complete seating had been achieved, firm finger pressure was applied by one operator (AA). Excess cement was cleaned around the margin, and light polymerization was performed with a light-emitting diode (LED) light (VALO; Ultradent Products Inc., South Jordan, UT, USA) for 20 s on each surface.

Then, 24 h following cementation, all models were sectioned at 3 specific levels representing coronal, middle, and apical thirds of the post. To ensure consistency in sectioning, all models were mounted in the same position and sectioned with a low-speed saw machine (Techcut 4; Allied high tech products Inc., Compton, CA, USA) using 0.3 mm thickness diamond saw blades (Covington engineering, Meridian, ID, USA). A total of 5 diamond saw blades were used for sectioning. Each blade was used for sectioning 9 samples, 3 from each group, in an ordered fashion in which a different group was sectioned with each new blade. Sections were created horizontally under water cooling at levels of 1, 4, and 7 mm from the resin model margin, dividing the resin model/post-and-core assembly into 4 sections, of which the middle 2 sections were of the same separation dimension of 3 mm. The first of the two sectioned resin model/post assembly was used for measurements of the coronal and middle sections, and the second for the apical section measurements (Figure 4). Blue dye (2% methylene blue, Polysciences, Inc., Warrington, PA, USA) was used to stain the sectioned model/post-and-core assembly for 1 min, and then each was dried carefully with absorbent paper.

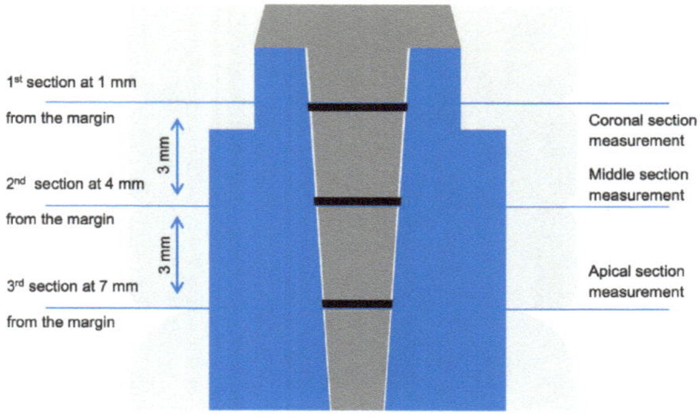

Figure 4. A scheme representing the resin model/post-and-core assembly and sectioning levels.

A light microscope (SteREO; ZEISS, Oberkochen, Germany) was used at 2.5× objective magnification and 10× eyepiece magnification, and images of these sections were captured using a digital single-lens reflex (DSLR) camera (EOS Rebel T6s; Canon, Tokyo, Japan). To ensure standardization, all microscope and camera settings were fixed. A stand was designed, 3D-printed (Form2, Formlabs Inc., Somerville, MA, USA) and then fixated on the microscope platform to ensure all sections were placed in the same position on the microscope platform and to keep the microscope settings the same throughout all sample measurements.

Measurements on the images of the cement surface area were made by a blinded examiner (SB) using image editing software (Photoshop; Adobe Systems Ltd., San Jose, CA, USA) by using the "pen tool" and "make path" options (Figures 5 and 6). The total surface area was marked, and the number of pixels recorded from the histogram option for each section. Descriptive statistics of the mean and standard deviation of the cement film thickness around posts, represented in pixel count, were calculated.

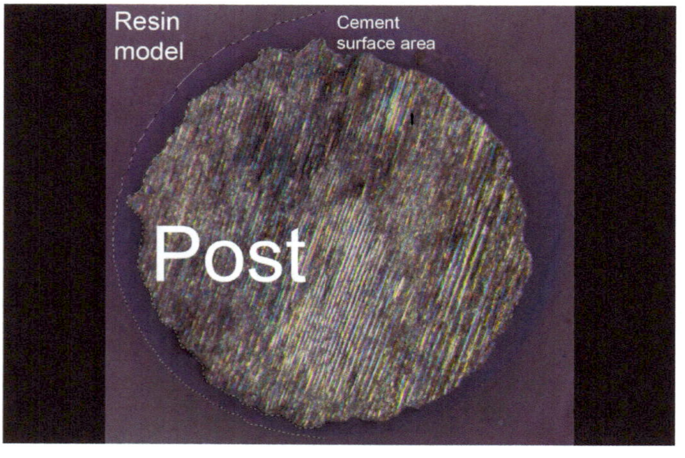

Figure 5. A representative sample of group B obtained using a light microscope at 2.5× objective magnification and 10× eyepiece magnification. The cement surface area was marked using image editing software (Photoshop; Adobe Systems Ltd., San Jose, CA, USA).

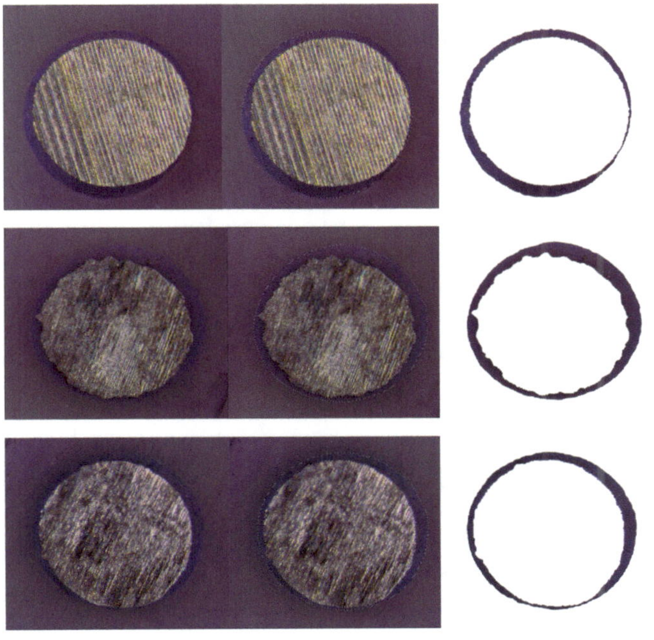

Figure 6. Microscopic images of the coronal section of representative samples of cast-gold CPC, 3D-printed Ti alloy CPC, and milled Ti alloy CPC, respectively, highlighting the cement space.

Statistical analysis was performed using Jamovi [26] and R Core [27] software. A Kruskal–Wallis test was used to compare the three groups. The Dwass–Steel–Critchlow–Fligner method was used for pairwise comparison. All tests were performed with a significance level of $\alpha = 0.05$.

3. Results

To achieve full CPC seating prior to cementation, group C required the greatest number of adjustments, ranging from 2 to 14 times. Group A adjustments ranged from nought to five times, and group B adjustments ranged from nought to three times. The mean and standard deviation values for CPC adaptation in each group for each section were calculated (Table 1). The Kruskal–Wallis test revealed a statistically significant difference among all groups in the coronal (χ^2 33.1, df 2, $p < 0.001$, ε^2 0.705), middle (χ^2 16.2, df 2, $p < 0.001$, ε^2 0.345), and apical (χ^2 35.4, df 2, $p < 0.001$, ε^2 0.753) sections.

Table 1. Means ± standard deviations of pixel count for all groups.

Groups	Group A *	Group B #	Group C ^
Coronal	10,451 ± 4701 [a]	26,044 ± 4464	13,992 ± 4344 [a]
Middle	11,412 ± 8164	30,458 ± 21,955 [b]	23,337 ± 8860 [b]
Apical	17,737 ± 6391	51,106 ± 5949	31,193 ± 9609

Data within rows followed by the same superscript letter are not significantly different ($p > 0.05$). * Group A: cast-gold custom post and core (CPC); # group B: 3D-printed Ti alloy CPC; ^ group C: milled Ti alloy CPC.

For the coronal section, pairwise comparisons revealed a statistically significant difference ($p < 0.001$) between groups A and B. Also, there was a statistically significant difference ($p < 0.001$) between groups B and C. No statistically significant differences were found ($p = 0.113$) between groups A and C (Table 1). For the middle sections, pairwise comparisons revealed statistically significant differences between groups A and B ($p < 0.001$)

and groups A and C ($p = 0.006$). There was no statistically significant difference ($p > 0.05$) between groups B and C (Table 1). For the apical section, pairwise comparisons revealed statistically significant differences ($p < 0.001$) among all groups (Table 1).

4. Discussion

Based on the findings of the present study, the null hypothesis was rejected. Neither 3D-printed nor milled Ti CPC could achieve comparable cement film thickness to cast-gold CPC in all three sections. Most in vitro post-and-core studies have used extracted teeth for the testing of examined variables, such as adaptation, retention, and fracture resistance [2–6]. However, to eliminate the added anatomical variations between extracted teeth, 3D-printed resin models were used in this study. This was possible through advancements in additive manufacturing and CAD/CAM technology capable of producing accurate and reliable models for dental workflow [28,29].

Other studies have used pixel counts for the calculation and comparison of surface areas of shapes with irregular configurations [30,31]. This method can produce accurate results under two conditions: first, the camera position and setting must be standardized; second, all samples must be positioned at a standardized location and distance from the camera. These conditions were uniformly applied in the present study.

For restoring endodontically treated teeth, CPC offers superior adaptation and fit [32], improved resistance to rotational forces [33], and higher success rates [5,6] when compared to prefabricated posts. Cast-gold CPC has been proven to have long-term success rates [5–7], high fracture resistance [34], high corrosion and tarnish resistance, biocompatibility [35], and casting predictability [36]. Hence, cast-gold CPC is still considered the "gold standard" for restoring endodontically treated teeth [5,6,36], and was chosen to serve as the control group in the current study. However, like any restorative material, cast-gold CPC has some disadvantages that might influence clinicians to seek alternative materials. These disadvantages include higher cost, increased fabrication time, limitations with translucent higher esthetic restorations, and unfavorable failure patterns [32].

When used as a restorative material for endodontically treated teeth, titanium has the following advantages: high corrosion resistance, very low allergenic potential, low toxicity, and high biocompatibility. All of these will eventually result in a favorable biological response [37,38]. Also, the modulus of elasticity of CPC manufactured from Ti alloys is lower than predominantly base metal alloys and zirconia CPC [35]. This can result in superior fracture resistance and a more favorable failure pattern for teeth restored with milled Ti alloy or 3D-printed CPC. Furthermore, titanium's color can be altered through anodization, which could be advantageous in esthetic situations [39].

The American Society for Testing and Materials (ASTM) [40] classifies commercially pure titanium (CP Ti) into four grades based on the concentration of impurities. Of these, grade I is the purest and grade IV is the least pure. As CP Ti impurity concentration increases, its mechanical properties will improve. However, due to its overall low mechanical properties, CP Ti bio-medical utilization is limited to situations where high strength is not required. Ti alloys were developed to overcome CP Ti mechanical properties and maintain their favorable biological response. Ti-6Al-4V alloy is the most widely used Ti alloy for medical and dental applications due to its superior mechanical properties and long-term success, and therefore it was used in the current study [35,41].

Results from this study showed that the mean for cement film thickness around cast-gold CPC was less than that of the other two groups. Also, the standard deviation was more uniform, which indicates the reproducibility of the cement film thickness between cast-gold CPC samples in all sections. When compared to 3D-printed Ti CPC, cast-gold CPC had significantly lower cement film thickness ($p < 0.05$), and therefore superior adaptation in all three sections examined. When compared to milled Ti CPC, cast-gold CPC showed significantly lower cement film thickness in the apical and middle sections ($p < 0.05$), but there was no significant difference in the coronal section ($p > 0.05$). However, the authors suggest interpreting this finding with caution due to the fact that milled Ti CPCs needed more adjustments to achieve full seating.

Milled Ti CPC had significantly lower cement film thickness ($p < 0.05$) in the coronal and apical sections than 3D-printed Ti CPC, and this can be interpreted to mean that milling is superior to 3D printing in the manufacturing of CPC, in terms of post adaptation, but it requires more adjustments to achieve full seating. This finding is highlighted in the apical section measurements, where 3D-printed Ti CPC showed very consistent and large cement film thickness results in all of its samples. This observation could offer a better understanding of the limitations of Ti 3D printing technology with finer detail production. In an attempt to reduce human errors and maximize the validity of this study, finishing and polishing was not performed for the samples since these procedures could improve the fit of CPC and influence the outcome of the study. The only necessary adjustments to achieve full seating were completed by a single examiner to eliminate inter-examiner variations. Among all groups, the milled Ti CPC group showed the worst initial fit for all samples, and required considerable adjustments, ranging from 2 to 14 times, to achieve full seating. In comparison, the cast-gold CPC group adjustment range was nought to five times, and the 3D-printed Ti CPC group adjustment range was nought to three times, suggesting that even though milled Ti CPC showed better results compared to 3D-printed Ti CPC, its clinical application might be less appealing due to the extended chairside time that will be required for necessary adjustments.

Liu et al. [42] performed a similar study comparing the internal adaptation of cobalt–chromium (Co-Cr) posts manufactured by conventional casting, milling, and 3D printing, and they concluded that milled and 3D-printed posts are a suitable replacement for conventionally casted posts. Their findings contrast with the findings of the present study, with the adaptation of 3D-printed and milled CPC. This might be attributed to casting inaccuracies that could be introduced to Co-Cr alloys, which have been reported to be greater than those of gold alloys [43]. Various studies have investigated the adaptation of dental restorations produced through CAD/CAM technology [44–49]. However, this is not the case for CPCs, mainly due to difficulties in evaluating the adaptation of CPC using conventional techniques and the complexity of production of conventional CPC as the comparison counterpart.

The limitations of this study include using one design for the resin models, the adaptation being evaluated in three sections only, and the samples, for the purpose of standardization, being 3D-printed models and not natural teeth. Also, the control group was fabricated through conventional methods while the experimental groups were fabricated fully digitally, which led to the involvement of different manufacturers to fabricate the samples. The authors recommend additional studies evaluating the adaptation of CPC using emerging technologies, micro-CT for instance, that could produce more accurate results. Furthermore, the authors recommend more comprehensive comparative research about the new available materials for prefabrication and CPC and the correlations between retention, fracture resistance, and adaptation. Finally, the authors suggest the utilization of a single facility to produce the examined materials, since this will allow for better control and provide reliable future study outcomes.

5. Conclusions

Within the limitations of the present study, gold-cast CPC showed a more consistent and reduced cement film thickness and, hence, superior internal adaptation, compared to 3D-printed and milled Ti CPC. Among all groups, milled Ti CPC had the lowest initial adaptation and 3D-printed Ti CPC had the lowest final adaptation. In vitro studies with even larger sample sizes are required to confirm and correlate conclusions from this study and to test its clinical relevance.

Author Contributions: Conceptualization, A.A.A. (Abdulaziz A. Alzaid), S.B., H.A. and A.A.A. (Abdulaziz A. AlHelal); methodology, A.A.A. (Abdulaziz A. Alzaid); software, A.A.A. (Abdulaziz A. Alzaid) and S.B.; validation, A.A.A. (Abdulaziz A. Alzaid), S.B. and H.A.; formal analysis, A.A.A. (Abdulaziz A. Alzaid), K.K.A., M.S.S. and R.J.; investigation, A.A.A. (Abdulaziz A. Alzaid), M.T.K. and K.K.A.; resources, A.A.A. (Abdulaziz A. Alzaid), M.T.K., R.J. and E.F.S.; data curation, A.A.A. (Abdulaziz A. Alzaid), R.J. and A.A.A. (Abdulaziz A. AlHelal); writing—original draft preparation, A.A.A. (Abdulaziz A. Alzaid), A.A.A. (Abdulaziz A. AlHelal), S.B., H.A. and M.T.K.; writing—review and editing, A.A.A. (Abdulaziz A. Alzaid), H.A. and M.T.K.; visualization, A.A.A. (Abdulaziz A. Alzaid), K.K.A. and E.F.S.; supervision, A.A.A. (Abdulaziz A. Alzaid) and M.T.K.; project administration, A.A.A. (Abdulaziz A. Alzaid), M.T.K., K.K.A., M.S.S. and A.A.A. (Abdulaziz A. AlHelal). All authors have read and agreed to the published version of the manuscript.

Funding: This research was approved by the Center for Dental Research and funded by the advanced specialty education program in prosthodontics at School of Dentistry, Loma Linda University.

Institutional Review Board Statement: Not applicable.

Informed Consent Statement: Not applicable.

Data Availability Statement: The original contributions presented in this study are included in the article; further inquiries can be directed to the corresponding author.

Acknowledgments: The authors thank Udochukwu Oyoyo, for his statistical analysis, his dedication, and all his efforts during this project.

Conflicts of Interest: The authors declare no conflicts of interest.

References

1. Rosenstiel, S.F.; Land, M.F.; Fujimoto, J. *Contemporary Fixed Prosthodontics*, 5th ed.; Mosby: St. Louis, MO, USA, 2015.
2. Ng, C.C.; Dumbrigue, H.B.; Al-Bayat, M.I.; Griggs, J.A.; Wakefield, C.H. Influence of remaining coronal tooth structure location on the fracture resistance of restored endodontically treated anterior teeth. *J. Prosthet. Dent.* **2006**, *95*, 290–296. [CrossRef]
3. Fokkinga, W.A.; Kreulen, C.M.; Le Bell-Rönnlöf, A.M.; Lassila, L.V.; Vallittu, P.K.; Creugers, N.H. In vitro fracture behavior of maxillary premolars with metal crowns and several post-and-core systems. *Eur. J. Oral Sci.* **2006**, *11*, 250–256. [CrossRef]
4. Hegde, J.; Bashetty, K.; Ramakrishna, S.; Lekha, C. An in vitro evaluation of fracture strength of endodontically treated teeth with simulated flared root canals restored with different post and core systems. *J. Conserv. Dent.* **2012**, *15*, 223–227. [CrossRef] [PubMed]
5. Bergman, B.; Lundquist, P.; Sjögren, U.; Sundquist, G. Restorative and endodontic results after treatment with cast posts and cores. *J. Prosthet. Dent.* **1989**, *61*, 10–15. [CrossRef]
6. Creugers, N.H.; Mentink, A.G.; Käyser, A.F. An analysis of durability data on post and core restorations. *J. Dent.* **1993**, *21*, 281–284. [CrossRef] [PubMed]
7. Raedel, M.; Fiedler, C.; Jacoby, S.; Boening, K.W. Survival of teeth treated with cast post and cores: A retrospective analysis over an observation period of up to 19. *5 years. J. Prosthet. Dent.* **2015**, *114*, 40–45. [CrossRef] [PubMed]
8. Linde, L.A. The use of composites as core material in root-filled teeth. II. Clinical investigation. *Swed. Dent. J.* **1984**, *8*, 209–216.
9. Worni, A.; Kolgeci, L.; Rentsch-Kollar, A.; Katsoulis, J.; Mericske-Stern, R. Zirconia-based screw-retained prostheses supported by implants: A Retrospective study on technical complications and failures. *Clin. Implant. Dent. Relat. Res.* **2015**, *17*, 1073–1081. [CrossRef]
10. Moscovitch, M. Consecutive case series of monolithic and minimally veneered zirconia restorations on teeth and implants: Up to 68 months. *Int. J. Periodontics Restor. Dent.* **2015**, *35*, 314–323. [CrossRef]
11. Maló, P.; de Araújo Nobre, M.; Borges, J.; Almeda, R. Retrievable metal ceramic implant-supported fixed prostheses with milled titanium frameworks and all-ceramic crowns: Retrospective clinical study with up to 10 years of follow-up. *J. Prosthodont.* **2012**, *21*, 256–264. [CrossRef]
12. Paniz, G.; Stellini, E.; Meneghello, R.; Cerardi, A.; Gobbato, E.A.; Bressan, E. The precision of fit of cast and milled full-arch implant-supported restorations. *Int. J. Oral Maxillofac. Implant.* **2013**, *28*, 687–693. [CrossRef] [PubMed]
13. Wang, Z.; Xiao, Z.; Huang, C.; Wen, L.; Zhang, W. Influence of ultrasonic surface rolling on microstructure and wear behavior of selective laser melted Ti-6Al-4V alloy. *Materials* **2017**, *10*, 1203–1217. [CrossRef] [PubMed]
14. Sun, J.; Zhang, F.Q. The application of rapid prototyping in prosthodontics. *J. Prosthodont.* **2012**, *21*, 641–644. [CrossRef] [PubMed]
15. McLean, J.W.; von Fraunhofer, J.A. The estimation of cement film thickness by an in vivo technique. *Br. Dent. J.* **1971**, *131*, 107–111. [CrossRef]
16. Ferrari, M.; Vichi, A.; Mannocci, F.; Mason, P.N. Retrospective study of the clinical performance of fiber posts. *Am. J. Dent.* **2000**, *13*, 9B–13B.
17. Ferrari, M.; Cagidiaco, M.C.; Goracci, C.; Vichi, A.; Mason, P.N.; Radovic, I.; Tay, F. Long-term retrospective study of the clinical performance of fiber posts. *Am. J. Dent.* **2007**, *20*, 287–291.

18. Grandini, S.; Goracci, C.; Monticelli, F.; Borracchini, A.; Ferrari, M. SEM evaluation of the cement layer thickness after luting two different posts. *J. Adhes. Dent.* **2005**, *7*, 235–240.
19. Sorensen, J.A.; Engelman, M.J. Effect of post adaptation on fracture resistance of endodontically treated teeth. *J. Prosthet. Dent.* **1990**, *64*, 419–424. [CrossRef]
20. Geramipanah, F.; Rezaei, S.M.; Sichani, S.F.; Sichani, B.F.; Sadighpour, L. Microleakage of different post systems and a custom adapted fiber post. *J. Dent.* **2013**, *10*, 94–102.
21. Musikant, B.L.; Deutsch, A.S. Endodontic posts and cores. Part II. Design of the Flexi-post. *J. Ala. Dent. Assoc.* **1985**, *69*, 42–46.
22. Duc, O.; Krejci, I. Effects of adhesive composite core systems on adaptation of adhesive post and cores under load. *J. Dent.* **2009**, *37*, 622–626. [CrossRef] [PubMed]
23. Alves, J.; Walton, R.; Drake, D. Coronal leakage: Endotoxin penetration from mixed bacterial communities through obturated, post-prepared root canals. *J. Endod.* **1998**, *24*, 587–591. [CrossRef]
24. Mannocci, F.; Ferrari, M.; Watson, T.F. Microleakage of endodontically treated teeth restored with fiber posts and composite cores after cyclic loading: A confocal microscopic study. *J. Prosthet. Dent.* **2001**, *85*, 284–291. [CrossRef] [PubMed]
25. Muttlib, N.A.; Azman, A.N.; Seng, Y.T.; Alawi, R.; Ariffin, Z. Intracanal Adaptation of a Fiber Reinforced Post System as Compared to a Cast Post-and-Core. *Acta Stomatol. Croat.* **2016**, *50*, 329–336. [CrossRef] [PubMed]
26. Jamovi. The Jamovi Project. (Version 1.2). 2020. Available online: https://www.jamovi.org (accessed on 9 January 2020).
27. R Core Team. R: A Language and Environment for Statistical Computing. (Version 3.6). 2019. Available online: https://cran.r-project.org/ (accessed on 9 January 2020).
28. Patzelt, S.B.; Bishti, S.; Stampf, S.; Att, W. Accuracy of computer-aided design/computer-aided manufacturing-generated dental casts based on intraoral scanner data. *J. Am. Dent. Assoc.* **2014**, *145*, 1133–1140. [CrossRef]
29. Revilla-León, M.; Gonzalez-Martín, Ó.; López, J.P.; Sánchez-Rubio, J.L.; Özcan, M. Position Accuracy of Implant Analogs on 3D Printed Polymer versus Conventional Dental Stone Casts Measured Using a Coordinate Measuring Machine. *J. Prosthodont.* **2018**, *27*, 560–567. [CrossRef]
30. Linkevicius, T.; Vindasiute, E.; Puisys, A.; Maslova, N.; Puriene, A. The influence of the cementation margin position on the amount of undetected cement. A prospective clinical study. *Clin. Oral Implant. Res.* **2013**, *24*, 71–76. [CrossRef]
31. Bukhari, S.A.; AlHelal, A.; Kattadiyil, M.T.; Wadhwani, C.P.K.; Taleb, A.; Dehom, S. An in vitro investigation comparing methods of minimizing excess luting agent for cement-retained implant-supported fixed partial dentures. *J. Prosthet. Dent.* **2020**, *124*, 706–715. [CrossRef]
32. Goto, Y.; Nicholls, J.I.; Phillips, K.M.; Junge, T. Fatigue resistance of endodontically treated teeth restored with three dowel-and-core systems. *J. Prosthet. Dent.* **2005**, *93*, 45–50. [CrossRef]
33. Fernandes, A.S.; Shetty, S.; Coutinho, I. Factors determining post selection: A literature review. *J. Prosthet. Dent.* **2003**, *90*, 556–562. [CrossRef] [PubMed]
34. Pontius, O.; Nathanson, D.; Giordano, R.; Hutter, J.W. Survival rate and fracture strength of incisors restored with different post and core systems and endodontically treated incisors without coronoradicular reinforcement. *J. Endod.* **2002**, *28*, 710–715. [CrossRef] [PubMed]
35. Anusavice, K.J.; Shen, C.; Rawls, H.R. *Phillips' Science of Dental Materials*, 12th ed.; Saunders: Philadelphia, PA, USA, 2013.
36. Knosp, H.; Holliday, R.J.; Corti, C.W. Gold in Dentistry: Alloys, Uses and Performance. *Gold Bull.* **2003**, *36*, 93–102. [CrossRef]
37. Niinomi, M. Mechanical biocompatibilities of titanium alloys for biomedical applications. *J. Mech. Behav. Biomed. Mater.* **2007**, *1*, 30–42. [CrossRef] [PubMed]
38. Steinemann, S. Titanium—The material of choice. *Periodontology 2000* **1998**, *17*, 7–21. [CrossRef] [PubMed]
39. Wadhwani, C.; Brindis, M.; Kattadiyil, M.T.; O'Brien, R.; Chung, K. Colorizing titanium-6aluminum-4vanadium alloy using electrochemical anodization: Developing a color chart. *J. Prosthet. Dent.* **2018**, *119*, 26–28. [CrossRef]
40. *ASTM Standard F67*; Standard Specification for Unalloyed Titanium, for Surgical Implant Applications. ASTM International: West Conshohocken, PA, USA, 2017. Available online: http://www.astm.org/ (accessed on 9 January 2020).
41. Koizumi, H.; Takeuchi, Y.; Imai, H.; Kawai, T.; Yoneyama, T. Application of titanium and titanium alloys to fixed dental prostheses. *J. Prosthodont. Res.* **2019**, *63*, 266–270. [CrossRef]
42. Liu, W.; Qing, H.; Pei, X.; Wang, J. Internal adaptation of cobalt-chromium posts fabricated by selective laser melting technology. *J. Prosthet. Dent.* **2019**, *121*, 455–460. [CrossRef]
43. Pulskamp, F.E. A comparison of the casting accuracy of base metal and gold alloys. *J. Prosthet. Dent.* **1979**, *41*, 272–276. [CrossRef]
44. Tamac, E.; Toksavul, S.; Toman, M. Clinical marginal and internal adaptation of CAD/CAM milling, laser sintering, and cast metal ceramic crowns. *J. Prosthet. Dent.* **2014**, *112*, 909–913. [CrossRef]
45. Kocaağaoğlu, H.; Kılınç, H.İ.; Albayrak, H.; Kara, M. In vitro evaluation of marginal, axial, and occlusal discrepancies in metal ceramic restorations produced with new technologies. *J. Prosthet. Dent.* **2016**, *116*, 368–374. [CrossRef]
46. Akçin, E.T.; Güncü, M.B.; Aktaş, G.; Aslan, Y. Effect of manufacturing techniques on the marginal and internal fit of cobalt-chromium implant-supported multiunit frameworks. *J. Prosthet. Dent.* **2018**, *120*, 715–720. [CrossRef] [PubMed]
47. Nesse, H.; Ulstein, D.M.; Vaage, M.M.; Øilo, M. Internal and marginal fit of cobalt-chromium fixed dental prostheses fabricated with 3 different techniques. *J. Prosthet. Dent.* **2015**, *114*, 686–692. [CrossRef] [PubMed]

48. Katsoulis, J.; Mericske-Stern, R.; Yates, D.M.; Izutani, N.; Enkling, N.; Blatz, M.B. In vitro precision of fit of computer-aided design and computer-aided manufacturing titanium and zirconium dioxide bars. *Dent. Mater.* **2013**, *29*, 945–953. [CrossRef] [PubMed]
49. Al-Rubaye, T.; Elsubeihi, E. The Accuracy of Custom-Made Milled Metal Posts as Compared to Conventional Cast Metal Posts. *Dent. J.* **2024**, *12*, 309. [CrossRef] [PubMed]

prosthesis

Article

The Effects of Different Chemical Disinfectants on the Strength, Surface, and Color Properties of Conventional and 3D-Printed Fabricated Denture Base Materials

Ahmed Fathey Elhagali [1,*], Mohamed Y. Sharaf [2,3], Mahmoud El-Said Ahmed Abd El-Aziz [4], Ali Sayed Ali Bayiumy [5], Mahmoud Abdellah Ahmed Refaei [4], Ahmed Hassan Al-Agamy [1], Ahmed Ali [4], Ahmed Elakel [6], Raand Altayyar [7], Riham Alzahrani [7], Mohammed M. Gad [8] and Mohamed Aboshama [1]

1 Department of Removable Prosthodontics, Faculty of Dental Medicine, Al-Azhar University, Assiut 71524, Egypt; ahmedelagmy2222@gmail.com (A.H.A.-A.); mohammedaboshama.19@azhar.edu.eg (M.A.)
2 Prosthetic Department, College of Dentistry, Misr University of Science and Technology, 6th October City 12585, Egypt; mohamedsharaf@dent.bsu.edu.eg
3 Prosthetic Department, College of Dentistry, National University of Science and Technology, Thi Qar 64011, Iraq
4 Dental Biomaterials, Faculty of Dental Medicine, Al-Azhar University (Assiut), Assiut Branch 71524, Egypt; mahmoudabdelziz.46@azhar.edu.eg (M.E.-S.A.A.E.-A.); mahmoudabdellah14@yahoo.com (M.A.A.R.); drahmedelfeky@gmail.com (A.A.)
5 Dental Biomaterials, Faculty of Dentistry, Assiut University, Assiut 71524, Egypt; alydent@aun.edu.eg
6 Department of Preventive Dental Sciences, College of Dentistry, Imam Abdulrahman Bin Faisal University, P.O. Box 1982, Dammam 31441, Saudi Arabia; amelakel@iau.edu.sa
7 College of Dentistry, Imam Abdulrahman Bin Faisal University, Dammam 31441, Saudi Arabia; 2180001780@iau.edu.sa (R.A.); 2180001557@iau.edu.sa (R.A.)
8 Department of Substitutive Dental Sciences, College of Dentistry, Imam Abdulrahman Bin Faisal University, P.O. Box 1982, Dammam 31441, Saudi Arabia; mmjad@iau.edu.sa
* Correspondence: ahmedelhagali86@gmail.com

Abstract: Objectives: The disinfection of fabricated prostheses is crucial to prevent cross-infection between dental laboratories and clinics. However, there is a lack of information about the effects of chemical disinfection on 3D-printed denture base resins. This study aimed to evaluate the impact of different disinfectants on the flexural strength, elastic modulus, micro-hardness, surface roughness (Ra), and change in color of 3D-printed and conventional heat-polymerized (HP) denture base resins (DBRs). Methods: A total of 240 specimens (80 bar-shaped (64 × 10 × 3.3 mm) and 160 disk-shaped (10 × 2 mm)) were made from HP and 3D-printed DBRs. For each resin, the specimens were divided into four groups ($n = 10$) according to the disinfectant solution. One remained in water without disinfection as a control group, while the other three groups were disinfected using 1% sodium hypochlorite, 2% glutaraldehyde, or 10% Micro 10+ for 30 min. The flexural strength, elastic modulus, micro-hardness, Ra, and color change were measured. The collected data were statistically analyzed using a two-way ANOVA and Tukey's post hoc test ($\alpha = 0.05$). Results: A significant decrease in flexural strength, elastic modulus, and hardness was found with sodium hypochlorite ($p < 0.05$). When comparing the resins per solution, the 3D-printed resin showed a significant decrease in flexural strength, elastic modulus, and hardness compared with PMMA ($p < 0.001$), while no change was found in the Ra of both resins with all disinfectants ($p > 0.05$). Disinfecting with sodium hypochlorite resulted in a significant increase in color change for both resins ($p < 0.05$); however, all the changes were within clinically acceptable limits. Sodium hypochlorite showed the highest color change, while 2% glutaraldehyde and 10% Micro 10+ showed no significant changes in the tested properties ($p > 0.05$). Conclusions: Neither resin showed a change in surface roughness with immersion in disinfectants. Sodium hypochlorite had an adverse effect on

 https://doi.org/10.3390/prosthesis7020024

the flexural properties, hardness, and change in color of the PMMA and 3D-printed DBRs, while the other disinfectants had no effect on the tested properties.

Keywords: 3D printing; disinfectants; mechanical testing; surface properties; denture base; acrylic resin

1. Introduction

Edentulism, or complete tooth loss, is a very challenging condition that has a negative impact on quality of life [1]. Complete denture (CD) fabrication is the conventional rehabilitation of edentulism and is still a feasible treatment strategy [1,2]. The material of choice for CD fabrication is polymethylmethacrylate (PMMA), and this is attributed to the following reasons: first, it has a lower cost compared to other options; second, it has superior physical and mechanical properties, with an excellent esthetic appearance; and finally, it affords better manipulation and handling for any dental technician [3]. On the other hand, this material reveals some dimensional and color changes, with a possibility of fracture in spite of its mechanical and physical properties; moreover, a CD, if not properly cleaned and maintained by the patient, can cause Candidal infection or even tissue abrasions and irritation [4]. One major factor that could really affect the mechanical and esthetic properties of this material is surface and subsurface voids [5]. Thus, paying attention to and obtaining superior surface characteristics during denture fabrication can pave the way for better esthetics and higher longevity of the denture base [2].

Digital technology has become increasingly popular in several dental specialties in recent years. Nevertheless, the use of digital tools, materials, and computer-aided design and computer-assisted manufacturing (CAD/CAM) in the design and production of dental prostheses has helped to lessen the workload for dental technicians and dentists [6]. Two methods have been designed for denture fabrication: the subtractive method, in which the denture is milled from a prefabricated resin disc, and the additive method, in which the denture is built up by three-dimensional printing (3D printing) using fluid resins [3]. The benefits of 3D-printed dentures include shorter production times, more accuracy, lower costs, fewer patient visits, and greater patient comfort [7]. The primary disadvantages of subtraction procedures are the high cost and waste of milling machines, burs, and restorative materials, as well as the device's restricted motion range to make complicated forms [8].

Different types of technologies for denture fabrication have been reported, such as stereolithography (SLA), digital light projection (DLP), and photopolymer jetting (PolyJet) [9]. Among these methods, DLP is considered to be superior when compared to the others because of its lower material consumption, faster speed, and greater precision [10]. In term of the fabrication method effect, 3D-printed DBRs show comparable properties to conventional PMMA [3], while 3D-printed DBRs show superior properties compared to PMMA regarding accuracy, fit, and adaptability; therefore, 3D printing is recommended for DBR fabrication [5]. According to a recent review report, printed dentures have problems with strength, color stability, and stainability; however, these issues could be resolved by using new materials and modifying existing technology [11]. With growing evidence supporting the benefits of 3D printing technologies, several studies have been conducted to assess the performance of dentures printed with different printing technologies. In terms of strength, 3D-printed dentures show low strength but are still within the ISO recommendations [12]; in addition, they also have poor surface properties [4,11,13]. To obtain the benefits from additive technologies with optimum properties, controlling the printing factors (pre-printing, printing, and/or post-printing parameters) is suggested [14]. By modifications to these parameters, improvements in strength, surface properties, and antimicrobial efficacy were achieved [8,14,15].

Prosthetics produced in dental laboratories are vulnerable to microbial contamination during the manufacturing process [16]. Consequently, various contamination sources are reported, including contact with contaminated hands, felt disks and pumice used in the polishing process, and machines and equipment used for denture base resin (DBR) fabrication. Additionally, the fabrication process involves several steps, most of which are conducted in a laboratory environment with the risk of prosthesis contamination [17,18]. Other sources of contamination include when the prosthesis is returned from the dental office for repair, relining, or rebasing after patient use [19]. According to the literature, significant microbiological cross-contamination can occur when transferring prostheses between dental offices and dental laboratories [18]. Therefore, less attention has been paid to disinfecting dentures; instead, the focus should be on preventing cross-contamination through infection-control procedures, including the barrier technique, sterilization, and disinfection of the dental office and its equipment. A dental prosthesis provides a conduit for organisms to be transferred from patients to laboratory and dental staff [19,20]. Dentures must be disinfected to prevent cross-contamination and enhance cleanliness. If proper disinfection measures are not followed, the dental office–prosthesis laboratory connection could be a cross-infection conduit [18].

Several disinfectants have been suggested at different concentrations and durations with variations in disinfection level [16]. However, suitable disinfection should be effective without deteriorating the prosthesis structures and properties [21]. Among the common disinfectants, sodium-hypochlorite- and glutaraldehyde-based disinfectants are often used in dentistry, and aldehyde-free disinfectants are commonly used for complete microbial elimination from disinfected prostheses [16,22–24].

Because the restorations are exposed to temperature changes and functional stress during their clinical service, reversible elastic deformation, irreversible plastic deformation, and fracture [9] are all possible outcomes of residual stresses created by these dynamic changes [25]. Furthermore, a full denture's color stability may be reduced by its surface roughness [8]. CD discoloration may be a sign of aging and material degradation [25], which may ultimately necessitate denture replacement [7].

Before considering 3D-printed dentures as a good substitute for traditional PMMA dentures, their mechanical, physical, and cosmetic qualities should be carefully examined. Therefore, the present in vitro study aimed to evaluate and compare the flexural strength, modulus of elasticity, micro-hardness, surface roughness, and change in color of heat-polymerized acrylic resin with that of 3D-printed resin after disinfections with various chemical disinfectants. The null hypothesis was that the disinfectants would not affect the tested properties of the tested DBRs.

2. Materials and Methods

2.1. Specimen Preparation

The sample size calculation revealed that $n = 10$ was sufficient to detect effect sizes for the main effects and pairwise comparisons, with the power level set at 80% and 95% confidence [26]. According to each test specification, specimens were fabricated with different dimensions; for flexural properties, rectangular specimens ($64 \times 10 \times 3.3 \pm 0.2$ mm), while surface and color changes were tested using disk-shaped (10×2 mm) specimens.

The test specimens were designed using a software program (SolidWorks version 2024, Dassault Systèmes SolidWorks Corp., Aix-en-Provence, France) and saved as a standard tessellation language (STL). The STL was used to mill the specimens from milling wax blocks (Duo Cad; FSM Dental, Ankara, Türkiye), which were placed inside a metal flask or (split stainless-steel mold with metal slots of required dimensions) produced to obtain PMMA (Temdent Classic; Schütz Dental GmbH, Rosbach, Germany) specimens [27]. As instructed by the manufacturer, the PMMA was hand-blended and packed into the mold followed by polymerization as a conventional method.

For 3D-printed resin, an STL file was used to print 3D-printed specimens. The specimens were printed vertically with 100 µm thick layers (z-direction angulated 90° to the printing direction) with a digital light processing printer (D30 II, Rapid Shape, Heimsheim, Germany) using a fluid resin (FREEPRINT denture, Detax, Ettlingen, Germany) [28]. Following printing, the specimens were post-cured from all sides for 20 min using ultraviolet light (385 nm) with UV-A type 3 in a light box (type E0202; Yizhet, Shenzhen, China) and cleaned with 99% isopropyl alcohol for 5 min as per the manufacturer's instructions. Silicon carbide paper grit P1200 (Paper SiC P1200; Struers GmbH, Rosbach, Germany) was then used to grind all the manufactured specimens to their final dimensions. A digital caliper (Digimatic Micrometer, Mitutoyo, Kanagawa, Japan) was used to validate the specimen dimensions to the closest ±0.02 mm after they had been ground. The approved specimens were then kept for 48 h in distilled water at 37 °C [29].

2.2. Specimen Disinfection Procedures

The prepared specimens (for each main group, conventional and 3D-printed resins) were randomly divided into four groups (n = 10). One was immersed in water as a control, while the other three groups were disinfected using 1% sodium hypochlorite, 2% glutaraldehyde, and 10% Micro 10+ for a specified time based on immersion protocols (Table 1).

Table 1. Immersion solutions.

Solution	Composition	Immersion Protocol
Glutaraldehyde	An organic compound with the formula $(CH_2)_3(CHO)_2$. The molecule consists of a five-carbon chain doubly terminated with formyl groups.	20 min immersion in at least a 2% solution of glutaraldehyde at room temperature.
Sodium Hypochlorite	0.5% Sodium hypochlorite solution, 1% active chlorine.	Solution of 5.25% sodium hypochlorite (1:5 dilution) diluted to obtain 1% sodium hypochlorite by adding 50 mL of sodium hypochlorite to 200 mL of water with iImmersion for 10 min at room temperature.
Micro 10+	Micro 10+ is an aldehyde-free concentrated solution. A total of 100 g of Micro 10+ contains 9 g of alkylbenzyldimethylammonium chloride, amphoteric and non-ionic surfactants, complexant, corrosion inhibitor, and additives.	Very economical 2% dilution. Contact time: 15 min.

2.3. Specimen Testing

The flexure strength and elastic modulus were assessed using a three-point bending test. A load cell with a 5 kN capacity and a 1 mm/min cross-head speed was used to apply the load until the specimen fractured using a universal testing machine (Instron Industrial Products, Model 3345). Stress–strain curves were generated using software (Instron® Bluehill Lite Software, Norwood, MA, USA). The limiting stress at which failure occurs is represented by the flexural strength (FS), calculated using Equation (1):

$$FS\ (б) = 3F\ (L)/2wh^2 \quad (1)$$

Here, F is the maximum load at the point of fracture, L is the span, w is the width of the sample, and h is its height. The modulus of elasticity was calculated mathematically from the stress–strain curve obtained during the flexural strength test. The modulus of elasticity (MPa) = stress/strain within the elastic portion [30].

Surface roughness (Ra) measurements were measured on the disk specimens, and a USB digital surface profile gauge (Elcometer 224/2, Elcometer Instruments, Manchester, England) was used to assess surface roughness. The roughness (Ra, μm) was generated as the arithmetic mean between the peaks and valleys recorded after the profilometer needle had scanned a span of 2 mm in length, with a cut-off of 0.25 mm, to optimize filtering and surface undulation. Each surface was read five times, starting from three different positions and always with the needle scanning the specimen's geometric center. The mean roughness of each specimen was determined as the average of the five values.

The micro-hardness (VHN) was assessed using a Vickers hardness tester (ZHU 2.5, Zwick/Roell GmbH, Ulm, Germany). The load was applied using a Vickers indenter with a speed of 1 mm/min at the contact point and a dwell time of 2 s at the loading point. For each specimen, three readings at different points on the specimen surfaces were conducted, followed by an average calculation per specimen.

For color changes (ΔE_{ab}), a reflective spectrophotometer (X-Rite, model RM200QC) was used to measure the colors of all the disk specimens after fabrication and after treatment. The aperture size was set at 4 mm, and the specimens were precisely positioned in relation to the device. Measurements were made against the CIE standard illuminant D65 against a white background using the CIE L*a*b* color space. Equation (2) was used to assess the specimen color changes:

$$\Delta E_{ab} = (\Delta L^{*2} + \Delta a^{*2} + \Delta b^{*2})\ \frac{1}{2} \quad (2)$$

Here, L* = lightness (0–100), a* = the red/green axis, and b* = the yellow/blue axis [31]. The National Bureau of Standards (NBS) was used as a reference for the color change comparison and calculated using the following equation: NBS = $\Delta E_{ab} \times 0.92$. A material is deemed esthetically and clinically acceptable when NBS units fall within the range of 3.7 NBS units. Differences exceeding 3.7 NBS units are classified as a "mismatch" and viewed as clinically unacceptable [31–33].

The data was statistically analyzed using SPSS v28 (IBM, Armonk, NY, USA). The Shapiro-Wilk test and histograms were used to evaluate the normal distribution of data. Data were presented as mean and standard deviation (SD) and were analyzed using a two-way ANOVA test (for the combined effect of disinfectant and material type) followed by a post hoc analysis (Tukey's) test. The significance level was set at $p < 0.05$.

3. Results

Table 2 summarizes the mean values, SD, and significance for the groups regarding flexural strength, elastic modulus, hardness, and surface roughness. For PMMA, a significant difference was found when comparing the disinfectants ($p = 0.04$). For the pairwise comparison, a significant decrease in flexural strength with sodium hypochlorite ($p < 0.05$) was found compared with the other solutions, while no significant difference was found between water, glutaraldehyde, and Micro 10+ ($p > 0.05$). For the 3D-printed DBRs, no significant differences were found between the tested groups ($p = 0.640$). When comparing resins per solution, the 3D-printed resin showed a significant decrease in flexural strength ($p < 0.001$), and the highest flexural strengths were recorded with water immersion (101.3 ± 4.00 and 89.3 ± 3.22 MPa for PMMA and 3D-printed resin, respectively).

For both resins, immersion in disinfectant did not produce a significant difference in the elastic modulus compared to water immersion ($p > 0.05$), except with sodium hypochlorite, which showed a significant decrease in the elastic modulus ($p < 0.05$). When comparing resin based on the immersion solution, the 3D-printed resin showed a significantly decreased elastic modulus compared with PMMA, and the highest elastic moduli were reported for water immersion (2954.3 ± 93.77 and 2208.5 ± 87.27 MPa for PMMA and 3D-printed DBRs, respectively).

Table 2. Mean value and significance between groups of both resins with different disinfectants, showing effects for flexural strength, elastic modulus, hardness, and surface roughness properties.

Tested Properties	Immersion Solution	Resins Mean ± SD		
		Heat-Polymerized	3D-Printed	p Value
Flexural strength (MPa)	Water	101.3 ± 4.00 [a]	89.3 ± 3.22 [a]	0.000 *
	Glutaraldehyde	99.4 ± 2.244 [a]	86.7 ± 3.76 [a]	0.000 *
	Sodium Hypochlorite	91.5 ± 2.87	82.7 ± 1.61	0.000 *
	Micro 10+	98.1 ± 2.91 [a]	86.9 ± 3.17 [a]	0.000 *
	p value	0.020 *	0.04 *	
Elastic modulus (MPa)	Water	2954.3 ± 93.77 [a]	2208.5 ± 87.27 [a]	0.000 *
	Glutaraldehyde	2920.8 ± 94.4 [a]	2182.5 ± 92.94 [a]	0.000 *
	Sodium Hypochlorite	2514.4 ± 98.71	1975.5 ± 69.12	0.000 *
	Micro 10+	2854.6 ± 123.99 [a]	2105.5 ± 143.06 [a]	0.000 *
	p value	0.032 *	0.010 *	
Hardness (VHN)	Water	17.38 ± 0.35 [a]	16.8 ± 0.44 [a]	0.73
	Glutaraldehyde	17.25 ± 0.33 [a]	16.58 ± 0.47 [a]	0.08
	Sodium Hypochlorite	14.02 ± 0.39	13.27 ± 0.31	0.10
	Micro 10+	17.11 ± 0.25 [a]	16.52 ± 0.41 [a]	0.53
	p value	0.03 *	0.04 *	
Roughness (Ra, µm)	Water	0.155 ± 0.035	0.399 ± 0.029	0.000 *
	Glutaraldehyde	0.159 ± 0.026	0.405 ± 0.035	0.000 *
	Sodium Hypochlorite	0.167 ± 0.014	0.414 ± 0.032	0.000 *
	Micro 10+	0.160 ± 0.015	0.411 ± 0.032	0.000 *
	p value	0.174	0.091	

* Significant at p value < 0.05. Same small letter vertically per column indicates insignificant difference pair-wise comparison.

For both resins, immersion in disinfectant showed no significant difference in hardness compared to water immersion ($p > 0.05$), except with sodium hypochlorite, which showed a significant decrease in hardness ($p < 0.05$), leading to the lowest hardnesses (14.02 ± 0.39 and 13.27 ± 0.31 VHN for PMMA and 3D-printed resin, respectively). When comparing the resins based on the immersion solution, the 3D-printed resin showed insignificantly low hardnesses compared with PMMA ($p > 0.05$).

The immersion of the PMMA and 3D-printed DBRs in disinfectant showed no significant difference in surface roughness ($p = 0.174$ and $p = 0.091$ for PMMA and 3D-printed DBRs, respectively). When comparing the resins based on the immersion solution, the 3D-printed resin showed a significantly increased surface roughness ($p < 0.001$).

Table 3 shows the mean values, SD, and significance of the color changes. Immersion of the specimens in sodium hypochlorite significantly increased the color change compared with glutaraldehyde and Micro 10+ ($p < 0.05$), with no significant difference in color change between glutaraldehyde and Micro 10+ ($p > 0.05$). When comparing the resins based on the immersion solution, no significant differences were found in the color change between the PMMA and 3D-printed DBRs ($p > 0.05$). The highest NBS value (1.04) was recorded with sodium hypochlorite, which was lower than 3.7, revealing that all changes were within the clinically acceptable value.

Table 3. Change in color (ΔE_{ab}) of tested resins after disinfectants.

Immersion Solutions	Resin and NBS Unit Mean ± SD				
	Heat-Polymerized	NBS	3D-Printed	NBS	*p* Value
Glutaraldehyde	0.61 ± 0.09 [a]	0.56	0.68 ± 0.10 [a]	0.62	0.30
Sodium Hypochlorite	1.13 ± 0.09	1.04	1.67 ± 0.06	1.53	0.09
Micro 10+	0.71 ± 0.06 [a]	0.65	0.78 ± 0.09 [a]	0.71	0.11
p value	0.001 *		0.001 *		

* $p < 0.05$ set as significant level. Same small letter vertically per column indicates insignificant difference pairwise comparison. The National Bureau of Standards (NBS) values deemed esthetically and clinically acceptable when falling within the range of 3.7, while values greater than 3.7 are classified as a mismatch (clinically unacceptable).

4. Discussion

This study was conducted to investigate the effect of chemical disinfectants on the flexural strength, elastic modulus, hardness, surface roughness, and change in color of 3D-printed DBRs compared with PMMA DBRs. The null hypothesis was partially rejected, as all the disinfectant solutions significantly impacted all the tested properties except roughness, which showed no significant change.

Denture disinfection is a mandatory process to avoid cross-contamination between dental offices and dental laboratories where the prosthesis is fabricated, adjusted, repaired, and rebased [16,34,35]. With advanced technology for removable prosthesis fabrication, no study has investigated the effect of the disinfection process on the properties of 3D-printed DBRs. Thus, one 3D-printed resin was selected for investigation in the present study compared with one conventional PMMA DBR. Two commonly used disinfectant solutions were selected, sodium hypochlorite and glutaraldehyde, as well as one aldehyde-free disinfectant (Micro 10+), and the three selected disinfectants have strong antimicrobial activities [36]. An infection-control procedure designed to avoid cross-contamination was assessed in preliminary studies [36,37]. The findings showed that 4% chlorhexidine gluconate solutions and sodium hypochlorite solutions reduced the microbial growth on the dentures in 10 min [36,37].

The literature reports significant fluctuation in the use of several disinfectants regarding concentration and duration [16,20]. The most appropriate disinfectant should meet most of the ideal agent's criteria while retaining the prosthesis structure [21]. Sodium hypochlorite is a commonly used disinfectant and has a wide range of activity within a short disinfectant period [17,18]. Rodrigues et al. and Salvia et al. reported immersion in sodium hypochlorite containing 2% active chloride for 30 min as the most efficient approach for disinfecting acrylic resin prostheses [38,39]. Furthermore, Chau et al. [23] reported strong disinfectant activities of 1% sodium hypochlorite in removing microbes from denture surfaces. Despite its effectiveness as a disinfectant, sodium hypochlorite has numerous disadvantages, including corrosion of metal surfaces and irritation [22]. Glutaraldehyde-based disinfectants are frequently used and recommended for instrument disinfection [21]. The literature describes glutaraldehyde's high antibacterial activity, and its efficiency varies with the exposure time [24]. However, glutaraldehyde-based solutions should be used for an appropriate time due to the reported toxicity with prolonged immersion [18].

Because acrylic resins are hydrophilic, they can absorb solvents and water, triggering hydrolysis and the emergence of acrylic regions with unique optical characteristics when the absorbed liquids diffuse into the polymer network [6,34]. DBR immersion in glutaraldehyde and Micro 10+ did not affect the flexural strength, while sodium hypochlorite decreased it. This is consistent with previous studies [40,41] reporting a decrease in the flexural strength of DBRs after immersion in 1% NaOCl. This decline is due to the sorption of the NaOCl aqueous solution and its active chlorine content. The absorbed solution acts as a plasticizing agent, which could be pivotal in altering the chemical structure [42]. Moreover, the pendant monomer solubility increased due to its active chlorine, which facilitated the increase in the leachability of the remaining monomer. This was replaced with greater solution sorption [43], which is regarded as the key factor controlling the strength and surface integrity of DBRs [1].

When comparing the PMMA and 3D-printed resins based on the immersion solution, the 3D-printed resin showed low strength. This is consistent with Prpic' et al. [3], who reported a high flexural strength of conventional compared with 3D-printed DBRs. The decreased strength of 3D-printed resins may be attributed to the painting method (layer-by-layer), with apparent weak bonding between successive printed layers [3,5,14,44]. In addition to the direction of the printed layer in relation to the direction of the load applied, which was parallel to the layer direction when the specimen was vertically printed [14], another reason was attributed to the degree of conversion of photo-polymerized 3D-printed resins, as reported in a previous study, compared with conventional PMMA. As the degree of conversion decreased, the residual monomer increased, adversely affecting the strength of the printed object [3,5,44,45]. Additionally, the poor strength of the 3D-printed resins after disinfectant immersion was exaggerated due to the chemical composition of the disinfectant [46]. With immersion in solution, the resin absorbs water and the absorbed water acts as a plasticizer, affecting the mechanical properties [14,44]. A previous study [6] compared the water sorption of 3D-printed DBRs with conventional ones and reported an increase in water sorption with 3D-printed DBRs. This could also explain why the flexural strength decreased after immersion in disinfectant solution.

According to ISO-20795-1:2013 [12], a flexural strength above 65 MPa and an elastic modulus above 2000 MPa are clinically acceptable. In this context, the elastic modulus was found to be a measure of the resins' rigidity or flexibility; a higher elastic modulus denotes a more rigid material [17]. To distribute the load evenly and reduce the danger of breakage, denture base materials should have flexibility and rigidity to withstand stresses [27]. In this case, the flexural strength was mirrored by the resins' elastic moduli, with the heat-polymerized (HP) resin having a higher elastic modulus than the 3D-printed resins. This outcome is consistent with the findings of Fouda et al. [46], who examined the elastic modulus of HP and 3D-printed resins and found that the former had a lower elastic modulus. With effervescent pills, the HP and 3D-printed resins' elastic moduli dropped; however, with NaOCl, they dropped precipitously. The water and chemical uptake during submersion in these disinfectant solutions may explain this observation. Comparisons with earlier research are problematic, since no studies have assessed the impact of denture cleaners on 3D-printed resins' elastic moduli. However, the low modulus may be interpreted similarly to the flexural strength because both were tested under the same load, direction, and conditions and were reported as flexural characteristics [34].

A material's hardness is a crucial characteristic. A DBR's surface hardness indicates how much the pressures involved in mastication can be resisted. Although hardness is evaluated in numerous ways, the most practical means to determine a material's hardness is measuring its resistance to indentation [47]. A logical definition of hardness might then be the "resistance of a material to indentation"; thus, greater indentation indicates softer material [47]. It has been reported that resin immersion in disinfectant solutions weakens the secondary bond between the polymeric chains of the acrylic resins [47]. Shen et al. [35] discovered that all the resins they tested had a soft surface after being subjected to a glutaraldehyde alkaline disinfection solution with a phenolic buffer for at least 2 h. In a subsequent study [48], a surprising change in hardness was discovered after immersing specimens in glutaraldehyde for 7 days. Ma et al. [49] discovered that a phenolic-based disinfectant induced surface weakening of resins after 30 min of immersion.

The hardnesses of both resins were significantly decreased during immersion in sodium hypochlorite. In addition to the effect of disinfectants on the surface, the absorbed fluid with effective components penetrated the polymeric chains, altering the bonding and acting as a plasticizer, affecting the mechanical properties of resins and material deformation under mechanical testing [1,26,41]. The negative effect of sodium hypochlorite on both denture resins is consistent with previous studies [26,41,50] reporting a significant decrease in the hardness of 3D-printed and heat-polymerized resins after immersion in sodium hypochlorite. This is consistent with Asad et al. [48], who discovered a decrease in hardness. During the polymerization step, various amounts of residual monomer persist in the acrylic resin [51] and may function as a plasticizer, reducing the mechanical characteristics of polymerized resins [52–54]. Simultaneously, acrylic resins absorb water molecules [52,55]. Likewise, the remaining monomer can progressively leak into storage solutions, reducing the hardness of acrylic resins [23,53] that additionally serve as plasticizers, diminishing the mechanical strength of DBRs [56]. Von Fraunhofer and Suchatlampong [57] investigated the indentation resistance of denture base polymers and discovered that storage in water caused mellowing of the surface in heat-polymerized acrylic resins.

Regarding roughness analysis after immersion in solutions, the current study showed no significant difference in surface roughness between the tested specimens compared to the control group. This is consistent with a review [58] of the effect of disinfectants on the roughness of DBRs, and most of the reviewed studies reported the same findings with different disinfectant solutions. Moreover, Fotovat et al. [34] found that disinfectants produced no change in the surface roughness of 3D-printed resins. Shen et al. [35] found that glutaraldehyde-based disinfectants caused no apparent surface alteration with the standard alkaline formulation.

The present study found statistically significant differences in surface roughness between the heat-polymerized and 3D-printed acrylic for all disinfection methods. The increased roughness of the 3D-printed resins may be attributed to the orientation of the printed layers producing a stepwise effect, being perpendicular to the profilometer scan. With thermal cycling, the temperature accelerated water sorption and increased water absorption. The absorbed water moved the layers apart, impacting the surface irregularity [15,44,59]. The type and curing method of acrylic resin have been consistently reported to significantly influence surface changes after chemical disinfection because component elution may directly impact Ra [58]. Photo-polymerized resins produce more elution than heat-polymerized resins. Heat-polymerized materials have higher monomer-to-polymer conversion rates and a lower residual monomer content [60,61]. The clinical threshold for microbial adhesion is 0.2 μm, and microbial adhesion increased above this threshold [44,62]. The PMMA values were less than the clinical threshold, while all 3D-printed DBRs exceeded this value.

Fotovat et al. compared conventional and 3D-printed resins after immersion in different disinfectants and reported a significant difference in color change between resins. They attributed that change to the difference in composition (filler contents) and printing technologies (layer-by-layer) [63]. Within the category of 3D-printed resins, this component may be pivotal in the observed rise in color change [34]. This finding is consistent with our finding that no significant difference existed between the resins based on disinfectant immersion. This difference in variation could be due to the immersion time and color calculation method, as well as the resin materials used. The change in color of both resins was significantly altered with immersion in sodium hypochlorite compared to glutaraldehyde and Micro 10+. The color change with sodium hypochlorite was consistent with Carvalho et al. [64] and Rocha et al. [25], who found a significant increase in the change in color of PMMA after immersion in sodium hypochlorite. This was consistent with previous studies [19,50,65] confirming the pronounced color change with sodium hypochlorite over other disinfectants. The color changes were attributed to disinfectant solvents permeating the polymer network, expanding the intermolecular spaces. Subsequently, this process resulted in exchanges of internal and external pigments, alterations in the polymer matrix, and the chemical degradation and dissolution of their compounds, leading to color changes [30,63,65]. A previous study [66] reported that sodium hypochlorite exhibited whitening of resins through the oxidation of resin surfaces, consistent with the finding of the present study. Conversely, another study [67] disagreed with the present study, finding that the color steadiness of DBRs was enhanced following immersion in 2% alkaline glutaraldehyde as well as 0.5% sodium hypochlorite solution. These variations in results are attributed to differences in material type and immersion time.

Despite the color changes with sodium hypochlorite, all changes fell within the clinically acceptable value for color changes based on the NBS value. Regarding the thresholds established in the study by Fotovat et al. [34], the color change observed in all three groups was deemed clinically acceptable. This may be due to the immersion time compared to other studies reporting noticeable color changes after immersion for a prolonged time [26,34,68]. Finally, the conventional denture bases outperformed the 3D-printed bases in color stability. Specifically, sodium hypochlorite caused substantially more color change than the other disinfectants. Among all groups, the conventional DBR samples immersed in glutaraldehyde and Micro 10+ showed the least color change.

Regarding disinfectant solutions, regardless of material type, sodium hypochlorite adversely affected the tested properties. Although sodium hypochlorite showed a strong antimicrobial effect, other disinfectants could be recommended as alternatives, as these showed antimicrobial effectiveness without adverse effects on the strength, surface, and color properties. Regarding resin type, although PMMA showed superior performances to the 3D-printed DBRs, all values were within the clinically acceptable range except the roughness of the 3D-printed resins, which exceeded the clinical threshold. Clinically, glutaraldehyde and Micro 10+ could be recommended as disinfectants in dental laboratories and dental offices to control infections and overcome cross-contamination.

The use of bar- and disk-shaped specimens rather than a denture configuration and the duration of immersion for each disinfectant solution are the two limitations of this study. Other limitations are this study's in vitro nature, the restriction to specimens fabricated in the laboratory, and the absence of aged specimens returned to the laboratory for repair or modification after patient use for a prolonged time. Therefore, different disinfectant effects on a real denture base with different immersion times should be investigated in future studies. Additionally, aging specimens using thermal cycling in chewing simulators representing the specimens' return from the clinic to the laboratory for repair or adjustment are required in future investigations.

5. Conclusions

The disinfectant type affects the properties of DBRs. Sodium hypochlorite has adverse effects on the flexural strength, elastic modulus, hardness, and change in color of PMMA and 3D-printed DBRs. Conversely, other disinfectants do not affect the tested properties. Based on the findings of the present study, glutaraldehyde and Micro 10+ could be recommended for DBR disinfection to overcome cross-contamination between dental laboratories and dental offices.

Author Contributions: Conceptualization, A.F.E. and M.Y.S.; methodology, M.E.-S.A.A.E.-A. and A.S.A.B.; software, M.A.A.R.; validation, A.H.A.-A., A.A. and A.E.; formal analysis, M.A. and M.M.G.; investigation, M.Y.S. and M.E.-S.A.A.E.-A.; resources, A.F.E. and M.Y.S.; data curation, A.F.E. and M.Y.S.; writing—original draft preparation, A.F.E., M.Y.S., M.E.-S.A.A.E.-A. and A.S.A.B., M.Y.S., M.E.-S.A.A.E.-A., M.A., M.M.G., R.A. (Raand Altayyar) and R.A. (Riham Alzahrani); writing—review and editing, A.F.E., M.Y.S., M.E.-S.A.A.E.-A., A.S.A.B., M.Y.S., M.E.-S.A.A.E.-A., M.A., M.M.G., A.H.A.-A., A.A., A.E., R.A. (Raand Altayyar) and R.A. (Riham Alzahrani); visualization, A.F.E. and M.Y.S.; supervision, A.H.A.-A., A.A. and A.E.; project administration, A.F.E., M.Y.S., M.E.-S.A.A.E.-A.; funding acquisition, A.H.A.-A., A.A., A.E., R.A. (Raand Altayyar) and R.A. (Riham Alzahrani). All authors have read and agreed to the published version of the manuscript.

Funding: This research received no external funding.

Institutional Review Board Statement: Not applicable.

Informed Consent Statement: Not applicable.

Data Availability Statement: Data are contained within the article.

Conflicts of Interest: The authors declare no conflicts of interest.

References

1. Alkaltham, N.S.; Aldhafiri, R.A.; Al-Thobity, A.M.; Alramadan, H.; Aljubran, H.; Ateeq, I.S.; Khan, S.Q.; Akhtar, S.; Gad, M.M. Effect of Denture Disinfectants on the Mechanical Performance of 3D-Printed Denture Base Materials. *Polymers* **2023**, *15*, 1175. [CrossRef]
2. Al-Thobity, A.M.; Gad, M.; ArRejaie, A.; Alnassar, T.; Al-Khalifa, K.S. Impact of denture cleansing solution immersion on some properties of different denture base materials: An in vitro study. *J. Prosthodont.* **2019**, *28*, 913–919. [CrossRef]
3. Prpić, V.; Schauperl, Z.; Ćatić, A.; Dulčić, N.; Čimić, S. Comparison of mechanical properties of 3D-printed, CAD/CAM, and conventional denture base materials. *J. Prosthodont.* **2020**, *29*, 524–528. [CrossRef]
4. Tuna, S.H.; Keyf, F.; Gumus, H.O.; Uzun, C. The evaluation of water sorption/solubility on various acrylic resins. *Eur. J. Dent.* **2008**, *2*, 191–197.6. [CrossRef]
5. Perea-Lowery, L.; Gibreel, M.; Vallittu, P.K.; Lassila, L.V. 3D-printed vs. heat-polymerizing and autopolymerizing denture base acrylic resins. *Materials* **2021**, *14*, 5781. [CrossRef]
6. Gad, M.M.; Alshehri, S.Z.; Alhamid, S.A.; Albarrak, A.; Khan, S.Q.; Alshahrani, F.A.; Alqarawi, F.K. Water Sorption, Solubility, and Translucency of 3D-Printed Denture Base Resins. *Dent. J.* **2022**, *10*, 42. [CrossRef]
7. Xu, Y.; Xepapadeas, A.B.; Koos, B.; Geis-Gerstorfer, J.; Li, P.; Spintzyk, S. Effect of post-rinsing time on the mechanical strength and cytotoxicity of a 3D printed orthodontic splint material. *Dent. Mater.* **2021**, *37*, e314–e327. [CrossRef]
8. Namala, B.B.; Hegde, V. Comparative evaluation of the effect of plant extract, Thymus vulgaris and commercially available denture cleanser on the flexural strength and surface roughness of denture base resin. *J. Indian Prosthodont. Soc.* **2019**, *19*, 261–265.
9. Hwangbo, N.K.; Nam, N.E.; Choi, J.H.; Kim, J.E. Effects of the Washing Time and Washing Solution on the Biocompatibility and Mechanical Properties of 3D Printed Dental Resin Materials. *Polymers* **2021**, *13*, 4410. [CrossRef]
10. Anjum, R.; Dhaded, S.V.; Joshi, S.; Sajjan, C.S.; Konin, P.; Reddy, Y. Effect of plant extract denture cleansing on heat-cured acrylic denture base resin: An in vitro study. *J. Indian Prosthodont. Soc.* **2017**, *17*, 401–405. [CrossRef]
11. Goodacre, B.J. 3D Printing of Complete Dentures: A Narrative Review. *Int. J. Prosthodont.* **2024**, *37*, 159–164. [CrossRef]
12. *ISO 2039-1:2001*; Plastics—Determination of Hardness—Part 1: Ball Indentation Method. International Organization for Standardization (ISO): Geneva, Switzerland, 2001. Available online: https://www.iso.org/standard/31264.html/ (accessed on 5 December 2023).
13. Kharat, S.; Dudhani, S.I.; Kouser, A.; Subramanian, P.; Bhattacharjee, D.; Jhamb, V. Exploring the Impact of 3D Printing Technology on Patient-Specific Prosthodontic Rehabilitation: A Comparative Study. *J. Pharm. Bioallied Sci.* **2024**, *16* (Suppl. S1), S423–S426. [CrossRef]
14. Gad, M.M.; Fouda, S.M. Factors affecting flexural strength of 3D-printed resins: A systematic review. *J. Prosthodont.* **2023**, *32*, 96–110. [CrossRef]
15. Shim, J.S.; Kim, J.; Jeong, S.H.; Choi, Y.J.; Ryu, J.J. Printing accuracy, mechanical properties, surface characteristics, and microbial adhesion of 3D-printed resins with various printing orientations. *J. Prosthet. Dent.* **2020**, *124*, 468–475. [CrossRef]

16. da Silva, F.C.; Kimpara, E.T.; Mancini, M.N.; Balducci, I.; Jorge, A.O.; Koga-Ito, C.Y. Effectiveness of six different disinfectants on removing five microbial species and effects on the topographic characteristics of acrylic resin. *J. Prosthodont.* **2008**, *17*, 627–633. [CrossRef]
17. Cotrim, L.E.F.; Santos, E.M.; Jorge, A.O.C. Procedimentos de Biossegurança Realizados por Cirurgiões-Dentistas e Laboratórios Durante a Confecção de Próteses Dentárias. *Rev. Odontol. UNESP* **2001**, *30*, 233–244.
18. Coordenaçao, N. *Controle de Infecções e a Prática Odontológica em Tempos de AIDS: Manual de Condutas*, 1st ed.; Ministério da Saúde: Brasília, Brazil, 2000.
19. Li, J.; Helmerhorst, E.J.; Leone, C.W.; Troxler, R.F.; Yaskell, T.; Haffajee, A.D.; Socransky, S.S.; Oppenheim, F.G. Identification of early microbial colonizers in human dental biofilm. *J. Appl. Microbiol.* **2004**, *97*, 1311–1318. [CrossRef]
20. Bhathal, M.K.; Kukreja, U.; Kukreja, N. Evaluation of Efficacy of Different Denture Disinfectants on Biofilms Formed on Acrylic Resin. *Dent. J. Adv. Stud.* **2018**, *6*, 20–27.
21. Guimaraes Junior, L. *Biossegurança e Controle de Infecção Cruzada em Consultórios Odontológicos*, 1st ed.; Santos: São Paulo, Brazil, 2001.
22. Bell, J.A.; Brockmann, M.S.; Feil, P.; Sackuvich, D. The effectiveness of two disinfectants on denture base acrylic resin with an organic load. *J. Prosthet. Dent.* **1989**, *61*, 581–589. [CrossRef]
23. Chau, V.B.; Saunders, T.R.; Pimsler, M.; Elfring, D.R. In-depth disinfection of acrylic resins. *J. Prosthet. Dent.* **1995**, *74*, 309–313. [CrossRef]
24. Angelillo, I.F.; Bianco, A.; Nobile, C.G.A.; Pavia, M. Evaluation of the efficacy of glutaraldehyde and peroxygen for disinfection of dental instruments. *Lett. Appl. Microbiol.* **1998**, *27*, 292–296. [CrossRef]
25. Rocha, M.M.; Carvalho, A.M.; Coimbra, F.C.; Arruda, C.N.; Oliveira, V.D.; Macedo, A.P.; Silva-Lovato, C.H.; Pagnano, V.O.; Paranhos, H.D. Complete denture hygiene solutions: Antibiofilm activity and effects on physical and mechanical properties of acrylic resin. *J. Appl. Oral Sci.* **2021**, *29*, e20200948. [CrossRef]
26. Alqanas, S.S.; Alfuhaid, R.A.; Alghamdi, S.F.; Al-Qarni, F.D.; Gad, M.M. Effect of denture cleansers on the surface properties and color stability of 3D printed denture base materials. *J. Dent.* **2022**, *120*, 104089. [CrossRef]
27. Chhabra, M.; Kumar, M.N.; RaghavendraSwamy, K.N.; Thippeswamy, H.M. Flexural strength and impact strength of heat-cured acrylic and 3D printed denture base resins-A comparative in vitro study. *J. Oral Biol. Craniofacial Res.* **2022**, *12*, 1–3. [CrossRef]
28. Abbasi, S.; Nikanjam, S.; Shishehian, A.; Khazaei, S.; Fotovat, F.; Pana, N.H. A comparative evaluation of wear resistance of three types of artificial acrylic teeth after removing the glaze layer. *Dent. Res. J.* **2022**, *19*, 59.
29. Campbell, S.D.; Cooper, L.; Craddock, H.; Hyde, T.P.; Nattress, B.; Pavitt, S.H.; Seymour, D.W. Removable partial dentures: The clinical need for innovation. *J. Prosthet. Dent.* **2017**, *118*, 273–280. [CrossRef]
30. Jain, S.; Sayed, M.; Ahmed, W.M.; Halawi, A.H.; Najmi, N.M.; Aggarwal, A.; Bhandi, S.; Patil, S. An in-vitro study to evaluate the effect of denture cleansing agents on color stability of denture bases fabricated using CAD/CAM milling, 3D-printing and conventional techniques. *Coatings* **2021**, *11*, 962. [CrossRef]
31. *ISO/TR-28642:2016(E)*; Dentistry—Guidance on Colour Measurements. International Organization for Standardization: Geneva, Switzerland, 2016.
32. Nimeroff, I. *Colorimetry National Bureau of Standards*; Monograph 104; U.S. Government Printing Office: Washington, DC, USA, 1968.
33. Porwal, A.; Khandelwal, M.; Punia, V.; Sharma, V. Effect of denture cleansers on color stability, surface roughness, and hardness of different denture base resins. *J. Indian Prosthodont. Soc.* **2017**, *17*, 61–67. [CrossRef]
34. Fotovat, F.; Abassi, S.; Nikanjam, S.; Alafchi, B.; Baghiat, M. Effects of various disinfectants on surface roughness and color stability of thermoset and 3D-printed acrylic resin. *Eur. J. Transl. Myol.* **2024**, *34*, 11701. [CrossRef]
35. Shen, C.; Javid, N.S.; Colaizzi, F.A. The effect of glutaraldehyde base disinfectants on denture base resins. *J. Prosthet. Dent.* **1989**, *61*, 583–589. [CrossRef]
36. Pavarina, A.C.; Pizzolitto, A.C.; Machado, A.L.; Vergani, C.E.; Giampaolo, E.T. An infection control protocol: Effectiveness of immersion solutions to reduce the microbial growth on dental prostheses. *J. Oral Rehabil.* **2003**, *30*, 532–536. [CrossRef]
37. Azevedo, A.; Machado, A.L.; Vergani, C.E.; Giampaolo, E.T.; Pavarina, A.C.; Magnani, R. Effect of disinfectants on the hardness and roughness of reline acrylic resins. *J. Prosthodont.* **2006**, *15*, 235–242. [CrossRef]
38. Salvia, A.C.; Matilde Fdos, S.; Rosa, F.C.; Kimpara, E.T.; Jorge, A.O.; Balducci, I.; Koga-Ito, C.Y. Disinfection protocols to prevent cross-contamination between dental offices and prosthetic laboratories. *J. Infect. Public Health* **2013**, *6*, 377–382. [CrossRef]
39. Rodrigues, E.A.; Reis, R.F.; Camargo, R.W. Eficácia de três desinfetantes em próteses totais à base de resina acrílica. *Odontol. Mod.* **1994**, *21*, 11–15.
40. Davi, L.R.; Peracini, A.; Ribeiro Nde, Q.; Soares, R.B.; da Silva, C.H.; de Freitas Oliveira Paranhos, H.; De Souza, R.F. Effect of the physical properties of acrylic resin of overnight immersion in sodium-hypochlorite solution. *Gerodontology* **2010**, *27*, 297–302. [CrossRef]

41. Kurt, A.; Erkose-Genc, G.; Uzun, M.; Sarı, T.; Isik-Ozkol, G. The Effect of Cleaning Solutions on a Denture Base Material: Elimination of Candida albicans and Alteration of Physical Properties. *J. Prosthodont.* **2018**, *27*, 577–583. [CrossRef]
42. Durkan, R.; Ayaz, E.A.; Bagis, B.; Gurbuz, A.; Ozturk, N.; Korkmaz, F.M. Comparative effects of denture cleansers on physical properties of polyamide and polymethyl methacrylate base polymers. *Dent. Mater. J.* **2013**, *32*, 367–375. [CrossRef]
43. Pisani, M.X.; Da Silva, C.L.; Paranhos, H.O. The effect of experimental denture cleanser solution Ricinus communis on acrylic resin properties. *Mater. Res.* **2010**, *13*, 369–373. [CrossRef]
44. Gad, M.M.; Fouda, S.M.; Abualsaud, R.; Alshahrani, F.A.; Al-Thobity, A.M.; Khan, S.Q.; Akhtar, S.; Ateeq, I.S.; Helal, M.A.; Al-Harbi, F.A. Strength and surface properties of a 3D-printed denture base polymer. *J. Prosthodont.* **2021**, *31*, 412–418. [CrossRef]
45. Kosti'c, M.; Kruni'c, N.; Nikoli'c, L.; Nikoli'c, V.; Najman, S.; Koci'c, J. Residual monomer content determination in some acrylic denture base materials and possibilities of its reduction. *Vojn. Pregl.* **2009**, *66*, 223–227. [CrossRef]
46. Fouda, S.M.; Gad, M.M.; Abualsaud, R.; Ellakany, P.; AlRumaih, H.S.; Akhtar, S.; Al-Qarni, F.D.; Al-Harbi, F.A. Flexural Properties and Hardness of CAD-CAM Denture Base Materials. *J. Prosthodont.* **2023**, *32*, 318–324. [CrossRef]
47. Asad, T.; Watkinson, A.C.; Huggett, R. The effects of various disinfectant solutions on the surface hardness of an acrylic resin denture base material. *Int. J. Prosthodont.* **1993**, *6*, 9–12.
48. Asad, T.; Watkinson, A.C.; Huggett, R. The effect of disinfection procedures on flexural properties of denture base acrylic resins. *J. Prosthet. Dent.* **1992**, *68*, 191–195. [CrossRef]
49. Ma, T.; Johnson, G.H.; Gordon, G.E. Effects of chemical disinfectants on the surface characteristics and color of denture resins. *J. Prosthet. Dent.* **1997**, *77*, 197–204. [CrossRef]
50. Atalay, S.; Çakmak, G.; Fonseca, M.; Schimmel, M.; Yilmaz, B. Effect of different disinfection protocols on the surface properties of CAD-CAM denture base materials. *J. Prosthet. Dent.* **2022**, *8*, 787–795. [CrossRef]
51. Vallittu PK: The effect of surface treatment of denture acrylic resin on the residual monomer content and its release into water. *Acta Odontol. Scand.* **1996**, *54*, 188–192. [CrossRef]
52. Dogan, A.; Bek, B.; Cevik, N.N.; Usanmaz, A. The effect of preparation conditions of acrylic denture base materials on the level of residual monomer, mechanical properties and water absorption. *J. Dent.* **1995**, *23*, 313–318. [CrossRef]
53. Braun, K.O.; Mello, J.A.; Rached, R.N.; Cury, A.A.D.B. Surface texture and some properties of acrylic resins submitted to chemical polishing. *J. Oral Rehabil.* **2003**, *30*, 91–98. [CrossRef]
54. Kwon, T.Y.; Imai, Y. Polymerization characteristics of ethyl methacrylate-based resin initiated by TBB. *Dent. Mater. J.* **2004**, *23*, 161–165. [CrossRef]
55. Cucci, A.L.; Vergani, C.E.; Giampaolo, E.T.; Afonso, M.C.S.F. Water sorption, solubility, and bond strength of two autopolymerizing acrylic resins and one heat-polymerizing acrylic resin. *J. Prosthet. Dent.* **1998**, *80*, 434–438. [CrossRef]
56. Takahashi, Y.; Chai, J.C.; Kawaguchi, M. Equilibrium strengths of denture polymers subjected to long-term water immersion. *Int. J. Prosthodont.* **1999**, *12*, 348–352.
57. Von Fraunhofer, J.A.; Suchatlampong, C. The surface characteristics of denture base polymers. *J. Dent.* **1975**, *3*, 105–109. [CrossRef]
58. Schwindling, F.S.; Rammelsberg, P.; Stober, T. Effect of chemical disinfection on the surface roughness of hard denture base materials: A systematic literature review. *Int. J. Prosthodont.* **2014**, *27*, 215–225. [CrossRef]
59. Lin, C.T.; Lee, S.Y.; Tsai, T.Y.; Dong, D.; Shih, Y. Degradation of repaired denture base material in simulated oral fluid. *J. Oral Rehabil.* **2000**, *27*, 190–198. [CrossRef]
60. Koda, T.; Tsuchiya, H.; Yamauchi, M.; Hoshino, Y.; Takagi, N.; Kawano, J. High-performance liquid chromatographic estimation of eluates from denture base polymers. *J. Dent.* **1989**, *17*, 84–89. [CrossRef]
61. Zissis, A.; Yannikakis, S.; Polyzois, G.; Harrison, A. A long term study on residual monomer release from denture materials. *Eur. J. Prosthodont. Restor. Dent.* **2008**, *16*, 81–84.
62. Bollen, C.M.; Lambrechts, P.; Quirynen, M. Comparison of surface roughness of oral hard materials to the threshold surface roughness for bacterial plaque retention: A review of the literature. *Dent. Mater.* **1997**, *13*, 258–269. [CrossRef]
63. Kessler, A.; Reichl, F.X.; Folwaczny, M.; Högg, C. Monomer release from surgical guide resins manufactured with different 3D printing devices. *Dent. Mater.* **2020**, *36*, 1486–1492. [CrossRef]
64. Carvalho, C.F.; Vanderlei, A.D.; Salazar Marocho, S.M.; Pereira, S.; Nogueira, L.; Arruda Paes-Junior, T.J. Effect of disinfectant solutions on a denture base acrylic resin. *Acta Odontol. Latinoam.* **2012**, *25*, 255–260.
65. Gad, M.M.; Abualsaud, R.; Fouda, S.M.; Rahoma, A.; AlThobity, A.M.; Khan, S.Q.; Akhtar, S.; Al-Abidi, K.S.; Ali, M.S.; Al-Harbi, F.A. Color Stability and Surface Properties of PMMA/ZrO_2 Nanocomposite Denture Base Material after Using Denture Cleanser. *Int. J. Biomater.* **2021**, *2021*, 6668577. [CrossRef]
66. Yiu, C.K.; King, N.M.; Pashley, D.H.; Suh, B.; Carvalho, R.; Carrilho, M.; Tay, F. Effect of resin hydrophilicity and water storage on resin strength. *Biomaterials* **2004**, *25*, 5789–5796. [CrossRef]

67. Almuraikhi, T. Evaluation of the Impact of Different Disinfectants on Color Stability of Denture Base Materials: A Comparative Study. *J. Contemp. Dent. Pract.* **2022**, *23*, 543–547. [CrossRef]
68. Piskin, B.; Sipahi, C.; Akin, H. Effect of different chemical disinfectants on color stability of acrylic denture teeth. *J. Prosthodont.* **2014**, *23*, 476–483. [CrossRef]

prosthesis

Case Report

Enhanced Retention of Mandibular Digital Complete Dentures Using an Intraoral Scanner: A Case Report

Edgar García [1,*] and Stephanie Jaramillo [2]

[1] Department of Prosthodontics, The University of Iowa, Iowa City, IA 52242, USA
[2] Independent Researcher, Guayaquil 090112, Ecuador; stephaniejaramillo.ch@gmail.com
* Correspondence: egarciazea@uiowa.edu

Abstract: Introduction: Mandibular complete dentures often pose challenges due to anatomical and functional limitations. Impression techniques, including functional, mucostatic, compressive, selective pressure, and neutral zone methods, play a crucial role in achieving stability and retention. In 1999, Abe introduced the Suction Effective Mandibular Complete Denture (SEMCD) technique, revolutionizing mandibular denture retention by incorporating functional extensions and achieving a peripheral seal even in the presence of mobile soft tissues. **Case report:** An 87-year-old male presented to a private dental clinic with the chief complaint that his current lower complete denture lacked retention and stability. Intraoral examination revealed a severely resorbed mandibular edentulous ridge with movable retromolar pads and a prominent spongy lingual area. This case report describes the integration of Abe's concepts into a digital workflow, using a single-step intraoral scanning technique and digital design software to fabricate a mandibular denture with enhanced retention and stability. **Conclusions:** This approach minimizes the number of clinical steps involved, improves patient comfort, and achieves predictable outcomes, highlighting the utility of digital technologies in modern prosthodontics.

Keywords: digital denture; suction denture; complete dentures; intraoral scans

1. Introduction

Achieving acceptable stability and retention with mandibular complete dentures presents significant challenges. Specific issues such as high tongue mobility, insufficient residual ridge height, movable retromolar pads, and anatomical structures that change size when the patient opens or closes their mouth complicate the use of mandibular dentures [1,2].

Definitive impressions for complete dentures play a crucial role in treatment success and can be classified into techniques such as neutral zone, mucostatic, mucocompressive, selective pressure, and functional. Among these, functional techniques aim to capture the dynamic relationship between soft tissues and the denture base, enhancing retention and stability during function [3–5].

In 1999, Abe introduced an innovative functional impression technique and new concepts regarding the correct extension of the prosthesis that revolutionized denture fitting by ensuring a peripheral seal even with mobile soft tissues at the denture's base. This method, known for its high predictability, enabled mandibular dentures to restore function and masticatory capacity, significantly improving patients' quality of life, all without requiring surgical procedures [6].

Advancements in technology have significantly transformed the fabrication of complete dentures, offering more efficient and precise methods than traditional approaches. The integration of computer-aided design (CAD) and computer-aided manufacturing (CAM) has enabled the creation of digital complete dentures with a high degree of accuracy and repeatability [7]. The use of intraoral scanners (IOS) has significantly improved the patient experience by eliminating the need for physical impressions, which can be uncomfortable and less accurate [8].

The purpose of this report is to describe a technique for fabricating mandibular dentures with increased retention, utilizing a single-step scanning technique combined with Abe's concepts for both the intraoral scanning process and the design of the prosthesis extension.

2. Materials and Methods

An 87-year-old male presented to a private dental clinic with the chief complaint that his current lower complete denture, which he had worn for approximately two years, lacked retention and stability, making it difficult for him to eat and speak. The patient's medical history was non-contributory, with no contraindications for dental treatment.

Intraoral examination revealed a severely resorbed mandibular edentulous ridge with movable retromolar pads and a prominent spongy lingual area (Figure 1). The existing denture exhibited excessive movement of the mandibular base during functional activities, with teeth positioned in suboptimal locations.

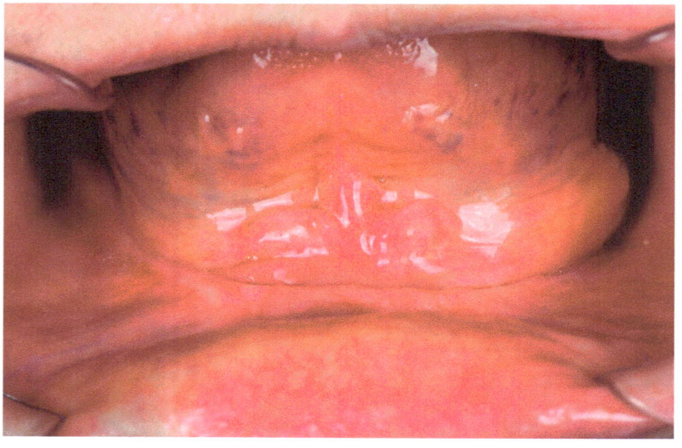

Figure 1. Intraoral view of the edentulous mandible.

The patient was presented with several treatment options, including implant overdentures, implant-supported fixed prostheses, and new complete dentures. However, he opted for a new mandibular complete denture, emphasizing his preference for a design closely resembling his current prosthesis and expressing his intention to use it temporarily while deciding whether to pursue implant treatment at a later date.

During the first appointment, an intraoral scanner (Aoralscan 3, Shining 3D, Hangzhou, China) was used to scan the patient's existing dentures in occlusion. Although the dentures lacked adequate stability, they provided a reference for the patient's vertical dimension of occlusion. This process not only facilitated the capture of the antagonist but also preserved the vertical dimension, eliminating the need for an additional appointment. In cases where the patient does not have existing dentures, it would be necessary to schedule an extra visit to design and print bases with wax rims to obtain an occlusal record.

To perform the mandibular scan, an intraoral retractor for edentulous arches (Lo Russo Retractors, Vallesaccarda, Italy) was employed (Figure 2). The patient was instructed to keep their mandible slightly closed, in a resting position, while the scan was performed following a specific strategy: starting posteriorly and proceeding along the occlusal aspect of the ridge to the opposite side, subsequently returning along the palatal or lingual aspect, and finally scanning the buccal aspect (Figure 3).

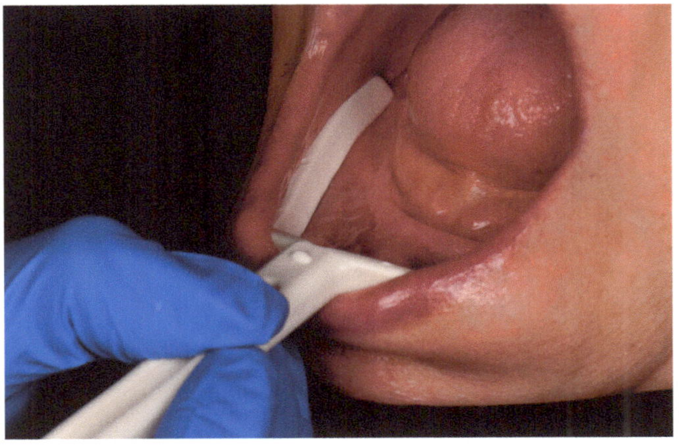

Figure 2. Intraoral retractor used for scanning edentulous arches.

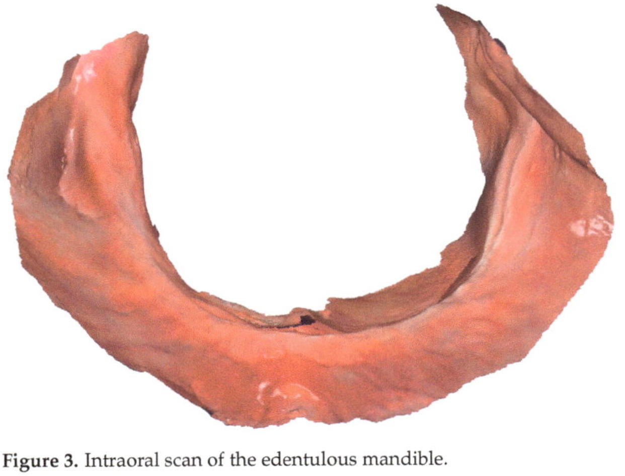

Figure 3. Intraoral scan of the edentulous mandible.

Once the mandibular scan was completed, the scanner software automatically aligned the scan with the reference of the patient's existing dentures due to similarities in their surface geometry. If automatic alignment is not feasible because the existing denture lacks sufficient extension, it may be necessary to reline the denture using an impression material or a relining material. This adjustment enables proper matching, which can be performed through various methods, either directly in the scanner software or later in the design software. The scan files were subsequently exported to design software (DentalCAD 3.2 Elefsina, Exocad GmbH, Darmstadt, Germany) to create the prosthesis.

For this design, the outline was carefully extended to fully cover the retromolar pad, avoid interference with Someya's sinew string, reach the deepest point of the buccal shelf, relieve pressure on the buccal frenum, and extend two millimeters beyond the mylohyoid ridge. The teeth were positioned based on the patient's request, using the previous denture as a reference (Figure 4).

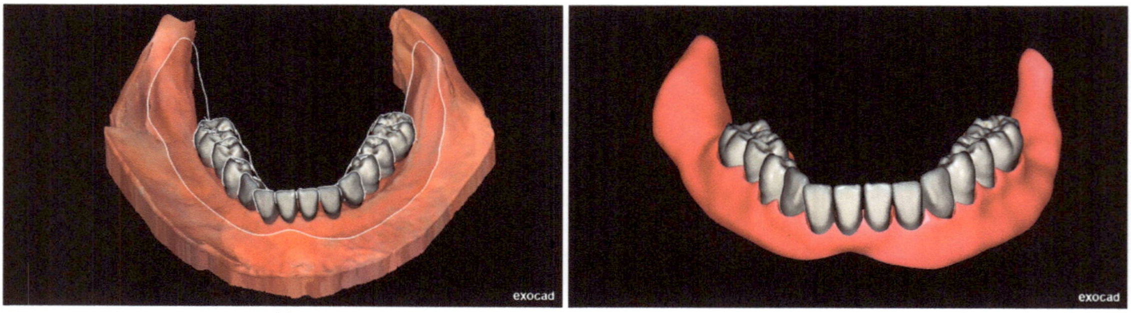

(**a**) (**b**)

Figure 4. Digital design of the denture: (**a**) extension of the denture covering the entirety of the retromolar papilla, avoiding Someya's sinew string, extending to the deepest point of the buccal shelf, relieving the buccal frenum, and extending two millimeters beyond the mylohyoid ridge; (**b**) final design of the denture.

The final denture base was printed using pink base resin (Denture Base Resin, Formlabs, Somerville, MA, USA), and the teeth were printed with denture teeth resin (Denture Teeth Resin A2, Formlabs, Somerville, MA, USA). The prosthesis was then assembled, finished, and polished (Figure 5).

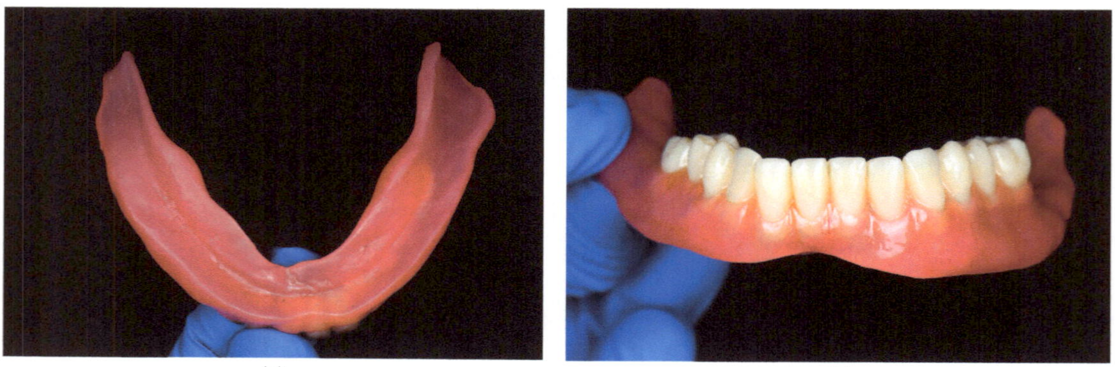

(**a**) (**b**)

Figure 5. The 3D-printed mandibular denture: (**a**) intaglio surface of the denture and (**b**) frontal view of the denture.

At the second appointment, the patient received the final denture, which fit well and restored his masticatory function. He reported satisfaction with the comfort, stability, and esthetics of the new denture (Figure 6). He was instructed to attend regular follow-up appointments to monitor and maintain the functionality and comfort of the prosthesis. At the first follow-up visit, a minor adjustment was made to a small area of the denture's intaglio surface to alleviate slight pressure. During follow-up appointments at one and four months, the patient reported high satisfaction with both function and esthetics.

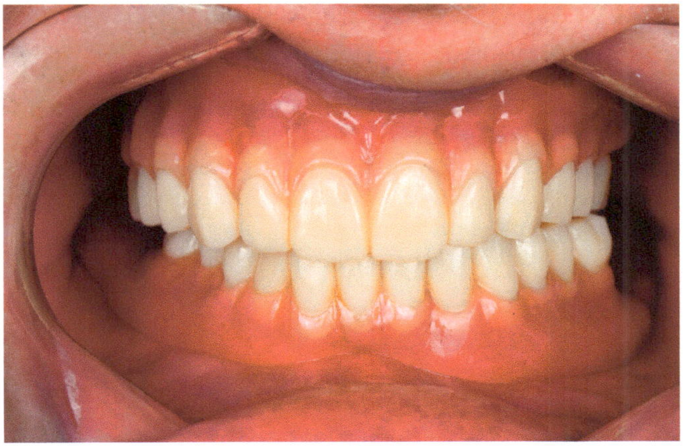

Figure 6. Intraoral view of the mandibular complete denture.

3. Discussion

The advent of digital technologies in dentistry has significantly transformed the fabrication of complete dentures, introducing more efficient and precise methods compared to traditional techniques [9]. Conventional impression methods have long been considered the gold standard for obtaining accurate anatomical records; however, they present limitations such as patient discomfort, potential for tissue distortion, and challenges in capturing dynamic anatomical structures, especially in the mandibular arch [10].

IOS have emerged as valuable alternatives, offering efficient and reproducible means of capturing oral anatomy while enhancing patient comfort. Studies have demonstrated that digital scans obtained via IOS can capture fine anatomical details without the distortion associated with conventional impression materials, a critical factor for achieving optimal fit in mandibular complete dentures, particularly in patients with compromised residual ridges or mobile soft tissues [11].

Despite these advantages, achieving an effective peripheral seal—a crucial element for the retention and stability of complete dentures—remains a challenge with digital impressions. The development of refined scanning protocols and techniques has addressed many of these obstacles, improving clinical predictability [12]. Utilizing IOS as a primary record before definitive impressions allows for precise planning and iterative adjustments of prosthetic designs, reducing clinical time and the number of appointments required [13]. The implementation of specific scanning strategies is essential to ensure accurate and functional digital impressions [14–17]. Techniques such as employing intraoral retractors to stabilize soft tissues and systematically capturing critical anatomical landmarks—such as the retromolar pad and buccal shelf—have enhanced the quality and clinical utility of digital records [18].

Jiro Abe's Suction Effective Mandibular Complete Denture (SEMCD) technique revolutionized the concept of achieving a peripheral seal in mandibular dentures by accounting for the dynamic nature of soft tissues. The key concepts of SEMCD involve capturing tissue dynamics by recognizing the dimensional changes in the retromolar area when the mouth opens or closes. Abe emphasized taking impressions with the mouth closed to ensure that the prosthesis aligns accurately with the tissues during function, as impressions taken with the mouth open can lead to discrepancies between the prosthesis and posterior tissues, compromising the peripheral seal and suction effectiveness. Additionally, extending the denture 2 to 3 mm beyond the mylohyoid ridge increases stability during functional movements by engaging additional tissue support. Relieving pressure in the area of Someya's sinew string—a tendinous structure located in the floor of the mouth near the retromolar pad—is crucial. This anatomical feature plays a key role in regulating the movement of the buccal mucosa and maintaining the posterior seal of the oral vestibule [19]. It appears as a sinew-like band of connective tissue that stabilizes the mucosal tissues during functional movements, such as mastication. Failure to account for this structure during prosthesis design can cause discomfort, impair tissue function, and lead to prosthesis dislodgement during mastication, compromising patient comfort and prosthetic stability (Figure 7).

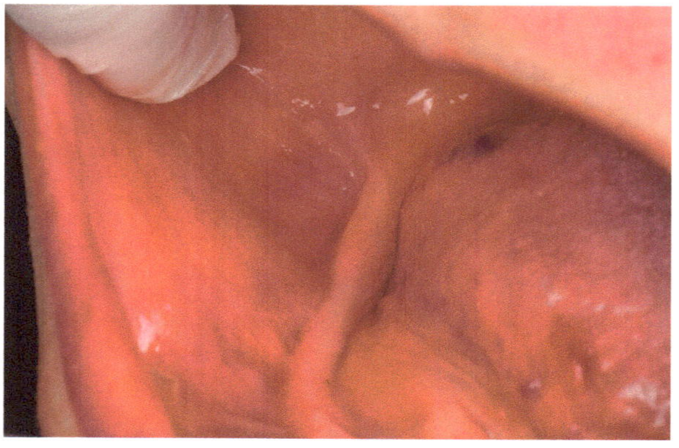

Figure 7. Intraoral view of Someya's sinew string on the right side of the mandibular arch.

In the present case, Abe's SEMCD principles were integrated into a digital workflow by considering three key elements: the use of an intraoral retractor to scan edentulous arches, performing the scan with the mandible in a proper resting position, and ensuring the ideal extension of the prosthesis. The intraoral retractor provided optimal soft tissue stabilization and visibility, enabling accurate execution of the scanning strategy. By performing the intraoral scan with the patient maintaining a mandibular resting position, the functional dimensions of the soft tissues were captured, improving the accuracy of the digital impression (Figure 8). Additionally, the digital prosthesis design was extended beyond the mylohyoid ridge, and pressure was relieved in the area of Someya's sinew string, replicating the functional extensions described by Abe and optimizing the stability and retention of the final denture [20].

Figure 8. Difference in scans of the retromolar area with the patient's mouth open and closed: (**a**) scan of the retromolar area with the mouth closed and (**b**) scan of the retromolar area with the mouth open.

Retention is a critical factor in the success of mandibular dentures. Studies comparing the force required to dislodge mandibular overdentures have shown that ball attachments require approximately 0.655 kg, implant bar attachments require 1.677 kg, and magnet attachments require 0.370 kg for removal [21]. In this case, dental floss and a dynamometer (Digital Force Gauge, Mxmoonfree, Hangzhou, China) were used to measure the retention of the denture. The dental floss was tied to the prosthesis, which was then inserted into the patient's mouth. The patient was in-structed to swallow to ensure proper seating of the denture. Afterward, the hook of the dynamometer was attached to the dental floss, and the patient was asked to open their mouth. Once the mouth was open, a vertical force was applied to measure the amount of force required to dislodge the prosthesis. The digitally fabricated mandibular complete denture required 0.31 kg of force for dislodgement (Figure 9). Although the retention force achieved is slightly lower than that of some attachment systems, this approach offers a viable treatment option for edentulous patients.

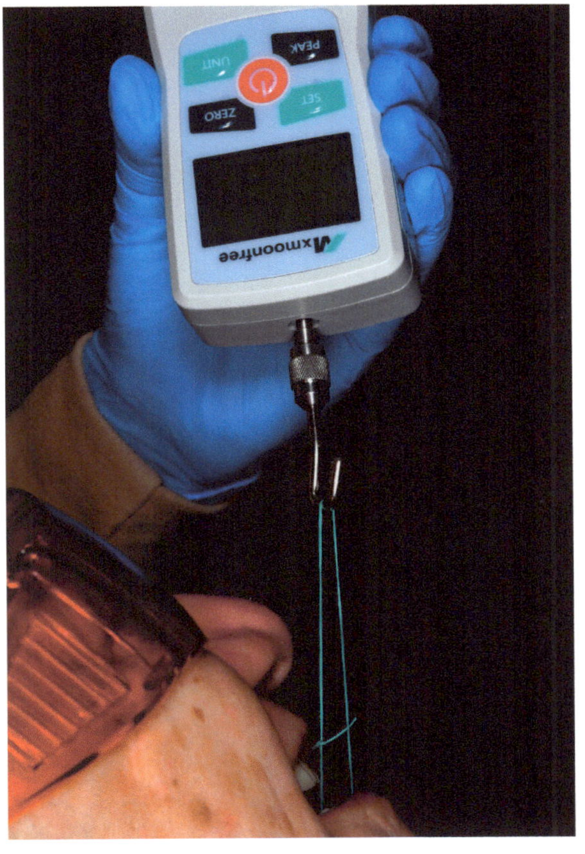

Figure 9. Maximum force required to remove the prosthesis using a dynamometer.

Incorporating SEMCD concepts into a digital workflow is both feasible and beneficial. The digital replication of functional extensions allows for high precision in prosthesis fabrication, potentially leading to improved clinical outcomes. Moreover, this method eliminates the need for physical impressions and models, streamlining the fabrication process and enhancing patient comfort. Due to the nature of the present article, further clinical studies are necessary to compare the impact of scanning edentulous mandibles with open or closed mouth techniques on long-term denture retention and stability.

4. Conclusions

The integration of Abe's principles into a digital workflow enhances the retention and stability of mandibular complete dentures while minimizing the number of clinical steps involved. The results of this study demonstrate that digital technologies can effectively replicate established functional concepts, providing a precise and efficient solution for the rehabilitation of edentulous patients.

Author Contributions: Conceptualization, E.G.; methodology, E.G. and S.J.; software, E.G.; validation, E.G. and S.J.; resources, E.G.; data curation, E.G.; writing—original draft preparation, E.G.; writing—review and editing, S.J.; visualization, E.G. and S.J.; supervision, E.G.; project administration, E.G.; funding acquisition, E.G. All authors have read and agreed to the published version of the manuscript.

Funding: This research received no external funding.

Institutional Review Board Statement: The patient provided written informed consent for the publication of their case, including the use of clinical data and images, in accordance with the definitions and regulations set forth by the U.S. Department of Health and Human Services, Office for Human Research Protections (45 CFR 46.102(l)), this article does not meet the regulatory definition of "research involving human subjects," as it is neither a systematic investigation nor intended to contribute to generalizable knowledge. Therefore, Institutional Review Board (IRB) approval was not required. The patient provided written informed consent for both treatment and publication of relevant clinical information.

Informed Consent Statement: Informed consent was obtained from the subject involved in this study.

Data Availability Statement: No new data were created or analyzed in this study. Data sharing is not applicable to this article.

Conflicts of Interest: The authors declare no conflicts of interest.

References

1. Jacobson, T.E.; Krol, A.J. A contemporary review of the factors involved in complete denture retention, stability, and support. Part I: Retention. *J. Prosthet. Dent.* **1983**, *49*, 5–15. [CrossRef]
2. Jacobson, T.E.; Krol, A.J. A contemporary review of the factors involved in complete dentures. Part II: Stability. *J. Prosthet. Dent.* **1983**, *49*, 165–172. [CrossRef] [PubMed]
3. Freeman, S.P. Impressions for complete dentures. *J. Am. Dent. Assoc.* **1969**, *79*, 1173–1178. [CrossRef] [PubMed]
4. Boucher, C.O. Complete denture impressions based upon the anatomy of the mouth. *J. Am. Dent. Assoc.* **1944**, *31*, 1174–1181. [CrossRef]
5. Boucher, C.O. A critical analysis of mid-century impression techniques for full dentures. *J. Prosthet. Dent.* **1951**, *1*, 472–491. [CrossRef] [PubMed]
6. Abe, J. Clinical denture fabrication to achieve an effective suction of the mandibular complete denture: Enhancement of the posterior end border seal around the retromolar pad. *Ann. Jpn. Prosthodont. Soc.* **2011**, *3*, 220–230. (In Japanese) [CrossRef]
7. Zandinejad, A.; Floriani, F.; Lin, W.S.; Naimi-Akbar, A. Clinical outcomes of milled, 3D-printed, and conventional complete dentures in edentulous patients: A systematic review and meta-analysis. *J. Prosthodont.* **2024**, *33*, 736–747. [CrossRef] [PubMed]
8. Lo Russo, L.; Salamini, A.; Troiano, G.; Guida, L. Digital dentures: A protocol based on intraoral scans. *J. Prosthet. Dent.* **2021**, *125*, 597–602. [CrossRef] [PubMed]
9. Lo Russo, L.; Salamini, A. Removable complete digital dentures: A workflow that integrates open technologies. *J. Prosthet. Dent.* **2018**, *119*, 727–732. [CrossRef] [PubMed]
10. Tripathi, A.; Singh, S.V.; Aggarwal, H.; Gupta, A. Effect of mucostatic and selective pressure impression techniques on residual ridge resorption in individuals with different bone mineral densities: A prospective clinical pilot study. *J. Prosthet. Dent.* **2019**, *121*, 90–94. [CrossRef] [PubMed]

11. LoRusso, L.; Caradonna, G.; Troiano, G.; Salamini, A.; Guida, L.; Ciavarella, D. Three-dimensional differences between intraoral scans and conven tional impressions of edentulous jaws: A clinical study. *J. Prosthet. Dent.* **2020**, *123*, 264–268. [CrossRef] [PubMed]
12. Thu, K.M.; Molinero-Mourelle, P.; Yeung, A.W.K.; Abou-Ayash, S.; Lam, W.Y.H. Which clinical and laboratory procedures should be used to fab ricate digital complete dentures? A systematic review. *J. Prosthet. Dent.* **2024**, *132*, 922–938. [CrossRef] [PubMed]
13. Lo Russo, L.; Guida, L.; Ronsivalle, V.; Ercoli, C. Digital denture with mucostatic base and functional borders: A cast-free digital technique. *J. Prosthodont.* **2024**. *epub ahead of print*. [CrossRef] [PubMed]
14. Jamjoom, F.Z.; Aldghim, A.; Aldibasi, O.; Yilmaz, B. Impact of intraoral scanner, scanning strategy, and scanned arch on the scan accuracy of edentulous arches: An in vitro study. *J. Prosthet. Dent.* **2024**, *131*, 1218–1225. [CrossRef] [PubMed]
15. Al Hamad, K.Q.; Al-Kaff, F.T. Trueness of intraoral scanning of edentulous arches: A comparative clinical study. *J. Prosthodont.* **2023**, *32*, 26–31. [CrossRef] [PubMed]
16. Zarone, F.; Ruggiero, G.; Ferrari, M.; Mangano, F.; Joda, T.; Sorrentino, R. Accuracy of a chairside intraoral scanner compared with a laboratory scanner for the completely edentulous maxilla: An in vitro 3-dimensional comparative analysis. *J. Prosthet. Dent.* **2020**, *124*, 761.e1–761.e7. [CrossRef] [PubMed]
17. Sorrentino, R.; Ruggiero, G.; Leone, R.; Irene Di Mauro, M.; Cagidiaco, E.F.; Joda, T.; Lo Russo, L.; Zarone, F. Influence of different palatal morphologies on the accuracy of intraoral scanning of the edentulous maxilla: A three-dimensional analysis. *J. Prosthodont. Res.* **2024**, *68*, 634–642. [CrossRef] [PubMed]
18. Lo Russo, L.; Sorrentino, R.; Esperouz, F.; Zarone, F.; Ercoli, C.; Guida, L. Assessment of distortion of intraoral scans of edentulous mandibular arch made with a 2-step scanning strategy: A clinical study. *J. Prosthet. Dent.* **2023**. *epub ahead of print*. [CrossRef] [PubMed]
19. Someya, S. The anatomical study of the sinew string observed on the buccal mucosa of mandibular second molar and posterior of retromolar pad. *J. Acad. Clin. Dent.* **2008**, *28*, 14–20. [CrossRef]
20. Abe, J.; Kokubo, K.; Sato, D. *Mandibular Suction-Effective Denture: The Professional*; Quintessence Publishing: Tokyo, Japan, 2012.
21. Naert, I.; Gizani, S.; Vuylsteke, M.; Van Steenberghe, D. A 5-year prospective randomized clinical trial on the influence of splinted and unsplinted oral implants retaining a mandibular overdenture: Prosthetic aspects and patient satisfaction. *J. Oral Rehabil.* **1999**, *26*, 195–202. [CrossRef] [PubMed]

prosthesis

Systematic Review

The Antimicrobial Effect of the Incorporation of Inorganic Substances into Heat-Cured Denture Base Resins—A Systematic Review

Mariana Lima [1], Helena Salgado [1,2], André Correia [1,2] and Patrícia Fonseca [1,2,*]

[1] Faculty of Dental Medicine (FMD), Universidade Católica Portuguesa, 3504-505 Viseu, Portugal; mariana.alexandra.lima@hotmail.com (M.L.); hsalgado@ucp.pt (H.S.); andrecorreia@ucp.pt (A.C.)
[2] Center for Interdisciplinary Research in Health (CIIS), Faculty of Dental Medicine (FMD), Universidade Católica Portuguesa, 3504-505 Viseu, Portugal
* Correspondence: pafonseca@ucp.pt

Abstract: Introduction: Polymethylmethacrylate (PMMA) is the most widely used denture base material due to its favourable properties. Several studies have tested the incorporation of anti-infective agents into PMMA as a strategy to prevent biofilm growth on the denture surface. This systematic review aims to evaluate the efficacy of incorporating inorganic antimicrobial particles into denture base resins in preventing antimicrobial growth, thereby identifying the most effective agents for enhancing PMMA's antimicrobial properties. Materials and methods: This systematic review followed the PRISMA guidelines, and the research protocol was registered in PROSPERO. The search was performed by using Medical Subject Headings and free text combined with Boolean operators in PubMed/Medline® and in Cochrane® and a free text combination in Web of Science® Core Collection. Data regarding the inorganic particles studied, their antimicrobial effect, and the type of samples produced were collected and analysed. Results: After screening, a total of fifteen studies were included in this review. Most samples were disk-shaped and of varying sizes, and the most tested microbial strain was *Candida albicans*. Silver was the most used antimicrobial particle, followed by gold, titanium, and copper. Conclusions: Overall, incorporating inorganic particles into PMMA has produced promising antimicrobial results, depending on the concentration. Due to the high heterogeneity observed in the samples, more studies are recommended, particularly clinical trials.

Keywords: anti-infective agents; nanoparticles; polymethyl methacrylate; denture bases; inorganic particles

1. Introduction

The oral microbiota is massively diverse, allowing different pathogenic species to establish metabolic communications, which is frequently observed between *Streptococcus mutans* (*S. mutans*) and *C. albicans* [1,2]. Several pathologies, such as denture stomatitis, tooth decay, and periodontal disease, are caused by fungi and bacteria that can severely affect the patient's health. *Candida albicans*, e.g., is regarded as the most prevalent fungus associated with the development of Candida-Associated Denture Stomatitis (CADS) in the palatal mucosa of denture wearers [1–4], while bacteria, such as streptococci and lactobacilli, are more associated with tooth decay [5]. Denture hygiene with antifungal disinfectants fails at removing *C. albicans* that has infiltrated the denture resin, thus allowing the permanence of biofilm [6]. Consequently, there is a growing need of measures to prevent biofilm formation.

Heat-cured PMMA requires heat energy to activate the initiator [2,7], being moulded into denture bases through a flask–pack–press technique [7]. Due to the usual presence of oral microbiota inside denture resin, several studies have tested the incorporation of antimicrobial particles. Natural products (specifically chitosan), chemical compounds (such as nystatin and chlorhexidine), organic extracts (namely, tea tree oil and thymoquinone), and inorganic particles present promising antimicrobial properties, as they causing cell death in microorganisms and preventing the adherence of bacteria and fungi [4,8,9]. Nanoparticles, such as silver (Ag), copper (Cu), and gold (Au), present antimicrobial properties that make them ideal inorganic particles to incorporate into denture base resins [8,9].

This systematic review aims to compare the antimicrobial properties of incorporating different inorganic antimicrobial agents into denture base heat-cured resin against conventional PMMA while assessing the resulting efficacy in preventing associated pathogenesis. Therefore, we intend to determine the inorganic agent that presents the best antimicrobial properties when incorporated into PMMA.

2. Materials and Methods

This systematic review follows the Preferred Reporting Items for Systematic Reviews and Meta-Analysis (PRISMA) guidelines [10,11], and the research question was formulated based on the PICO template (Population, Intervention, Comparison, and Outcome) [12]: "Does the incorporation of inorganic antimicrobial particles (I) into heat-cured denture base resins (P) result in enhanced antimicrobial properties (O) compared to conventional heat-cured resins (C)?" The research protocol was registered in PROSPERO (International Prospective Register of Systematic Reviews) on 3 January 2024 (ID CRD42024496013).

2.1. Information Sources and Search Strategy

The search, which included studies published from 2019 until January 2024, was conducted within the databases PubMed/Medline®, Cochrane®, and Web of Science® Core Collection. The research equation in PubMed/Medline® was determined by using MeSH terms (Medical Subject Headings) and free text terms, namely, dentures, acrylic resins, and antimicrobials, connected through Boolean operators such as "AND" and "OR". Therefore, the following equation was obtained:

((("Dentures" [MeSH]) OR ("Denture Bases" [MeSH]) OR ("Dental Prosthesis" [Mesh]) OR (dental prostheses) OR (denture) OR (denture base) OR (Prostheses, Dental) OR (Dental Prostheses) OR (Prosthesis, Dental)) AND (("Acrylic Resins" [MeSH]) OR (resin) OR (pmma) OR (polymethylmethacrylate) OR (denture base material)) AND (("Anti-Infective Agents" [MeSH]) OR ("Cariostatic Agents" [MeSH]) OR (anti-cariogenic) OR (cariostatic) OR (antibiotic) OR (antimicrobial) OR (antibacterial) OR (antifungal) OR (anti-infective))).

In Cochrane®, the research strategy started by searching for each MeSH term ("Dentures", "Denture Bases", "Dental Prosthesis", "Acrylic Resins", "Anti-Infective Agents", and "Cariostatic Agents") individually. Following this step, the MeSH terms were combined with related free text terms, obtaining three different search lines: "Dentures" OR "Denture Bases" OR "Dental Prosthesis" OR (dental prostheses) OR (denture) OR (denture base) OR (Prostheses, Dental) OR (Dental Prostheses) OR (Prosthesis, Dental); "Acrylic Resins" OR (resin) OR (pmma) OR (polymethylmethacrylate) OR (denture base material); "Anti-Infective Agents" OR "Cariostatic Agents" OR (anti-cariogenic) OR (cariostatic) OR (antibiotic) OR (antimicrobial) OR (antibacterial) OR (antifungal) OR (anti-infective). Finally, a final search line was used to combine the three searches by using the Boolean operator AND.

Also, the Web of Science® database was used in this research. Free text terms combined with Boolean operators were used withing the following equation: ALL = (((Dentures) OR (Denture Bases) OR (Dental Prosthesis) OR (dental prostheses) OR (denture) OR (denture base) OR (Prostheses, Dental) OR (Dental Prostheses) OR (Prosthesis, Dental)) AND ((Acrylic Resins) OR (resin) OR (pmma) OR (polymethylmethacrylate) OR (denture base material)) AND ((Anti-Infective Agents) OR (cardiostatic Agents) OR (anti-cariogenic) OR (cardiostatic) OR (antibiotic) OR (antimicrobial) OR (antibacterial) OR (antifungal) OR (anti-infective))).

2.2. Study Selection

Inclusion criteria were defined, aiming to include in vitro studies published after 2019 that focused on denture materials, specifically heat-cured PMMA, modified with inorganic antimicrobials. The studies were required to present a minimum of 5 samples to be included in the review. Database filters such as Books and documents, Meta-analysis, Review, and Systematic review were used for exclusion of results in PubMed/Medline®; in the Web of Science® Core Collection, the filters Article and Early Access were used for inclusion and Review Article and Proceeding Paper for exclusion of results, while the filter Trials was used in Cochrane®. The remaining studies were imported into a Microsoft© Excel spreadsheet, where duplicates and studies published before 2019 were removed. The screening was conducted by two independent investigators (M.L. and P.F.), beginning with a selection based on the title and then by abstract. A final screening, based on full-text reading, was performed, where all exclusions were justified. To evaluate inter-rater reliability during screening, Cohen's kappa statistic was adopted and determined for each step in the selection.

2.3. Data Extraction and Quality Assessment

A Microsoft© Excel version 2408 with Office 365 preadsheet was prepared, where the data extracted from each study were recorded. These variables included information such as title, author, year, country, journal, resin brand, resin processing, sample number, shape and size, inorganic antimicrobial incorporated, concentrations of antimicrobial, control group, microorganisms tested, results, and conclusions.

The Checklist for Quasi-Experimental Studies (Non-Randomized Experimental Studies) from the Joanna Briggs Institute was used to evaluate the methodology and determine the quality of the selected studies. This checklist is composed of the following nine questions, with four possible answers, i.e., "yes", "no", "unclear", and "not applicable":

1. Is it clear in the study what is the "cause" and what is the "effect" (i.e., there is no confusion about which variable comes first)?
2. Were the participants included in any comparisons similar?
3. Were the participants included in any comparisons receiving similar treatment/care, other than the exposure or intervention of interest?
4. Was there a control group?
5. Were there multiple measurements of the outcome both pre- and post-intervention/exposure?
6. Was follow-up complete and if not, were differences between groups in terms of their follow-up adequately described and analysed?
7. Were the outcomes of participants included in any comparisons measured in the same way?
8. Were outcomes measured in a reliable way?
9. Was appropriate statistical analysis used?

The risk of bias was evaluated and analysed by counting the total number of affirmative answers and dividing by the total number of questions. The resulting fraction was then converted into a percentage and interpreted as seen in the article by Paes et al. [13], who considered that a percentage between 0% and 49% represented a high risk of bias. Following this author, a value ranging from 50% to 69% could be interpreted as a moderate risk of bias, and a percentage superior to or equal to 70% was defined as a low risk of bias.

3. Results

As shown in the PRISMA flow diagram depicted in Figure 1, the initial database search resulted in 2645 articles, of which 1821 were removed with database filters. The remaining articles were imported into a Microsoft© Excel spreadsheet, where studies with a publication year before 2019 and duplicates were removed, leaving 246 articles. The screening was performed, firstly based on the title and then on abstract reading, obtaining almost perfect agreement, with calculated k values of approximately 0.93 and 0.96, respectively. The full texts of twenty-five studies were read, leading to the exclusion of ten articles due to study characteristics and sample preparation, resulting in an almost perfect inter-rater agreement and a value of k reaching 0.92.

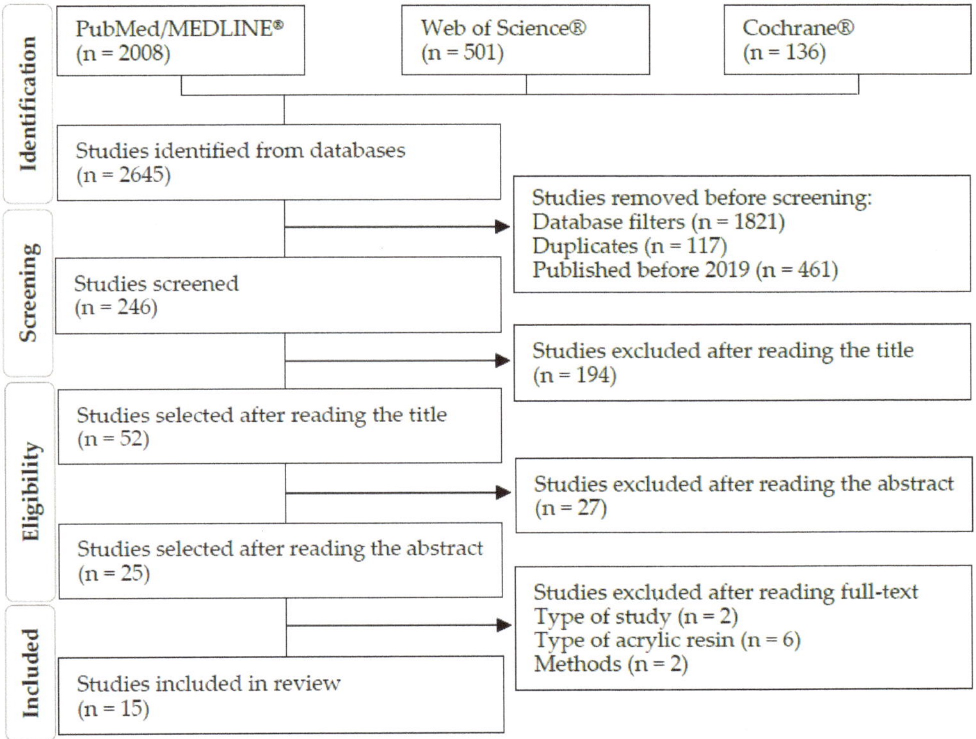

Figure 1. PRISMA flow diagram.

3.1. Study Characteristics

Fifteen studies, identified and detailed in Table 1, were selected for inclusion in the current review. Following quality evaluation, it was determined that fourteen of the fifteen included studies present a low risk of bias, while a moderate risk of bias was calculated for one study [14]. While most questions received affirmative responses, the question "Was follow-up complete and if not, were differences between groups in terms of their follow-up adequately described and analysed?" was deemed not applicable for all the included studies. On the other hand, the questions "Were there multiple measurements of the outcome both pre- and post-intervention/exposure?" and "Was appropriate statistical analysis used?" were marked as unclear in three and one article, respectively.

Most publications were published in 2021, followed by 2023. Considering the regional distribution of the studies, there were different prevalence rates of each continent, with a higher frequency of publications from Asia, followed by South America and Europe.

Table 2 depicts information about the acrylic resins tested and the sample preparation employed in each study, including the different sizes, particle concentrations, and microorganisms tested. The methodology employed while preparing the samples varied depending on the microbial strain the authors aimed to neutralise and the antimicrobial particle tested. Therefore, the samples prepared in each study presented different sizes and shapes, being moulded as disks, the most frequent, or coupons. While there was a high diversity in the inorganic particles incorporated, with sizes ranging from 5 nm to 150 nm, silver was the most common antimicrobial tested. By using pure PMMA as a control group, different concentrations were analysed, spanning from 0.015% to 20%, with a higher frequency of samples modified with a particle concentration of 5%. The antimicrobial tests were performed against various microbial strains, fungi, and bacteria, with higher incidence of *Candida albicans*.

Table 1. Included studies.

No.	Country	1st Author, Year	Title	Journal
1 [15]	Italy	De Matteis V, 2019	Silver Nanoparticles Addition in Poly(Methyl Methacrylate) Dental Matrix: Topographic and Antimycotic Studies	*Int J Mol Sci*
2 [16]	Brazil	Souza Neto FN, 2019	Effect of synthetic colloidal nanoparticles in acrylic resin of dental use	*Eur Polym J*
3 [17]	India	Gopalakrishnan S, 2020	Development of biocompatible and biofilm-resistant silver-poly(methylmethacrylate) nanocomposites for stomatognathic rehabilitation	*Int J Polym Mater Polym Biomater*
4 [18]	Iran	Giti R, 2021	Antimicrobial Activity of Thermocycled Polymethyl Methacrylate Resin Reinforced with Titanium Dioxide and Copper Oxide Nanoparticles	*Int J Dent*
5 [19]	Brazil	Pinheiro MCR, 2021	Thermopolymerized Acrylic Resin Immersed or Incorporated with Silver Nanoparticle: Microbiological, Cytotoxic and Mechanical Effect	*Mat Res*
6 [20]	Saudi Arabia	Alzayyat ST, 2021	Antifungal Efficacy and Physical Properties of Poly(methylmethacrylate) Denture Base Material Reinforced with SiO(2)Nanoparticles	*J Prosthodont*
7 [21]	Brazil	Takamiya AS, 2021	Biocompatible silver nanoparticles incorporated in acrylic resin for dental application inhibit *Candida albicans* biofilm	*Mater Sci Eng C Mater Biol Appl*
8 [22]	Saudi Arabia	Fouda SM, 2021	Effect of Low Nanodiamond Concentrations and Polymerization Techniques on Physical Properties and Antifungal Activities of Denture Base Resin	*Polymers (Basel)*
9 [23]	Iraq	Hazim RH, 2021	The Effect of Tellurium Oxide Micro Particles Incorporation into PMMA on *Candida albicans* Adherence	*J Res Med Dent Sci*
10 [24]	Serbia	Ivanovic V, 2022	Unraveling the Antibiofilm Activity of a New Nanogold Resin for Dentures and Epithesis	*Pharmaceutics*
11 [25]	Serbia	Gligorijevic N, 2022	Antimicrobial Properties of Silver-Modified Denture Base Resins	*Nanomaterials (Basel)*
12 [26]	Saudi Arabia	Ismaeil MA, 2023	Antifungal Effect of Acrylic Resin Denture Base Containing Different Types of Nanomaterials: A Comparative Study	*J Int Oral Health*
13 [27]	Brazil	Teixeira ABV, 2023	Adhesion of biofilm, surface characteristics, and mechanical properties of antimicrobial denture base resin	*J Adv Prosthodont*
14 [14]	Slovenia	Marić I, 2023	Antifungal Effect of Polymethyl Methacrylate Resin Base with Embedded Au Nanoparticles	*Nanomaterials (Basel)*
15 [28]	Chile	Correa S, 2024	Development of novel antimicrobial acrylic denture modified with copper nanoparticles	*J Prosthodont Res*

Figure 2 illustrates the prevalence and size (in nanometres) of nanoparticles incorporated into PMMA for each study. As observed in Figure 2a, while a great diversity in the antimicrobials tested in each study is evident, some particles were more prevalent than others. Over half of the studies [15–17,19,21,25–27] tested the incorporation into heat-cured PMMA of a compound containing silver, such as silver vanadate [27] or silver chloride [25]. Following silver, gold [14,24], titanium [18,26], and copper [18,28] were the most common inorganic antimicrobials, being tested in two articles each. Figure 2b illustrates the nanoparticle sizes incorporated into PMMA for each study, with bars indicating the ranges and lines representing the mean and exact values specified in the articles. To improve clarity and readability, the dimensions of silver chloride in the study by Gligorijevic et al. [25] were not represented in the chart, due to a significative deviation from the average range, while the two studies that did not specify the sizes [22,23] were also excluded from the graph. Therefore, a total of thirteen studies are illustrated in the chart, representing a total of fifteen inorganic compounds.

Table 2. Heat-cured PMMA samples used in antimicrobial tests.

No.	PMMA	Samples (mm)	Particle Size	Concentration (%)	Microorganism
1	Paladon® 65 (Kulzer) Gemany	Disk Ø ≈ 20	Ag 20 nm ± 3	3; 3.5	*C. albicans*
2	Lucitone® 550 (Dentsply® Ind. e Com. Ltd.a.) Brazil	Coupon 60 × 10 × 3	Ag 7.6 nm ± 2.3	0.05; 0.5; 5	*C. glabrata*
3	Alfa Aesar. USA	Not specified	Ag <100 nm	1; 2; 5;10	*S. mutans; C. albicans*
4	SR Triplex Hot (Ivoclar Vivadent®) Liechtenstein	150 (30 per group) disk 10 × 2	CuO 40 nm TiO$_2$ 17 nm	2.5; 7.5	*C. albicans; C. dubliniensis; S. mutans; S. sobrinus; S. salivarius; S. sanguis*
5	Vipicril (Vipi® Ind. e Com. Ltd.a.) Brazil	108 (27 per group) disk 15 × 2	Ag 50 nm	1; 2.5; 5	*C. albicans*
6	Major Base 20 (Major Prodotti Dentari SPA®) Italy	50 (10 per group) disk 15 × 2	SiO$_2$ 15 nm	0.05; 0.25; 0.5; 1	*C. albicans*
7	Lucitone® 550 (Dentsply® Ind. e Com. Ltd.a.) Brazil	63 (9 per group) disk 10 × 3	Ag 5/10 nm	0.05; 0.5; 5	*C. albicans*
8	Major base 20 (Major Prodotti Dentari SPA®) Italy	80 (20 per group) disk 15 × 2	ND	0.1; 0.25; 0.5	*C. albicans*
9	Not identified	25 (5 per group) disk 10 × 2	TeO	1; 3; 5; 7	*C. albicans*
10	PMMA Biogal® (Galenika) Serbia	48- 24 (6 per species) 24 (control) disk 5 × 2	Au 69.4 nm ± 12.42	2	*S. aureus; E. coli; C. albicans; S. mitis*
11	SR Triplex Hot (Ivoclar Vivadent®) Liechtenstein	375 (75 per group) disk 10 × 2	Ag <100 nm AgCl 1 µm	2; 5; 10 10	*S. aureus; C. albicans*
12	Major base, Trevalon/Universal Clear (Dentsply® Ind. e Com. Ltd.a.) Germany	100 (20 per group) disk 10 × 2	Ag 40 nm TiO$_2$ 50 nm	0.5; 1	*C. albicans*
13	Classic Dental Articles Ltd.a. Brazil	9 Disk 9 × 1	AgVO$_3$ Wires: Ø = 150 nm Particles: 25 nm	2.5; 5; 10	*C. albicans; C. glabrata; S. mutans*
14	Ivoclar Vivadent® Liechtenstein	Coupon 10 × 10 × 3	Au 11	20	*C. albicans*
15	Acryl BH (GDF) Germany	Disk 10 × 4	Cu 30 to 150	0.015; 0.045; 0.055; 0.06; 0.068	*C. albicans; S. mutans; A. actinomycetemcomitans; S. aureus*

Ag—silver; AgCl—silver chloride; AgVO$_3$—silver vanadate; Au—gold; Cu—copper; CuO—copper oxide; ND—nanodiamond; SiO$_2$—silicon dioxide; TeO—tellurium oxide; TiO$_2$—titanium dioxide.

The selected studies aimed to test the antimicrobial properties of incorporating different inorganic particles in PMMA. Therefore, each study employed various tests, attempting to achieve the most credible results, with the colony-forming unit (CFU) assay being the most frequent. Table 3 displays the assays and time intervals used to attest the effectiveness of the different concentrations tested and the results obtained.

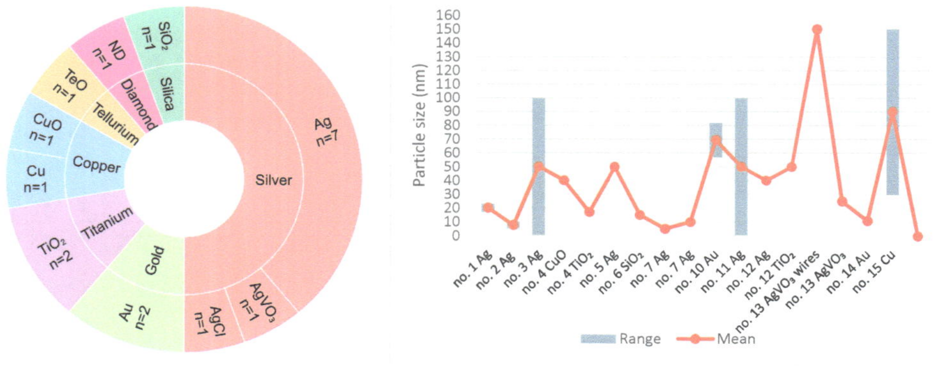

(a) (b)

Figure 2. Antimicrobial particles: (**a**) inorganic antimicrobials tested and respective incidence; (**b**) range of tested particle sizes per study and antimicrobial (nm).

Table 3. Effectiveness of different PMMA modifications.

No.	Particle	Tests	Microorganism	Best Effect	Worst Effect
1	Silver (Ag)	Viability CFU assay	*C. albicans*	3.5% Ag	3% Ag
		Circularity SEM assay		3.5% Ag	- Control
		Area covered Colonization assay		3.5% Ag	- Control
2	Silver (Ag)	CFU assay	*C. glabrata*	No statistically relevant difference	
		Biomass reduction CV assay		0.05% Ag	0.5% Ag
		Metabolic activity reduction XTT assay		0.05% Ag	5% Ag
		Micrographs of biofilms		0.05% Ag 0.5% Ag	5% Ag
3	Silver (Ag)	Cell count	*S. mutans*	10% Ag	1% Ag
		CFU assay	*C. albicans*	10% Ag	1% Ag
		Fluorescent microscopy	*S. mutans*	PMMA/Ag	- Control
4	Copper oxide (CuO) Titanium dioxide (TiO_2)	Optical density	*C. albicans*	7.5% TiO_2	2.5% TiO_2
			C. dubliniensis	7.5% CuO	2.5% TiO_2
			S. mutans		
			S. sobrinus	7.5% CuO	2.5% CuO
			S. salivarius	7.5% TiO_2	2.5% TiO_2
			S. sanguis	7.5% CuO	2.5% TiO_2 2.5% CuO
		Biofilm inhibition	*C. albicans*	7.5% TiO_2	2.5% TiO_2
			C. dubliniensis	7.5% CuO	2.5% TiO_2
			S. mutans	7.5% CuO	7.5% TiO_2
			S. sobrinus	7.5% CuO	2.5% CuO
			S. salivarius	7.5% TiO_2	2.5% TiO_2
			S. sanguis	7.5% CuO	2.5% CuO
5	Silver (Ag)	Viability assay Absorbance	*C. albicans*	1% Ag	2.5% Ag 5% Ag
6	Silicon dioxide (SiO_2)	Direct culture	*C. albicans*	1% SiO_2	0.05% SiO_2 0.25% SiO_2
		Slide count (CFU/mL)		1% SiO_2	0.05% SiO_2

Table 3. *Cont.*

No.	Particle	Tests	Microorganism	Best Effect	Worst Effect
7	Silver (Ag)	CFU assay	*C. albicans*	0.5% Ag 0.05% Ag	5% Ag
8	Diamond (ND)	CFU assay	*C. albicans*	0.5% ND	0.1% ND
9	Tellurium oxide (TeO)	Adherence test	*C. albicans*	5% TeO 7% TeO	1% TeO
10	Gold (Au)	CFU assay on discs	*S. aureus*	2% Au	- Control
			E. coli		
			C. albicans		
			S. mitis		
		MTT assay	*S. aureus*		
			E. coli		
			C. albicans		
			S. mitis		
		SEM assay	*S. aureus*		
			E. col		
			C. albicans		
			S. mitis		
		CFU assay in the surrounding medium	*S. aureus*	No statistically relevant difference	
			E. coli		
			C. albicans		
			S. mitis		
11	Silver (Ag)	Inhibition zone	*S. aureus*	10% Ag 10% AgCl	2% Ag
			C. albicans	10% AgCl 10% Ag	2% Ag 5% Ag
		CFU assay	*S. aureus*	10% Ag 10% AgCl 5% Ag	2% Ag
			C. albicans		
		Microdilution method Minimum inhibitory concentrations	*S. aureus*	10% AgCl	10% Ag
			C. albicans		
		Microdilution method Minimum microbicidal concentrations	*S. aureus*		
			C. albicans		
12	Silver (Ag) Titanium dioxide (TiO_2)	Disc diffusion Antifungal activity	*C. albicans*	1% Ag 1% TiO_2 0.5% Ag	0.5% TiO_2
		Elution test Colony counts		1% Ag 0.5% Ag 1% TiO_2	0.5% TiO_2
13	Silver vanadate ($AgVO_3$)	CFU assay	*C. albicans* In multispecies biofilm	- Control	10% $AgVO_3$ 5% $AgVO_3$ 2.5% $AgVO_3$
			C. glabrata In multispecies biofilm	10% $AgVO_3$	2.5% $AgVO_3$
			S. mutans In multispecies biofilm		
		Metabolic activity	Multispecies biofilm: *C. albicans* *C. glabrata* *S. mutans*	- Control	10% $AgVO_3$
14	Gold (Au)	Yeast adhesion	*C. albicans*	20% Au	- Control
15	Copper (Cu)	CFU assay	*C. albicans*	0.045% Cu	0.068% Cu
		SEM assay		0.045% Cu	- Control
		Surface inhibitory capacity	*A. actinomycetemcomitans*	0.045% Cu	-
			S. aureus		
			C. albicans		
			S. mutans		

3.1.1. Fungi

Candida albicans was cultured in fourteen out of the fifteen included studies, excluding the study by Souza Neto et al. [16]. Silver vanadate, proposed by Teixeira et al. [27], was discovered to favour *C. albicans* growth. Ivanovic et al. [24] and Marić et al. [14] concluded that gold provided favourable results against yeast, while titanium dioxide was compared with silver and copper in the studies by Ismaeil and Ebrahim [26] and Giti et al. [18]. The incorporation of nanodiamond (ND) (proposed by Fouda et al. [22]), silicon dioxide (SiO_2) (by Alzayyat et al. [20]), and tellurium oxide (TeO) (by Hazim and Fatalla [23]) led to improved antimicrobial properties at higher concentrations of inorganic antimicrobial. Overall, four studies [15,17,25,26] reported higher antimicrobial activity of Ag nanoparticles against *C. albicans* when incorporated at higher concentrations, while two studies [19,21] reported the inverse, with lower concentrations presenting the best antifungal effects.

For *Candida glabrata*, Souza Neto et al. [16] found that although there were no statistically relevant differences in the CFU results across the three concentrations, PMMA modified with lower concentrations of silver, specifically 0.5% in the micrographs of biofilm and 0.05% in the three assays, presented better antimicrobial effects. In contrast, using higher concentrations of silver, particularly 0.5% in the CV assay and 5% in the other tests, to modify resin proved to be less effective in preventing cell growth and biofilm formation. Teixeira et al. [27] reported an increase in the effectivity of silver vanadate at the highest concentration. For *Candida dubliniensis*, however, copper was proven to be more effective than TiO_2, with concentration-dependent efficacy, according to Giti et al. [18].

3.1.2. Bacteria

Four studies cultured *Streptococcus mutans*, obtaining an increase in antibacterial activity when higher concentrations of silver [17] or silver vanadate [27] were incorporated, as opposed to lower concentrations, which, in the case of $AgVO_3$, tended to favour growth. While the incorporation of gold resulted in the observation of small bacterial conglomerates [24], Correa et al. [28] concluded that copper presented favourable antimicrobial properties against *S. mutans* and *Aggregatibacter actinomycetemcomitans* [28].

Modifying PMMA with 2% Au reduced *Staphylococcus aureus* colonies to small conglomerates [24], with promising results in surface inhibitory capacity [28]. A higher antibacterial effect was detected at higher concentrations of Ag, as observed in the study by Gligorijevic et al. [25]. Higher antimicrobial efficacy of AgCl over Ag particles against both *C. albicans* and *S. aureus* was observed in the microdilution tests [25]. The incorporation of gold resulted in dispersed cells and chains of *Streptococcus mitis* and *Escherichia coli* in the modified PMMA [24]. Giti et al. [18] reported that 7.5% CuO was the most effective modification against both *Streptococcus sobrinus* and *Streptococcus sanguis*, while 7.5% TiO_2 provided the best antibacterial properties against *Streptococcus salivarius*.

4. Discussion

Several authors have studied the effect of incorporating chemicals and particles in PMMA and the resulting antimicrobial properties of the modified resin. Therefore, this systematic review aimed to determine which particle provided the best results in preventing the adherence of microorganisms to the denture and subsequent pathologies.

Six of the fifteen studies included in this review were published in 2021, while a single article was published in 2020 and one in 2024. The studies were published in Asia, South America, and Europe, with a noticeable absence of studies published in North America, Africa, and Oceania, potentially resulting from lack of research within these regions or concerns regarding cytotoxicity. There was a great diversity in sample sizes and shapes, with disk-shaped samples being the most frequent, most of which presented a thickness of 2 mm and varying diameters.

Antimicrobial particles varied in size from 5 nm to 150 nm, a broader range than the 10 nm to 100 nm observed in the study by Garcia et al. [29]. Silver was the most prevalent inorganic particle in antimicrobial tests, a trend also noted by Garcia et al. 2021 [29] and

An et al. 2023 [30]. The concentrations of nanoparticles tested ranged from 0% to 30% in the study by Garcia et al. 2021 [29]. Similarly, the studies in this review used control groups with unmodified PMMA, while modified samples had concentrations from 0.015% to 20%. Methods of incorporating particles into the denture resin varied, with some authors mixing particles with the monomer and others with the polymer, underscoring the need for a cautious comparison of the results.

Similar to what was observed by Garcia et al. [29], the CFU assay was the antimicrobial test performed at the highest frequency. However, CFU assays do not detect dead, culturable, or inactive cells, only measuring culturable live cells [31]. *C. albicans* was the strain tested in most of the studies, a tendency also observed by An et al. [30], which can be explained by this species being the most prevalent fungi in the oral cavity, presenting higher incidence in the palatal mucosa of denture wearers [32], often resulting in the development of Candida-Associated Denture Stomatitis due to the constant friction against the denture base [1,3]. Bacterial species such as *Streptococcus mutans* and *Lactobacillus acidophilus* are associated with tooth decay, due to bacteria-induced enamel demineralization [5].

The effect of pure silver particles against *C. albicans* was tested in four studies [15,17,19,21], resulting in different outcomes. De Matteis et al. [15] and Gopalakrishnan et al. [17] concluded that the antimicrobial properties were enhanced with higher concentrations of Ag, a tendency also observed by Adam and Khan [33], who found a correlation between higher concentrations of Ag nanoparticles and the lower values obtained in the CFU assays performed. On the other hand, Pinheiro et al. [19] and Takamiya et al. [21] reached an opposing conclusion, obtaining better antimicrobial results at lower silver concentrations, while the highest concentrations proved ineffective in some tests [21]. Most of the studies that prepared silver particles measured between 20 nm and 100 observed a concentration-dependent increase in effectiveness. The exception was the study by Pinheiro et al. [19], where the results did not present a statistically significant difference, which the author justified as a possible result from the bigger size of the particles and the incorporation into PMMA preventing the release of silver into the environment.

As evidenced in the studies by Fouda et al. [22] and Hazim and Fatalla [23], nanodiamond and tellurium provided favourable results against yeast. Similarly, Alzayyat et al. [20] observed that silicon dioxide provided concentration-dependant effectiveness against *C. albicans*. Ivanovic et al. [24] and Marić et al. [14] concluded that gold presented promising antimicrobial properties, although Ivanovic et al. [24] did not find a statistically significant difference in the cell count of the surrounding medium, indicating a possible lack of antimicrobial release. Giti et al. [18] demonstrated that copper oxide provided better properties against *Candida dubliniensis* when compared with titanium dioxide.

Two authors, Souza Neto et al. [16] and Teixeira et al. [27], tested the effectiveness of silver particles and silver vanadate against *Candida glabrata* strains. Souza Neto et al. [16] concluded that higher concentrations of silver resulted in worse antimicrobial properties against the strain, while the CFU assay resulted in no statistically significant difference for any of the concentrations. The author justified these findings by explaining that instead of forming a homogeneous dispersion, higher concentrations of silver formed agglomerates when mixed with PMMA, a tendency observed by An et al. [30] and Yudaev et al. [34]. Gopalakrishnan et al. [17] concluded that PMMA modified with silver nanoparticles provided higher antibacterial activity, while the incorporation of copper particles, as evident in the studies by Giti et al. [18] and Correa et al. [28], led to favourable antimicrobial properties.

Ivanovic et al. [24] obtained good antimicrobial properties against both *Streptococcus mitis* and *Staphylococcus aureus* when incorporating gold into PMMA, although no relevant effect was observed in the medium surrounding the modified samples. Copper proved effective against *S. aureus* in the study by Correa et al. [28], while silver and silver chloride presented similarly favourable antibacterial properties in the CFU and inhibition zone assays performed by Gligorijevic et al. [25]

Similarly to what was observed in the other bacterial and fungal strains, the study by Ivanovic et al. [24] obtained different CFU results on the discs modified with gold, which proved to be effective in reducing *Escherichia coli* cell count, and on the surrounding medium, where the difference was not statistically relevant. Correa et al. [28] examined the difference in surface inhibitory capacity between PMMA modified with copper and pure PMMA against a strain of *Aggregatibacter actinomycetemcomitans*, obtaining a favourable result in the inhibition tests with the modified resin.

Despite the filters and criteria applied, there was a great difference in the methodology and the morphology of the samples tested in the various studies, which was also noted in the studies by Garcia et al. [29] and Adam and Khan [33]. This lack of homogeneity complicates the organization of the findings and the elaboration of an effective comparison, as the differences observed may be a result of the different methodologies applied during sample preparation instead of the concentrations tested.

While performing the initial research, a notably low number of clinical studies were observed. The great variation between oral and in vitro conditions further increases the need for in vivo and clinical tests, as the laboratorial environment is incapable of perfectly replicating the oral conditions and possible interactions with oral microbiota. Additionally, as observed in the article by De Matteis et al. [15], colour changes may occur. Therefore, it is necessary to evaluate the effect of the incorporation of particles such as silver on the physical properties and aesthetics.

Even though this systematic review focused on the antimicrobial properties, other properties are also essential to providing the best functionality and biocompatibility, as stated by Garcia et al. [29] and Bangera et al. [35]. Adam and Khan [33] also found a low number of clinical trials, stating that clinical studies are required to assess the effect of modified PMMA on Candida-Associated Denture Stomatitis. Due to concerns regarding the toxicity of the particles, namely, silver, as acknowledged by Garcia et al. [29], biocompatibility tests must be performed to guarantee the safety of incorporating these particles.

5. Conclusions

Almost all the inorganic antimicrobials presented promising properties against the tested strains. Therefore, to answer the main question that prompted this review, modified PMMA did exhibit better antimicrobial properties than pure PMMA. However, it is not possible to declare a particle as the most efficient, due to the high heterogeneity in the samples and antimicrobial tests performed in the studies. Additionally, few authors compare different particles in the same test, which would have facilitated a comparison of the effects. More tests are necessary, especially in vivo studies and clinical trials, as the oral environment and microbiota are vastly different from in vitro strains. Likewise, biocompatibility tests are needed to evaluate the possibility of contact allergies and toxicity resulting from the incorporation of metallic particles into denture bases. It is important to analyse the effect of PMMA modifications on multispecies biofilms, to assess the interactions between different pathogenic strains in the presence of inorganic antimicrobials and their effect on the effectiveness of the modified resin.

Author Contributions: Conceptualization, H.S. and P.F.; methodology, H.S. and P.F.; validation, M.L. and P.F.; formal analysis, P.F. and A.C.; investigation, M.L. and P.F.; resources, A.C.; data curation, M.L.; writing—original draft preparation, M.L.; writing—review and editing, P.F. and A.C.; supervision, H.S. and P.F.; project administration, H.S. and P.F.; funding acquisition, A.C. All authors have read and agreed to the published version of the manuscript.

Funding: This work was financially supported by National Funds through FCT—Fundação para a Ciência e a Tecnologia—I.P., under project UIDP/04279/2020.

Institutional Review Board Statement: Not applicable.

Data Availability Statement: Not applicable.

Conflicts of Interest: The authors declare no conflicts of interest.

References

1. Vila, T.; Sultan, A.S.; Montelongo-Jauregui, D.; Jabra-Rizk, M.A. Oral Candidiasis: A Disease of Opportunity. *J. Fungi* **2020**, *6*, 15. [CrossRef] [PubMed]
2. Manikandan, S.; Vinesh, E.; Selvi, D.T.; Kannan, R.K.; Jayakumar, A.; Dinakaran, J. Prevalence of Candida among Denture Wearers and Nondenture Wearers. *J. Pharm. Bioallied Sci.* **2022**, *14* (Suppl. S1), S702–S705. [CrossRef] [PubMed]
3. Farid, D.A.M.; Zahari, N.A.H.; Said, Z.; Ghazali, M.I.M.; Hao-Ern, L.; Zol, S.M.; Aldhuwayhi, S.; Alauddin, M.S. Modification of Polymer Based Dentures on Biological Properties: Current Update, Status, and Findings. *Int. J. Mol. Sci.* **2022**, *23*, 10426. [CrossRef]
4. Walczak, K.; Schierz, G.; Basche, S.; Petto, C.; Boening, K.; Wieckiewicz, M. Antifungal and Surface Properties of Chitosan-Salts Modified PMMA Denture Base Material. *Molecules* **2020**, *25*, 5899. [CrossRef]
5. Mathur, V.P.; Dhillon, J.K. Dental Caries: A Disease Which Needs Attention. *Indian J. Pediatr.* **2018**, *85*, 202–206. [CrossRef] [PubMed]
6. Taebunpakul, P.; Jirawechwongsakul, P. Palatal Inflammation and the Presence of Candida in Denture-Wearing Patients. *J. Int. Soc. Prev. Community Dent.* **2021**, *11*, 272–280. [CrossRef]
7. Zafar, M.S. Prosthodontic Applications of Polymethyl Methacrylate (PMMA): An Update. *Polymers* **2020**, *12*, 2299. [CrossRef]
8. Saidin, S.; Jumat, M.A.; Mohd Amin, N.A.A.; Saleh Al-Hammadi, A.S. Organic and inorganic antibacterial approaches in combating bacterial infection for biomedical application. *Mater. Sci. Eng. C Mater. Biol. Appl.* **2021**, *118*, 111382. [CrossRef]
9. Spirescu, V.A.; Chircov, C.; Grumezescu, A.M.; Vasile, B.Ș.; Andronescu, E. Inorganic Nanoparticles and Composite Films for Antimicrobial Therapies. *Int. J. Mol. Sci.* **2021**, *22*, 4595. [CrossRef]
10. Moher, D.; Liberati, A.; Tetzlaff, J.; Altman, D.G.; PRISMA Group. Preferred reporting items for systematic reviews and meta-analyses: The PRISMA statement. *PLoS Med.* **2009**, *6*, e1000097. [CrossRef]
11. Page, M.J.; McKenzie, J.E.; Bossuyt, P.M.; Boutron, I.; Hoffmann, T.C.; Mulrow, C.D.; Shamseer, L.; Tetzlaff, J.M.; Akl, E.A.; Brennan, S.E.; et al. The PRISMA 2020 statement: An updated guideline for reporting systematic reviews. *BMJ* **2021**, *372*, n71. [CrossRef] [PubMed]
12. Schardt, C.; Adams, M.B.; Owens, T.; Keitz, S.; Fontelo, P. Utilization of the PICO framework to improve searching PubMed for clinical questions. *BMC Med. Inform. Decis. Mak.* **2007**, *7*, 16. [CrossRef]
13. Paes, S.M.; Pupo, Y.M.; Cavenago, B.C.; Fonseca-Silva, T.; Santos, C.C.O. Cryopreservation of mesenchymal stem cells derived from dental pulp: A systematic review. *Restor. Dent. Endod.* **2021**, *46*, e26. [CrossRef] [PubMed]
14. De Matteis, V.; Cascione, M.; Toma, C.C.; Albanese, G.; De Giorgi, M.L.; Corsalini, M.; Rinaldi, R. Silver Nanoparticles Addition in Poly(Methyl Methacrylate) Dental Matrix: Topographic and Antimycotic Studies. *Int. J. Mol. Sci.* **2019**, *20*, 4691. [CrossRef] [PubMed]
15. Neto, F.N.d.S.; Sala, R.L.; Fernandes, R.A.; Xavier, T.P.O.; Cruz, S.A.; Paranhos, C.M.; Monteiro, D.R.; Barbosa, D.B.; Delbem, A.C.B.; de Camargo, E.R. Effect of synthetic colloidal nanoparticles in acrylic resin of dental use. *Eur. Polym. J.* **2019**, *112*, 531–538. [CrossRef]
16. Gopalakrishnan, S.; Mathew, T.A.; Mozetič, M.; Jose, J.; Thomas, S.; Kalarikkal, N. Development of biocompatible and biofilm resistant silver-poly(methylmethacrylate) nanocomposites for stomatognathic rehabilitation. *Int. J. Polym. Mater. Polym. Biomater.* **2019**, *69*, 186–199. [CrossRef]
17. Giti, R.; Zomorodian, K.; Firouzmandi, M.; Zareshahrabadi, Z.; Rahmannasab, S. Antimicrobial Activity of Thermocycled Polymethyl Methacrylate Resin Reinforced with Titanium Dioxide and Copper Oxide Nanoparticles. *Int. J. Dent.* **2021**, *2021*, 6690806. [CrossRef]
18. Pinheiro, M.C.R.; Carneiro, J.A.O.; Pithon, M.M.; Martinez, E.F. Thermopolymerized Acrylic Resin Immersed or Incorporated with Silver Nanoparticle: Microbiological, Cytotoxic and Mechanical Effect. *Mat. Res.* **2021**, *24*, e20200115. [CrossRef]
19. Alzayyat, S.T.; Almutiri, G.A.; Aljandan, J.K.; Algarzai, R.M.; Khan, S.Q.; Akhtar, S.; Matin, A.; Gad, M.M. Antifungal Efficacy and Physical Properties of Poly (methylmethacrylate) Denture Base Material Reinforced with SiO2 Nanoparticles. *J. Prosthodont.* **2021**, *30*, 500–508. [CrossRef]
20. Takamiya, A.S.; Monteiro, D.R.; Gorup, L.F.; Silva, E.A.; Camargo, E.R.; Gomes-Filho, J.E.; De Oliveira, S.H.P.; Barbosa, D.B. Biocompatible silver nanoparticles incorporated in acrylic resin for dental application inhibit *Candida albicans* biofilm. *Mater. Sci. Eng. C Mater. Biol. Appl.* **2021**, *118*, 111341. [CrossRef]
21. Fouda, S.M.; Gad, M.M.; Ellakany, P.; Al Ghamdi, M.A.; Khan, S.Q.; Akhtar, S.; Al Eraky, D.M.; Al-Harbi, F.A. Effect of Low Nanodiamond Concentrations and Polymerization Techniques on Physical Properties and Antifungal Activities of Denture Base Resin. *Polymers* **2021**, *13*, 4331. [CrossRef] [PubMed]
22. Hazim, R.H.; Fatalla, A.A. The Effect of Tellurium Oxide Micro Particles Incorporation into PMMA on *Candida albicans* Adherence. *J. Res. Med. Dent. Sci.* **2021**, *9*, 129–135.
23. Ivanovic, V.; Popovic, D.; Petrovic, S.; Rudolf, R.; Majerič, P.; Lazarevic, M.; Djordjevic, I.; Lazic, V.; Radunovic, M. Unraveling the Antibiofilm Activity of a New Nanogold Resin for Dentures and Epithesis. *Pharmaceutics* **2022**, *14*, 1513. [CrossRef] [PubMed]
24. Gligorijević, N.; Mihajlov-Krstev, T.; Kostić, M.; Nikolić, L.; Stanković, N.; Nikolić, V.; Dinić, A.; Igić, M.; Bernstein, N. Antimicrobial Properties of Silver-Modified Denture Base Resins. *Nanomaterials* **2022**, *12*, 2453. [CrossRef]
25. Ismaeil, M.A.; Ebrahim, M. Antifungal effect of acrylic resin denture base containing different types of nanomaterials: A comparative study. *J. Int. Oral. Health* **2023**, *15*, 78–83. [CrossRef]

26. Teixeira, A.B.V.; Valente, M.L.D.C.; Sessa, J.P.N.; Gubitoso, B.; Schiavon, M.A.; Dos Reis, A.C. Adhesion of biofilm, surface characteristics, and mechanical properties of antimicrobial denture base resin. *J. Adv. Prosthodont.* **2023**, *15*, 80–92. [CrossRef] [PubMed]
27. Marić, I.; Zore, A.; Rojko, F.; Škapin, A.S.; Štukelj, R.; Učakar, A.; Vidrih, R.; Veselinović, V.; Gotić, M.; Bohinc, K. Antifungal Effect of Polymethyl Methacrylate Resin Base with Embedded Au Nanoparticles. *Nanomaterials* **2023**, *13*, 2128. [CrossRef]
28. Correa, S.; Matamala, L.; González, J.P.; de la Fuente, M.; Miranda, H.; Olivares, B.; Maureira, M.; Agüero, A.; Gómez, L.; Lee, X.; et al. Development of novel antimicrobial acrylic denture modified with copper nanoparticles. *J. Prosthodont. Res.* **2024**, *68*, 156–165. [CrossRef]
29. Garcia, A.A.M.N.; Sugio, C.Y.C.; de Azevedo-Silva, L.J.; Gomes, A.C.G.; Batista, A.U.D.; Porto, V.C.; Soares, S.; Neppelenbroek, K.H. Nanoparticle-modified PMMA to prevent denture stomatitis: A systematic review. *Arch. Microbiol.* **2021**, *204*, 75. [CrossRef]
30. An, S.; Evans, J.L.; Hamlet, S.; Love, R.M. Overview of incorporation of inorganic antimicrobial materials in denture base resin: A scoping review. *J. Prosthet. Dent.* **2023**, *130*, 202–211. [CrossRef]
31. Pan, H.; Zhang, Y.; He, G.X.; Katagori, N.; Chen, H. A comparison of conventional methods for the quantification of bacterial cells after exposure to metal oxide nanoparticles. *BMC Microbiol.* **2014**, *14*, 222. [CrossRef] [PubMed]
32. Qiu, J.; Roza, M.P.; Colli, K.G.; Dalben, Y.R.; Maifrede, S.B.; Valiatti, T.B.; Novo, V.M.; Cayô, R.; Grão-Velloso, T.R.; Gonçalves, S.S. Candida-associated denture stomatitis: Clinical, epidemiological, and microbiological features. *Braz. J. Microbiol.* **2023**, *54*, 841–848. [CrossRef] [PubMed]
33. Adam, R.Z.; Khan, S.B. Antimicrobial Efficacy of Silver Nanoparticles against Candida Albicans. *Materials* **2022**, *15*, 5666. [CrossRef] [PubMed]
34. Yudaev, P.; Chuev, V.; Klyukin, B.; Kuskov, A.; Mezhuev, Y.; Chistyakov, E. Polymeric Dental Nanomaterials: Antimicrobial Action. *Polymers* **2022**, *14*, 864. [CrossRef]
35. Bangera, M.K.; Kotian, R.; Madhyastha, P. Effects of silver nanoparticle-based antimicrobial formulations on the properties of denture polymer: A systematic review and meta-analysis of in vitro studies. *J. Prosthet. Dent.* **2023**, *129*, 310–321. [CrossRef]

prosthesis

MDPI

Case Report

A Milled-Provisional Crown with Attachment: A Novel Prosthodontic Design to Facilitate Orthodontic Treatment

Abdulaziz Alamri [1] and Yousif A. Al-Dulaijan [2,*]

[1] Department of Preventive Dental Sciences, College of Dentistry, Imam Abdulrahman Bin Faisal University, P.O. Box 1982, Dammam 31441, Saudi Arabia; absalamri@iau.edu.sa
[2] Department of Substitutive Dental Sciences, College of Dentistry, Imam Abdulrahman Bin Faisal University, P.O. Box 1982, Dammam 31441, Saudi Arabia
* Correspondence: yaaldulaijan@iau.edu.sa

Abstract: Background: This case report addresses the challenges of integrating orthodontic and prosthodontic treatment, particularly in clear aligner cases. The study introduces a novel milled-provisional crown with attachment (M-PCA) technique designed to enhance treatment efficacy and reduce orthodontic attachment debonding, a common issue in clear aligner therapy. **Case Report**: This is a case report that presents a 49-year-old female patient seeking orthodontic treatment for Class III malocclusion along with periodontal and prosthodontic challenges. The treatment plan involved a multidisciplinary approach, including using M-PCA for temporization during clear aligner therapy. **Conclusions**: The M-PCA approach demonstrated promising results, with no reported complications such as orthodontic attachment debonding throughout the treatment period. This innovation offers a significant advantage in managing orthodontic cases requiring provisional crowns, ensuring retention, and facilitating orthodontic treatment.

Keywords: orthodontic tooth movement; provisional crown; clear aligners; attachments; multidisciplinary treatment

1. Introduction

Orthodontic treatment conservatively enhances functionality, occlusion, and appearance [1,2]. Clear aligners are a new generation of orthodontic treatment and were designed to treat mild to moderate crowding [3]. Orthodontists claim it has achieved successful outcomes with more complex malocclusions [4]. Nowadays, it is one of the most popular and frequently used treatment options among orthodontists [3]. Compared to conventional orthodontics, clear aligners have shown superior esthetics, lower demineralization, and minimal soft tissue irritation [5,6]. Another advantage of the clear aligner option is digital planning. The software can let patients view the simulated smile and final proposed treatment during treatment planning [7]. It has been shown that clear aligners are more commonly used in adult patients for their esthetic advantage over fixed orthodontic appliances [8]. It was also reported that it is a better option for the gingiva, as clear aligners have fewer periodontal indices measurements compared to other appliances [9]. With the increase in periodontal diseases reaching 50% in several countries [10], clear aligners can be a better choice in some cases.

On the other hand, clear aligners may initially cause speech articulation issues, discomfort, and increased salivation [11]. In some situations, clear aligners may fail to achieve buccal torque of the posterior teeth, significant rotation correction, or significant vertical movement [11]. In addition, the bodily movement of the teeth can be a significant limitation in clear aligner extraction cases [11].

Like conventional orthodontic appliances, clear aligners' success depends on several factors, such as the duration of treatment. On average, the treatment duration of clear aligners for mild crowding takes 13 months to resolve, moderate crowding takes 15 months, and severe crowding takes around 17 months [12]. Clear aligners incorporate composite attachments during that time to increase their efficacy and predictability [13]. Unfortunately, there are reports of increased loss of these attachments at regular intervals during orthodontic treatment, reaching a prevalence of more than 60% [14]. Orthodontic attachment debonding can be related to several contributing factors, such as the frequency of aligner removal, occlusal forces introduced by mastication, and aligner wearing time [14]. In addition to their efficacy, some of these attachments are commonly used to deliver force to a specific part of the tooth to cause the orthodontic tooth movement necessary and to provide a better-controlled tooth movement depending on the software algorithm [15]. Other attachments are designed to retain the aligners on the teeth [15].

Provisional crowns are encountered in orthodontic cases that involve multidisciplinary approaches [16]. Fixed Partial Dentures (FPDs) and crowns are commonly found in adult patients who seek orthodontic treatment [17]. In some clinical situations, the existing fixed prostheses must be replaced due to biological, mechanical, or esthetic complications. The replacement of the defective prosthesis usually occurs during the first phase of treatment, in which the abutment tooth is temporized by provisional restoration. With clear aligner appliances, bonding attachments to provisional crowns come with challenges [18]. Therefore, this novel approach was developed to combine the attachment and provisional crown as one milled unit to overcome these challenges. This case report aims to present the workflow for the use of a Milled-Provisional Crown with Attachment (M-PCA) and its associated clinical outcomes.

2. Case Report

2.1. Case Presentation

A 49-year-old Saudi female patient came to the orthodontic faculty clinics at Imam Abdulrahman bin Faisal University Dental Hospital with a complaint: "My teeth are crooked, and I would like to correct them". Her medical history showed no significant findings. Her dental history showed multiple restorations and missing teeth. Upon clinical examination, multiple caries lesions, generalized calculus deposition, generalized gingival recession, a missing tooth (#3), two 3-unit FPDs on teeth #2, X, 4, & 13, X, 15, crowns on teeth #12, 19, 20, and 30, Class I malocclusion, an overjet range from 0 to 1 mm, a shallow overbite of 0–1 mm, a shifted lower midline of 1 mm, 3 mm upper crowding, and 1 mm lower crowding were found. The lateral cephalometric radiograph suggested a normodivergent pattern, with a Class III skeletal relation complicated by the retrognathic maxilla and proclined upper and lower incisors (Figures 1A–E and 2A,B).

The orthodontic diagnosis was Class III malocclusion due to retrognathic maxilla complicated with mild lower crowding and reverse overjet.

In Phase I, a standard treatment planning protocol was used to fabricate M-PCA prostheses. This was followed by a free gingival graft on the maxillary anterior teeth and orthodontic treatment using a clear aligner (comprehensive Invisalign® treatment).

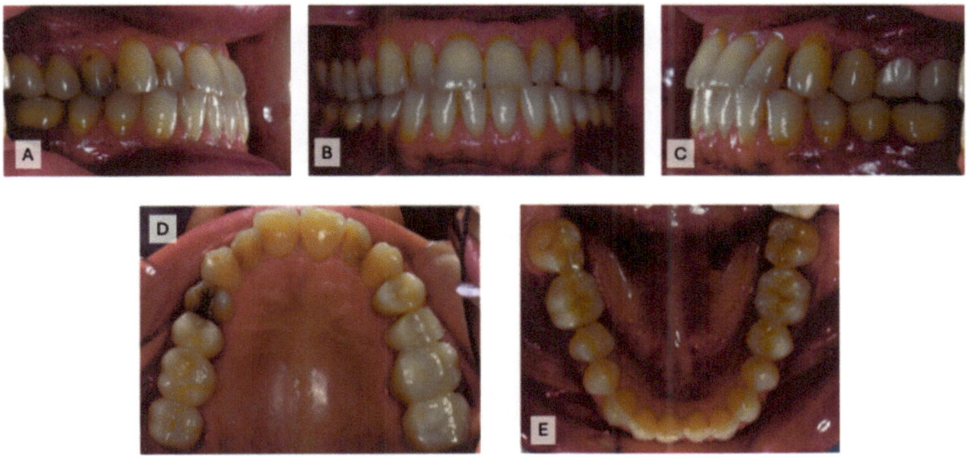

Figure 1. The initial intraoral photographs before restorability assessments: (**A**) right occlusion; (**B**) front occlusion; (**C**) left occlusion; (**D**) maxillary occlusion; (**E**) mandibular occlusion.

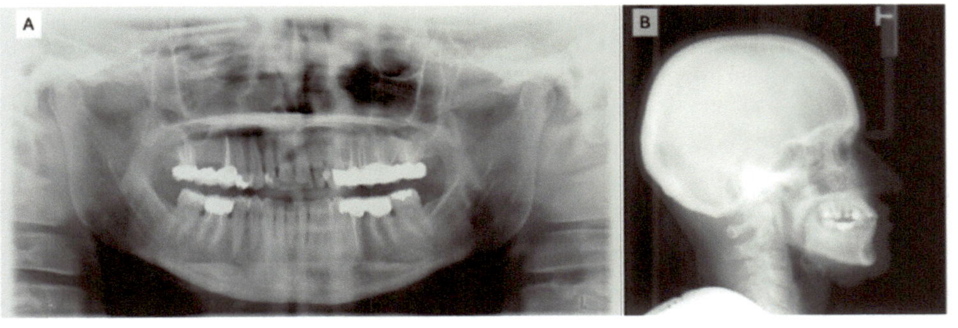

Figure 2. Radiographs taken before orthodontic treatment: (**A**) orthopantomography (OPG) taken during the initial visit and (**B**) a lateral cephalometric radiograph before the start of treatment.

2.2. M-PCA Design and Technique Description

The patient was referred to the prosthodontics faculty clinics to assess crowns, FPDs, and caries control. After the prosthodontic evaluation, tooth #4 was unrestorable, defective FPDs on #2, X, 4, and #13, X, 15, and several defective crowns needed to be removed. Abutment teeth #2, 13, 14, 15, 18, and 19 were selected for M-PCA units. The technique used for the fabrication of the M-PCA can be summarized as follows:

1. Obtain a digital impression of the abutment teeth using an intraoral scanner (Trios 3®, 3Shape, Copenhagen, Denmark) (Figure 3A,B), and then export the impression as standard tessellation language (STL) files.
2. Open a new job order in the CAD software program (Exocad DentalCAD Version 3.1, Exocad GmbH, Darmstadt, Germany), and then select the abutment teeth and their antagonist (Figure 3C).

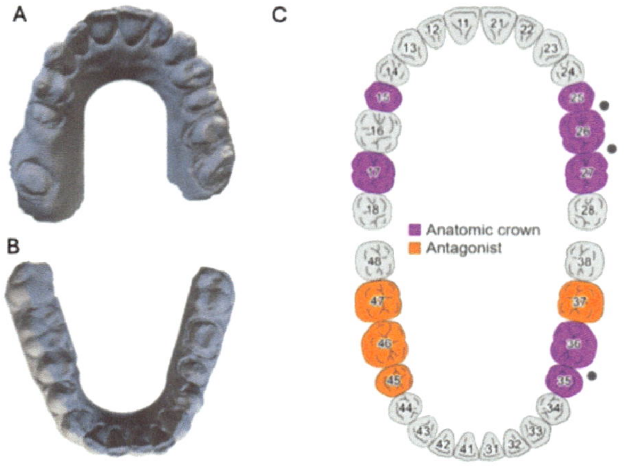

Figure 3. The process for fabricating M-PCA utilizing digital dentistry: (**A**) maxillary digital model; (**B**) mandibular digital model; (**C**) selecting the abutment teeth that will receive M-PCA and their antagonist.

3. After that, use the anatomical crown option under the crowns and copings for material configuration, select acrylic/PMMA for the material option, and select the other options and parameters, as seen in Figure 4.

Figure 4. A screenshot of the Exocad software shows the proper options for M-PCA fabrication.

4. Import the STL scan files of the maxillary and mandibular arches (Figure 5A), and then adjust the scan data orientation for them (Figure 5B).
5. Use the margin line detection feature to detect the margins of all abutment teeth (Figure 5C).

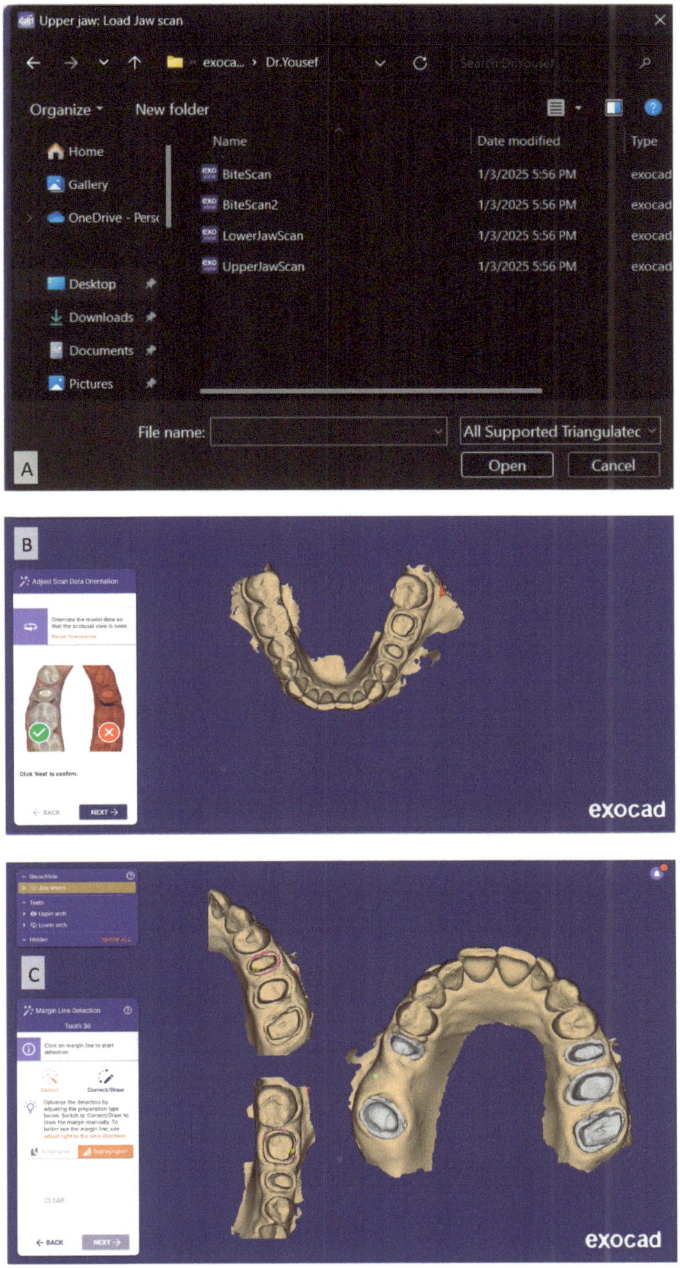

Figure 5. Screenshots of the Exocad software show (**A**) importing of digital impressions; (**B**) scan data orientation; and (**C**) marginal line detection.

6. Modify the cement gap from the crown bottoms menu and set it to "no cement gap" for the marginal area and "0.1 mm" for the remaining structure (Figure 6A).

7. From the tooth placement option, perform digital teeth wax-up using a digital library (Figure 6B), and then adjust the axial contour, proximal contact, and occlusion using the free-forming feature (Figure 6C).

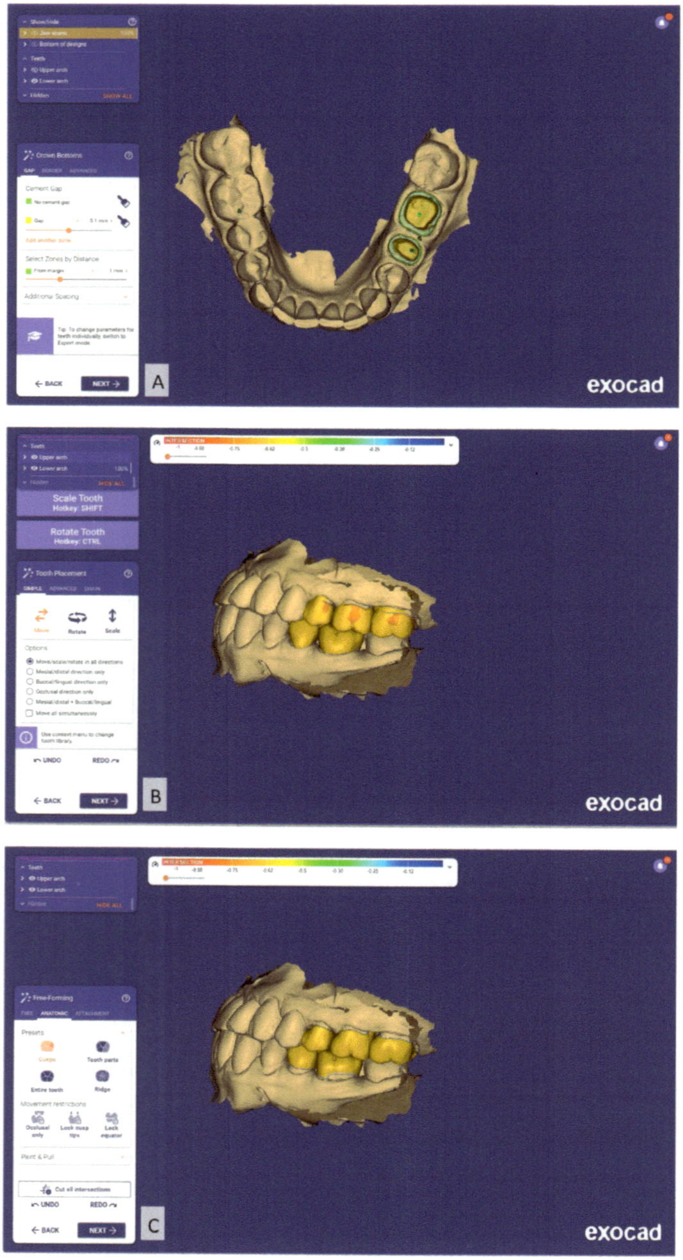

Figure 6. Screenshots of the Exocad software show: (**A**) cement gap setting; (**B**) digital teeth wax-up; (**C**) modification of the teeth wax-up.

8. From the drop menu, select add/remove mesh (Figure 7A), and then select the attachment tab (Figure 7B). Add the proper dimensions of the attachment (2 mm × 3 mm), as shown in Figure 7B.

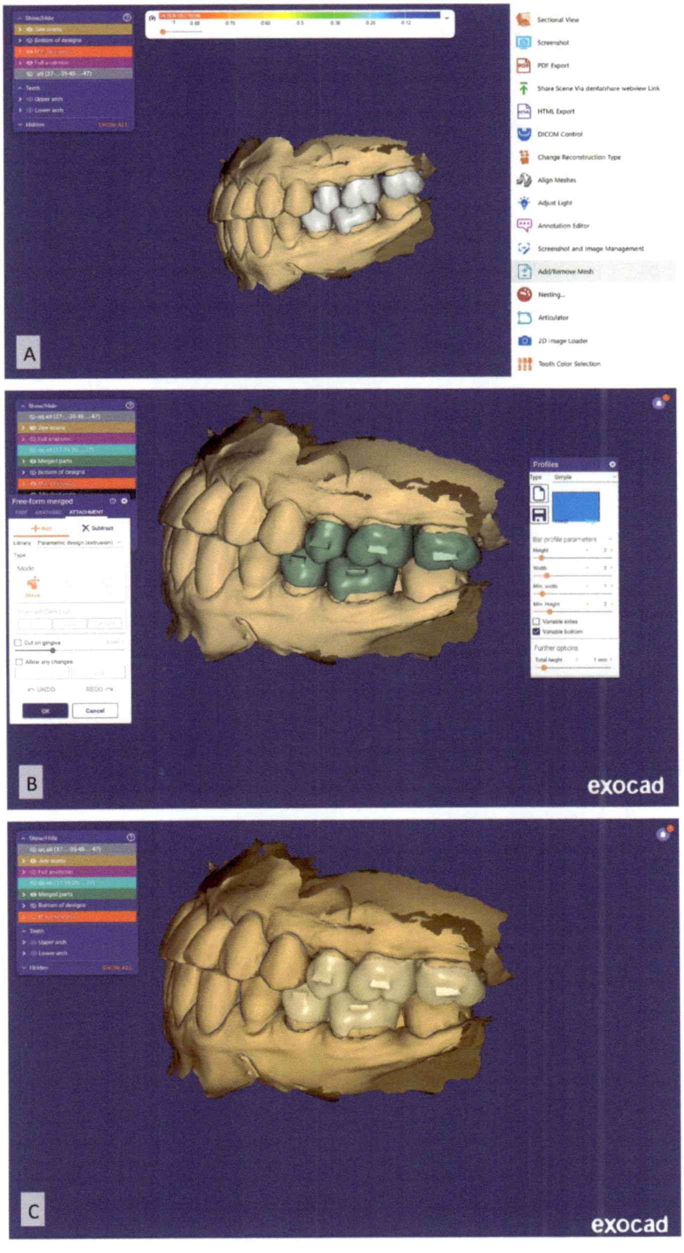

Figure 7. Screenshots of the Exocad software show: (**A**) option used to add the attachments for the teeth wax-up; (**B**) parameters and measurements of the attachment; (**C**) final M-PCA designs.

9. Rotate the attachment towards the gingival area to mimic the horizontal gingivally beveled attachment (Figure 7B).
10. At the end, check the final M-PCA design to ensure that the proper contour, proximal contact, occlusion, and shape and location of the attachment have been achieved (Figure 7C).
11. Transfer the virtual design to a milling machine (Zenotec T1, Wieland, Germany), and then use a monolayer Poly-(methyl methacrylate) (PMMA) disk to fabricate the M-PCA units.
12. After the clinical evaluation of the M-PCA units, perform an air abrasion treatment of the intaglio (inner) surface, and then cement the M-PCA units with an adhesive resin cement (RelyX Unicem, 3M ESPE, St. Paul, MN, USA).

2.3. Periodontic and Orthodontic Treatment Progress

After cementation of the M-PCA units, the patient was referred to the periodontics faculty clinics to evaluate the periodontium, extract tooth #4, and manage the recession with a free gingival graft (Figure 8A–E).

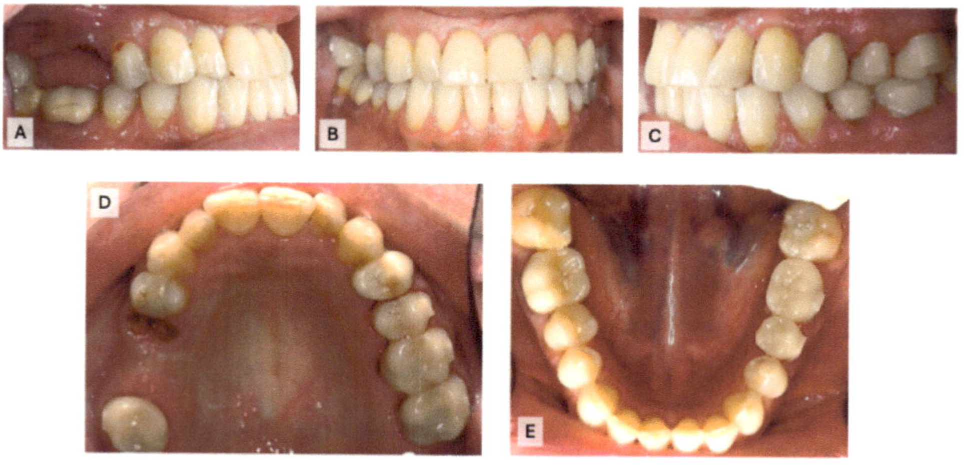

Figure 8. Intraoral photographs after caries and periodontal clearance with M-PCA cemented to abutment teeth: (**A**) right occlusion; (**B**) front occlusion; (**C**) left occlusion; (**D**) maxillary occlusion; (**E**) mandibular occlusion.

Three months later, the patient was scanned for Invisalign® treatment (Align Technology, Santa Clara, CA, USA) (Figure 9A–E).

The patient was seen for visit 1 for transitional aligners 1–3 (3 weeks). At visit 2, an IPR of 0.3 for mandibular anterior teeth was performed, and aligners 4–15 (12 weeks) were provided. In visit 3, aligners 16–27 (12 weeks) were given, followed by 28–39 (12 weeks) at visit 4. The patient was instructed to change aligners weekly. Teeth tracking was satisfactory at each visit, with no emergency incidents or attachment breakage, and aligners adapted well to the M-PCA (Figure 10A–E).

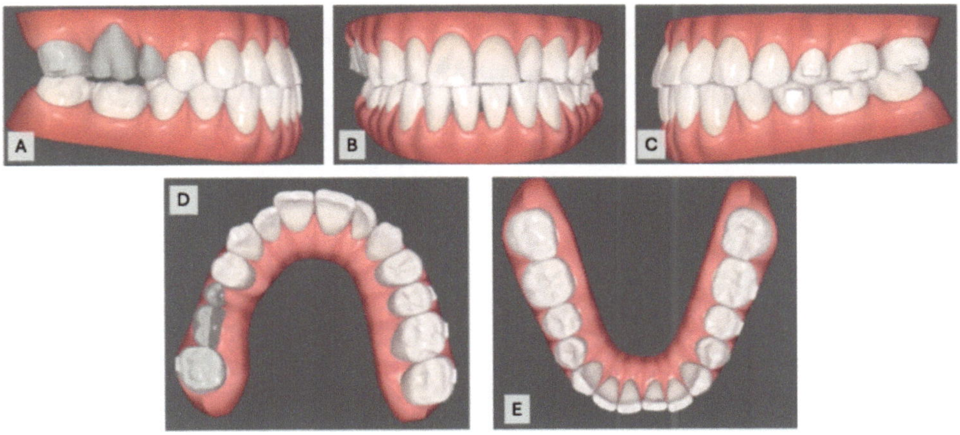

Figure 9. Intraoral scans for clear aligner: (**A**) right occlusion; (**B**) front occlusion; (**C**) left occlusion; (**D**) maxillary occlusion; (**E**) mandibular occlusion.

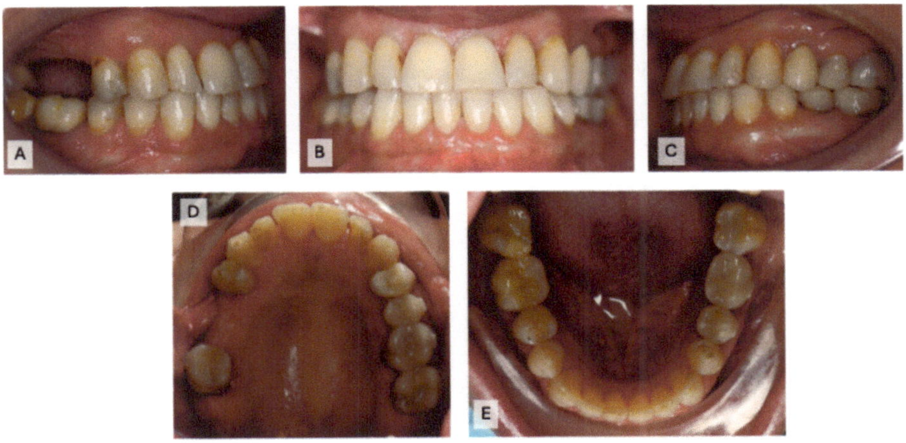

Figure 10. Intraoral photographs after orthodontic treatment: (**A**) right occlusion; (**B**) front occlusion; (**C**) left occlusion; (**D**) maxillary occlusion; (**E**) mandibular occlusion.

3. Discussion

The demand for adult orthodontic treatment utilizing clear aligner therapy is increasing [3]. However, multiple restorations, such as ceramic crowns and FPDs, pose difficulties for orthodontists regarding this age group. Currently, various esthetic restorations and bonding procedures are available [19,20]. Therefore, orthodontists struggle to find the optimal surface conditioning and bonding techniques for ceramics that produce strong attachment bonds without affecting the ceramic's surface after debonding [21,22]. One of the critical factors for the success of any orthodontic treatment is the retention of the orthodontic appliances on the teeth throughout the treatment. Increased incidence of breakage or detachment of the orthodontic appliance can negatively affect the overall success [23].

In clear aligner therapy, attachments are considered a crucial auxiliary device to deliver forces from the aligner to the tooth crown and root. These attachments are a force transducer that help improve clear aligners' biomechanics [24]. They are made of direct composite resin attached to the tooth surface [24]. With the aid of a computer program, attachments are automatically positioned in specific locations on teeth. They regulate the force's application point, direction, and amount applied [25]. Aligner attachments come in different forms that help retain teeth and enable greater control of specific tooth movements [26]. Thus, the primary factors determining the attachment's effectiveness and aligner fitting are its positioning and configuration [26]. Nevertheless, the material used plays a crucial role, since it should remain in the patient's mouth throughout orthodontic therapy and must maintain its features over time. Indubitably, these auxiliary components' mechanical and physical properties are significantly impacted by the material selection [27].

This case report presents a novel dental technique that outlines a method for fabricating M-PCA as a single unit. The attachment design aims to replicate retentive, gingivally beveled attachments as well as vertical and horizontal attachments. The selection of attachments was based on the necessity for tooth movement and to create/manage a force system. In this case report, a horizontal gingivally beveled attachment measuring (2×3 mm) was utilized in all M-PCA.

Digital dental workflows are user- and patient-friendly, delivering predictable, accurate, and effective treatment modalities [28,29]. This case report highlighted the opportunity to use Computer-Aided Design and Computer-Aided Manufacturing (CAD-CAM) to fabricate the M-PCA for clear aligner therapy. The results clearly show that the accuracy and adaptation of the M-PCA units used were clinically acceptable. This report provided information that may help the practitioner overcome the difficulties and challenges during bonding attachments on provisional crowns. The bonding time was also reduced, as this step was entirely eliminated during bonding at the attachments visit. Furthermore, the extra flush, discrepancy, and ill-fitting issues were eliminated using this technique.

In clinical settings, several factors contribute to early attachment failure in terms of debonding, shape changes, and loss of aligner fit [14]. Orthodontic attachment debonding can result in significant clinical issues that may increase the likelihood of treatment failure, lengthen treatment times, and require more follow-up visits [30]. Orthodontic attachment debonding can occur due to bond failure or patient negligence, influenced by operator-related, patient-related, and clinical factors [14]. Operator factors involve the bonding materials and protocols, while patient factors include the frequency of aligner removal [14]. Clinical variables encompass the attachment's location, number, and shape, with molar attachments being the most prone to loss due to increased forces during aligner use [14].

Several investigations evaluated the bond strength of conventional brackets to different restorative materials, such as dental ceramic [31], composite [32], amalgam [33], gold alloy [34], and PMMA [35]. It was reported that the bond strength of conventional stainless steel brackets to provisional crowns can tolerate orthodontic tooth movement using mechanical surface treatment [35]. For that reason, it was recommended to perform surface treatments, such as sandblasting of provisional crowns, to achieve clinically successful bonding and reduce bracket breakage during orthodontic therapy [35].

In clear aligner therapy, efforts have been conducted to improve the bonding of direct composite attachment [21,36]. Alsaud et al. evaluated the effect of different surface treatments of composite attachments bonded to lithium disilicate ceramics [21]. When the dislocation force outweighs the bonding force, the attachment will quickly be lost from the tooth surface. The dislocation force on the attachments will increase each time the patient places or removes the aligner [37].

A recent in vitro study by Shahin et al. [38] evaluated 3D-printed provisional crowns with attachments as one unit to overcome the debonding of orthodontic attachments. In this case report, milled-provisional PMMA crowns were used; with this novel technique, there was no emergency or breakdown of the crown attachment throughout the treatment. Moreover, the aligners were perfectly adapted to the M-PCA.

Since composite attachments are a crucial part of clear aligner therapy, it is vital to meticulously replicate them to preserve their integrity throughout the treatment [24]. The curing process can significantly impact the morphology and resistance of attachments, as in all other polymerization phases [39]. Nevertheless, the polymerization shrinkage factor is eliminated in this case by using M-PCA. After one year of bonding, the attachment deterioration rate was reported to be 14.79% for flowable composite material, while 9.70% was reported for packable composite [40]. In this case, all M-PCA had no damage or change in the shape and size during the "10 months" treatment time.

As reported, clear aligner patients have higher esthetic demands [41]. In addition to orthodontic attachment debonding, attachment discoloration can cause dissatisfaction during orthodontic treatment [42]. Studies have shown that surface roughness influences the staining of resin composites [43–46]. In this case report, M-PCA use showed good color stability throughout the treatment.

A systematic review confirmed that treatment with clear aligners results in better periodontal health [9]. In this case, the patient's periodontal health was optimal throughout the orthodontic treatment, and no periodontal breakdown was noticed, especially around the M-PCA abutments.

In summary, this report defined the areas of new methods that can be adopted for many clear aligner providers, especially in adult patients. However, some limitations are associated, such as the use of only one attachment design and dimension and one type of milled provisional restoration. Future in vitro and in vivo studies should be conducted to evaluate this concept using different attachment designs, dimensions, and materials. In addition, these types of attachments' performance (wear, color change, bacterial accumulation) should be investigated to simulate prolonged clinical usage.

4. Conclusions

This case showed a substantial potential for attachment modality in provisional crowns during orthodontic treatment. M-PCA can be used as a type of provisional crown where clear aligners are to be used. Using this method, the M-PCA was able to overcome most of the challenges encountered during orthodontic treatment. The emergency visit was eliminated, and aligners were found to be tracking exceptionally at each visit. The color stability and attachment durability (shape and size) were well maintained. Moreover, this paper defines and describes new and unique methods that can be adopted by many clear aligner providers, especially in adult patients. However, the M-PCA concept needs to be investigated further.

Author Contributions: Conceptualization, A.A. and Y.A.A.-D.; methodology, A.A. and Y.A.A.-D.; investigation, A.A. and Y.A.A.-D.; resources, A.A. and Y.A.A.-D.; writing—original draft preparation, A.A. and Y.A.A.-D.; writing—review and editing, A.A. and Y.A.A.-D. All authors have read and agreed to the published version of the manuscript.

Funding: This research received no external funding.

Institutional Review Board Statement: This study was conducted according to the guidelines of the Declaration of Helsinki and approved by the Institutional Review Board of Imam Abdulrahman Bin Faisal University (IRB: 2023-02-600, 21 December 2023).

Informed Consent Statement: Informed consent was obtained from the subject involved in the study.

Data Availability Statement: Data are available upon request.

Conflicts of Interest: The authors declare no conflicts of interest.

References

1. Swanson, K.; Hermanides, L. Biomechanics and Function: Altering Paradigms to Treat a Patient's Esthetic Disability Conservatively. *Compend. Contin. Educ. Dent.* **2020**, *41*, 284–289.
2. Guzman-Perez, G.; Jurado, C.A.; Alshahib, A.; Afrashtehfar, K.I. An Immediate Implant Approach to Replace Failing Maxillary Anterior Dentition Due to Orthodontically Induced Severe Root Resorption. *Int. J. Oral Implantol.* **2023**, *16*, 339–348.
3. Rossini, G.; Parrini, S.; Castroflorio, T.; Deregibus, A.; Debernardi, C.L. Efficacy of Clear Aligners in Controlling Orthodontic Tooth Movement: A Systematic Review. *Angle Orthod.* **2015**, *85*, 881–889. [CrossRef]
4. Yassir, Y.A.; Nabbat, S.A.; McIntyre, G.T.; Bearn, D.R. Clinical Effectiveness of Clear Aligner Treatment Compared to Fixed Appliance Treatment: An Overview of Systematic Reviews. *Clin. Oral Investig.* **2022**, *26*, 2353–2370. [CrossRef]
5. Patel, D.; Mehta, F.; Mehta, N. Aesthetic Orthodontics: An Overview. *Orthod. J. Nepal* **2014**, *4*, 38–43. [CrossRef]
6. Alsaeed, S.; Afrashtehfar, K.I.; Alharbi, M.H.; Alfarraj, S.S.; Alluhaydan, S.A.; Abahussain, F.A.; Alotaibi, G.M.; Awawdeh, M.A. Impact of Orthodontic Appliances on Hiring Prospects in Saudi Arabia: A Cross-Sectional Study. *Cureus* **2023**, *15*, e40173. [CrossRef]
7. Weir, T. Clear Aligners in Orthodontic Treatment. *Aust. Dent. J.* **2017**, *62*, 58–62. [CrossRef]
8. Phan, X.; Ling, P.H. Clinical Limitations of Invisalign. *J. Can. Dent. Assoc.* **2007**, *73*, 263–266.
9. Rossini, G.; Parrini, S.; Castroflorio, T.; Deregibus, A.; Debernardi, C.L. Periodontal Health during Clear Aligners Treatment: A Systematic Review. *Eur. J. Orthod.* **2014**, *37*, 539–543. [CrossRef]
10. Nazir, M.A. Prevalence of Periodontal Disease, Its Association with Systemic Diseases and Prevention. *Int. J. Health Sci.* **2017**, *11*, 72–80.
11. Lemos, C.R.; Fadel, M.A.V.; Polmann, H.; de Oliveira, J.M.D.; Pauletto, P.; Stefani, C.M.; Flores-Mir, C.; Canto, G.D.L. Clear Aligner's Adverse Effects: A Systematic Review Protocol. *PLoS ONE* **2024**, *19*, e0302049. [CrossRef]
12. Galan-Lopez, L.; Barcia-Gonzalez, J.; Plasencia, E. A Systematic Review of the Accuracy and Efficiency of Dental Movements with Invisalign®. *Korean J. Orthod.* **2019**, *49*, 140–149. [CrossRef] [PubMed]
13. Nucera, R.; Dolci, C.; Bellocchio, A.M.; Costa, S.; Barbera, S.; Rustico, L.; Farronato, M.; Militi, A.; Portelli, M. Effects of Composite Attachments on Orthodontic Clear Aligners Therapy: A Systematic Review. *Materials* **2022**, *15*, 533. [CrossRef]
14. Yaosen, C.; Mohamed, A.; Jinbo, W.; Ziwei, Z.; Al-Balaa, M.; Yan, Y. Risk Factors of Composite Attachment Loss in Orthodontic Patients during Orthodontic Clear Aligner Therapy: A Prospective Study. *BioMed Res. Int.* **2021**, *2021*, 6620377. [CrossRef]
15. El-Bialy, T.; Galante, D.; Daher, S. *Orthodontic Biomechanics: Treatment of Complex Cases Using Clear Aligner*; Bentham Science Publishers: Sharjah, United Arab Emirates, 2016; Volume 1.
16. Harrison, J.E.; Bowden, D.E. The Orthodontic/Restorative Interface. Restorative Procedures to Aid Orthodontic Treatment. *Br. J. Orthod.* **1992**, *19*, 143–152. [CrossRef]
17. Andreasen, G.F.; Stieg, M.A. Bonding and Debonding Brackets to Porcelain and Gold. *Am. J. Orthod. Dentofac. Orthop.* **1988**, *93*, 341–345. [CrossRef]
18. Barreda, G.J.; Dzierewianko, E.A.; Muñoz, K.A.; Piccoli, G.I. Surface Wear of Resin Composites Used for Invisalign® Attachments. *Acta Odontol. Latinoam. AOL* **2017**, *30*, 90–95. [PubMed]
19. Makhija, S.K.; Lawson, N.C.; Gilbert, G.H.; Litaker, M.S.; McClelland, J.A.; Louis, D.R.; Gordan, V.V.; Pihlstrom, D.J.; Meyerowitz, C.; Mungia, R.; et al. Dentist Material Selection for Single-Unit Crowns: Findings from the National Dental Practice-Based Research Network. *J. Dent.* **2016**, *55*, 40–47. [CrossRef]
20. Zarone, F.; Ferrari, M.; Mangano, F.G.; Leone, R.; Sorrentino, R. "Digitally Oriented Materials": Focus on Lithium Disilicate Ceramics. *Int. J. Dent.* **2016**, *2016*, 9840594. [CrossRef]
21. Alsaud, B.A.; Hajjaj, M.S.; Masoud, A.I.; Abou Neel, E.A.; Abuelenain, D.A.; Linjawi, A.I. Bonding of Clear Aligner Composite Attachments to Ceramic Materials: An in Vitro Study. *Materials* **2022**, *15*, 4145. [CrossRef]
22. Bayoumi, R.E.; El-Kabbany, S.M.; Gad, N. Effect of Different Surface Treatment Modalities on Surface Roughness and Shear Bond Strength of Orthodontic Molar Tubes to Lithium Disilicate Ceramics. *Egypt. Dent. J.* **2019**, *65*, 641–656. [CrossRef]
23. Borzangy, S. Impact of Surface Treatment Methods on Bond Strength of Orthodontic Brackets to Indirect Composite Provisional Restorations. *J. Contemp. Dent. Pract.* **2019**, *20*, 1412–1416. [CrossRef] [PubMed]
24. Garino, F.; Castroflorio, T.; Daher, S.; Ravera, S.; Rossini, G.; Cugliari, G.; Deregibus, A. Effectiveness of Composite Attachments in Controlling Upper-Molar Movement with Aligners. *J. Clin. Orthod.* **2016**, *50*, 341–347.
25. Abela, S. Interceptive Orthodontics: A Practical Guide to Occlusal Management. *Eur. J. Orthod.* **2017**, *39*, 342. [CrossRef]
26. Savignano, R.; Valentino, R.; Razionale, A.; Michelotti, A.; Barone, S.; D'anto, V. Biomechanical Effects of Different Auxiliary-Aligner Designs for the Extrusion of an Upper Central Incisor: A Finite Element Analysis. *J. Healthc. Eng.* **2019**, *2019*, 9687127. [CrossRef] [PubMed]

27. Mantovani, E.; Castroflorio, E.; Rossini, G.; Garino, F.; Cugliari, G.; Deregibus, A.; Castroflorio, T. Scanning Electron Microscopy Analysis of Aligner Fitting on Anchorage Attachments. *J. Orofac. Orthop.* **2019**, *80*, 79–87. [CrossRef] [PubMed]
28. Blatz, M.B.; Conejo, J. The Current State of Chairside Digital Dentistry and Materials. *Dent. Clin. N. Am.* **2019**, *63*, 175–197. [CrossRef] [PubMed]
29. Garaicoa, J.; Jurado, C.A.; Afrashtehfar, K.I.; Alhotan, A.; Fischer, N.G. Digital Full-Mouth Reconstruction Assisted by Facial and Intraoral Scanners: A Case Report and Technique Description. *Appl. Sci.* **2023**, *13*, 1917. [CrossRef]
30. Dasy, H.; Dasy, A.; Asatrian, G.; Rózsa, N.; Lee, H.-F.; Kwak, J.H. Effects of Variable Attachment Shapes and Aligner Material on Aligner Retention. *Angle Orthod.* **2015**, *85*, 934–940. [CrossRef] [PubMed]
31. Grewal Bach, G.K.; Torrealba, Y.; Lagravère, M.O. Orthodontic Bonding to Porcelain: A Systematic Review. *Angle Orthod.* **2013**, *84*, 555–560. [CrossRef] [PubMed]
32. Eslamian, L.; Borzabadi-Farahani, A.; Mousavi, N.; Ghasemi, A. The Effects of Various Surface Treatments on the Shear Bond Strengths of Stainless Steel Brackets to Artificially-Aged Composite Restorations. *Australas. Orthod. J.* **2011**, *27*, 28–32. [CrossRef]
33. Zachrisson, B.U.; Büyükyilmaz, T.; Zachrisson, Y. Improving Orthodontic Bonding to Silver Amalgam. *Angle Orthod.* **1995**, *65*, 35–42. [CrossRef]
34. Büyükyilmaz, T.; Zachrisson, Y.Ø.; Zachrisson, B.U. Improving Orthodontic Bonding to Gold Alloy. *Am. J. Orthod. Dentofac. Orthop.* **1995**, *108*, 510–518. [CrossRef] [PubMed]
35. Shahin, S.Y.; Abu Showmi, T.H.; Alzaghran, S.H.; Albaqawi, H.; Alrashoudi, L.; Gad, M.M. Bond Strength of Orthodontic Brackets to Temporary Crowns: In Vitro Effects of Surface Treatment. *Int. J. Dent.* **2021**, *2021*, 9999933. [CrossRef] [PubMed]
36. Chen, W.; Qian, L.; Qian, Y.; Zhang, Z.; Wen, X. Comparative Study of Three Composite Materials in Bonding Attachments for Clear Aligners. *Orthod. Craniofacial Res.* **2021**, *24*, 520–527. [CrossRef] [PubMed]
37. Skaik, A.; Wei, X.L.; Abusamak, I.; Iddi, I. Effects of Time and Clear Aligner Removal Frequency on the Force Delivered by Different Polyethylene Terephthalate Glycol-Modified Materials Determined with Thin-Film Pressure Sensors. *Am. J. Orthod. Dentofac. Orthop.* **2019**, *155*, 98–107. [CrossRef] [PubMed]
38. Shahin, S.Y.; Nassar, E.A.; Gad, M.M. 3D-Printed One-Piece versus Two-Piece Orthodontic Clear Aligner Attachments Bonded with Provisional Crowns: A Proof of Concept. *J. Dent.* **2024**, *151*, 105406. [CrossRef]
39. Gazzani, F.; Bellisario, D.; Quadrini, F.; Parrinello, F.; Pavoni, C.; Cozza, P.; Lione, R. Comparison Between Different Composite Resins Used for Clear Aligner Attachments: An In-Vitro Study. *Front. Mater.* **2022**, *8*, 789143. [CrossRef]
40. Lin, S.; Huang, L.; Li, J.; Wen, J.; Mei, L.; Xu, H.; Zhang, L.; Li, H. Assessment of Preparation Time and 1-Year Invisalign Aligner Attachment Survival Using Flowable and Packable Composites: A Split-Mouth Clinical Study. *Angle Orthod.* **2021**, *91*, 583–589. [CrossRef]
41. Alansari, R.A.; Faydhi, D.A.; Ashour, B.S.; Alsaggaf, D.H.; Shuman, M.T.; Ghoneim, S.H.; Linjawi, A.I.; Marghalani, H.Y.; Dause, R.R. Adult Perceptions of Different Orthodontic Appliances. *Patient Prefer. Adherence* **2019**, *13*, 2119–2128. [CrossRef]
42. Ardu, S.; Duc, O.; Di Bella, E.; Krejci, I. Color Stability of Recent Composite Resins. *Odontology* **2017**, *105*, 29–35. [CrossRef] [PubMed]
43. Lu, H.; Roeder, L.B.; Lei, L.; Powers, J.M. Effect of Surface Roughness on Stain Resistance of Dental Resin Composites. *J. Esthet. Restor. Dent.* **2005**, *17*, 102–108. [CrossRef] [PubMed]
44. Ghinea, R.; Ugarte-Alvan, L.; Yebra, A.; Pecho, O.E.; Paravina, R.D.; Perez, M. del M. Influence of Surface Roughness on the Color of Dental-Resin Composites. *J. Zhejiang Univ. Sci. B* **2011**, *12*, 552–562. [CrossRef] [PubMed]
45. Tavangar, M.; Bagheri, R.; Kwon, T.-Y.; Mese, A.; Manton, D.J. Influence of Beverages and Surface Roughness on the Color Change of Resin Composites. *J. Investig. Clin. Dent.* **2018**, *9*, e12333. [CrossRef] [PubMed]
46. Duc, O.; Di Bella, E.; Krejci, I.; Betrisey, E.; Abdelaziz, M.; Ardu, S. Staining Susceptibility of Resin Composite Materials. *Am. J. Dent.* **2019**, *32*, 39–42. [PubMed]

prosthesis

MDPI

Systematic Review

Microbial Adhesion to Poly Methyl Methacrylate (PMMA) Denture Base Resins Containing Zinc Oxide (ZnO) Nanostructures: A Systematic Review of In Vitro Studies

Nawal M. Majrashi [1], Mohammed S. Al Qattan [1], Noor S. AlMubarak [1], Kawther Zahar Alzahir [1] and Mohammed M. Gad [2,*]

1 College of Dentistry, Imam Abdulrahman Bin Faisal University, P.O. Box 1982, Dammam 31441, Saudi Arabia; 2160004637@iau.edu.sa (N.M.M.); 2180005539@iau.edu.sa (M.S.A.Q.); 2170002766@iau.edu.sa (N.S.A.); 2190004355@iau.edu.sa (K.Z.A.)

2 Department of Substitutive Dental Sciences, College of Dentistry, Imam Abdulrahman Bin Faisal University, P.O. Box 1982, Dammam 31441, Saudi Arabia

* Correspondence: mmjad@iau.edu.sa; Tel.: +966-592502080

Abstract: Background: Denture stomatitis is an inflammatory condition involving swelling and redness of the oral mucosa beneath a denture. Among various available treatments, zinc oxide nanoparticles (ZnONPs) and nano-wire nanostructures have been suggested as potential future therapies. However, there is a lack of information in the literature about the effectiveness of ZnONPs regarding microbial adhesion to different denture base resins. Here, we review studies on the effect of ZnONP use on microbial adhesion to denture base resins to answer the following study question: "Does incorporating ZnONPs into denture base resins reduce microbial adhesion?" Methods: Following the Preferred Reporting Items for Systematic Reviews and Meta-Analyses (PRISMA) guidelines, an electronic and manual search ranging from Jan 2000 to May 2024 was performed using PubMed, Web of Science, and Scopus databases to answer the study question. All full-length English-language articles investigating the effects of ZnO nanostructures on *Candida albicans* adhesion to polymethyl methacrylate (PMMA) denture base resins were included. The extracted data were tabulated for qualitative and quantitative analysis of the included studies. Results: Of the 479 studies reviewed, 7 studies successfully met the eligibility criteria. All included studies utilized PMMA as the denture base material with different polymerization methods. *C. albicans* was the most extensively studied microbial species, with various count methods used. Six studies concluded a statistically significant impact of ZnONPs on decreasing *C. albicans* adhesion to the denture base. However, one study reported the opposite. Conclusions: Incorporating ZnONPs into PMMA denture base resin has a positive impact on reducing *C. albicans* adherence and could be recommended for denture stomatitis treatment. However, further studies are needed to cover the notable gap in data regarding the safety and effectiveness of ZnO nanostructures.

Keywords: complete denture; microbial adhesion; PRISMA 2020; zinc oxide nanoparticle

1. Introduction

The denture base is that part of the prosthesis that carries the artificial teeth and rests on the mucous-bone support. Typically either resin- or metal-based, the denture base should have adequate physical, mechanical, esthetic, and biocompatibility properties [1]. Polymethyl methacrylate (PMMA) is the most used and recommended material for prosthetic fabrication [1,2]. PMMA offers a good esthetic, resembling the natural gingiva, and it is lightweight for patient comfort, cost-effective, easy to fabricate or repair, and highly biocompatible. However, limitations include low strength, risking fractures if dropped or subjected to heavy forces; fluid absorption, causing bad odor, discoloration, and bacterial overgrowth; polymerization-related shrinkage, affecting retention and stability; and, finally, a lack of thermal sensitivity compared to natural oral structures [3–5]. Different modification and fabrication techniques are recommended to overcome the disadvantages of traditional acrylic resins. To improve strength, high-impact acrylic (PMMA with additional rubber or fibers) can be used. Incorporating nanoparticles can enhance denture strength and esthetics and reduce bulkiness [2]. New fabrication methods, such as digital fabrication using CAD/CAM (computer-aided design/computer-aided manufacturing) techniques, improve control over denture design; this can lead to enhanced fit, durability, strength, and density while reducing production time and error [1,3].

The surface properties of denture base materials affect the esthetics and success of dentures, ultimately impacting the quality of life of the patient [3–5]. Surface roughness is a contributing factor to stomatitis, alongside poor hygiene, medications, and poor autoimmune resistance of the patient [3]. Inflamed and red oral mucosa beneath the denture is called denture stomatitis, and it is usually painless and asymptomatic. However, mucosal bleeding, taste alteration, and a burning sensation can develop [6]. The disease is multi-factorial, and it can develop from poor fit of the denture, poor denture hygiene, and wearing the denture at night [7,8]. *Candida albicans* is the principal pathogen causing denture stomatitis. Treatment options include anti-fungal agents, such as nystatin or miconazole, and laser therapy [9,10]. Several measures have been suggested to reduce the occurrence of denture stomatitis, including improved fabrication of dentures, improved oral hygiene, and the removal of dentures overnight, as well as the modification of denture bases with antifungal agents [11,12].

Nanotechnology is an area of science concerned with developing and producing extremely small tools and machines through the arrangement of atoms. Nanostructures include nanoparticles (NPs) and nanorods, and they are promising agents for antimicrobial applications [13]. They are mostly considered biocompatible due to the phagocytic activity of human cells against the particles. Nanotechnology could be used to deliver biocompatible therapeutic agents while reducing the development of resistance against regular anti-fungal therapies, which is the major drawback of common antimicrobial agents [14].

The nanostructure of zinc oxide (ZnO) is known for its ability to continuously release metal ions for up to two months; these ions are capable of reducing *C. albicans* growth [4]. Several studies have demonstrated the anti-fungal activity of ZnO nanoparticles (ZnONPs) added to PMMA in reducing the adherence of *C. albicans* [15,16]. ZnONPs increase the contact angle when incorporated into acrylic resin, thus increasing the hydrophobicity by increasing the roughness and modifying the surface energy. A hydrophobic surface is beneficial in applications requiring water resistance and in reducing microbial formation and supporting self-cleaning properties [17].

Although previous studies have investigated the capabilities of ZnO nanostructures, there is still a gap regarding the implementation of these nanocomposites as denture base resins, their antimicrobial properties, and their potential for reducing denture stomatitis. In addition, no previous review has been conducted to assess the performance of denture bases containing ZnO nanostructures. There remains a lack of information in the literature concerning the effectiveness of ZnONPs on microbial adhesion to different denture base resins. This review therefore focused on the effect of ZnONPs on microbial adhesion to denture base resins to address the question "Does incorporating ZnONPs into denture base resins reduce denture stomatitis"?

2. Materials and Methods

A systematic review was conducted according to the Preferred Reporting Items for Systematic Reviews and Meta-Analyses (PRISMA) guidelines (Figure 1) and structured according to the PICO method: the population consisted of PMMA denture base resins, the intervention was the addition of ZnONPs, the control was unmodified resins, and the outcome was microbial adhesion. This approach produced the following study question: "Does incorporating ZnONPs into denture base resins reduce microbial adhesion"?

To address the study question, the following keywords were used: PMMA, denture base, denture stomatitis, microbial adhesion, *C. albicans* adhesion and/or ZnONP, nanoparticles, nano-size, or nanostructure. The keywords were used to search the databases according to the inclusion and exclusion criteria. The inclusion criteria were in vitro studies written in English, full-length original articles, use of denture base resins, investigation of microbial adhesion, and use of ZnONPs. The exclusion criteria included documents where only the abstract was available, review articles, and studies not investigating denture base resins (such as those using soft liners).

An electronic search in databases (Web of Science and Scopus) was performed followed by manual searches for any additional articles meeting the inclusion criteria published from January 2000 to May 2024. Data extraction was conducted in three stages: (1) review of titles; (2) review of abstracts; and (3) review of full texts. Two reviewers (N.M.M. and N.S.A.) independently reviewed and analyzed the articles according to the inclusion criteria. Any discrepancies were analyzed by other reviewers (M.S.A. and M.M.G.) to resolve the issue. Data were extracted and tabulated according to items detailed in Table 1. Due to the variations in the included studies in terms of methodology, specimen shape, ZnONP content, salinization processes, testing methods, and aging procedures, it was not appropriate to conduct a meta-analysis. Therefore, the included articles were qualitatively analyzed and described.

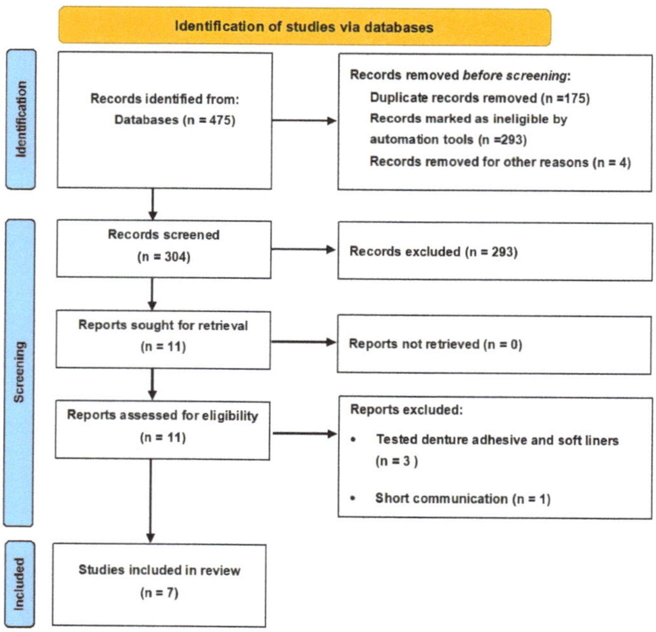

Figure 1. PRISMA flow chart of the study selection process.

Quality Assessment and Risk of Bias of Included Studies

The included studies were analyzed for quality assessment and risk of bias by two independent reviewers (N.M. and M.A.). Each article was evaluated using a risk of bias tool (the modified Consolidated Standards of Reporting Trials (CONSORT)) consisting of seven parameters and items, described in Table 2. A "yes" was assigned when the parameter was reported in the text, and a "no" if the information was absent or unclear. The risk of bias was classified according to the sum of "yes" marks received as follows: 1–3, high; 4–5, medium; 6–7, low risk of bias [18].

Table 1. Content of included studies.

Ref	Denture Base Resin Type/ Brand Name	ZnO Size and Brand Name	ZnO%	ZnO Treatment	Addition Method Monomer/ Polymer	Specimen Shape and Dimensions/ Aging	Sample Size	Microbial Species	Tested Properties	Assessment Method	Results and Outcome
Cierech M, et al. [19]	HP (PMMA)/ Superacryl Plus, Spofa Dental, Czech Republic		0%wt 2%wt	Sedimented (washed with deionized water, centrifuged, and dried)	Monomer	10 × 10 × 2 mm NS	N = 30 n = 15	*C. albicans*, ATCC 14055	Density MIC Roughness	CFU	ZnONPs display fungistatic or fungicidal activity and reduce C. albicans adhesion
Cierech M, et al. [20]	HP (PMMA)/ Superacryl Plus Spofa Dental, Czech Republic	25–30 nm CHEMPUR, piekary śląskie Poland 25–30 nm	0%wt 2%wt	Salinization	Monomer	10 × 10 × 2 mm NS	N = 30 n = 16	*C. albicans*, ATCC 14056	Anti-fungal morphology of cells	CFU	Significantly decreased *C. albicans* adhesion
Cierech M, et al. [21]	HP (PMMA)/Superacryl Plus Spofa Dental, Jicin, Czech Republic		0% 2.5% 5% 7.5%	Nanowires are produced	Monomer	13 × 13 × 2 mm NS	N = 20 n = 5	*C. albicans*, ATCC 14057	Zinc ions release Cytotoxicity to human cells	Optical emission spectrometry in inductively coupled plasma	ZnONPs prevent adhesion and biofilm development by *C. albican*
Kamonkhantikul K, et al. [22]	HP PMMA/Ivoclar Vivadent, Schaan, Liechtenstein	20–40 nm Nano Materials Technology Co.Led. Chonburi, Thailand	0% 1.25, 2.5, 5% wt	NS	Monomer	12 × 2 mm water storage time in 37C deionized water for 48 h or 1 month before testing	N = 98 n = 8	*C. albicans*, ATCC 90028	Antifungal, optical, and mechanical properties	CFU	ZnONPs cause reduction in *C. albicans* adherence
Apip C, et al. [23]	HP PMMA (MelioDent Heat Cure, Heareus Kulzer, Hanau, Germany)	25–45 nm NS	0 ppm Study: 250, 500, 1000 ppm	NS	ZnO-NWs were suspended in liquid monomer	10 × 5 × 3 mm Water storage not mentioned	N = 80	*C. albicans*, ATCC 10231	Anti-biofilm activity	Transmission electron microscopy Raman mapping images and spectra	*C. albicans* adherence and biofilm formation considerably decreased with increasing ZnO-NWs concentrations in PMMA-ZnO-NW
Anwander M, et al. [24]	AP (Palapress vario, Heraeus Kulzer GmbH, Hanau, Germany)	<100 nm Sigma-Aldrich Co., St. Louis, MO, USA	0% 0.1, 0.2, 0.4, and 0.8 wt	NS	Monomer	7 × 1.5 mm Stored in distilled water for 7 d prior to conducting the experiments	N = 60 n = 15	*C. albicans*, ATCC 10232	Roughness Biofilm formation Biomass	Energy-dispersive X-ray spectroscopy (EDX)	No statistically significant impact of available ZnO material on decreasing biofilm adhesion proven
Raj I, et al. [25]	PMMA of analytical-grade quality was procured from Alfa Aeser, Haverhill, Massachusetts	60 nm Sigma Aldrich, St. Luis, MO, USA	0 1, 2, 5, 10, and 15 by wt	NS	Polymer	Film, 12 cm length, 8 cm width, and 2 mm thickness Thermal and water storage	Not mentioned 4 experiments and 5 groups	*C. albicans*, ATCC 10233	Cytotoxicity Crystalline and morphological changes Density and abrasion resistance	Microscopically	ZnONP groups shows decrease in *C. albicans* adhesion and colonization

(PMMA) Poly methyl methacrylate; (CFU) colony-forming unit; (HP) heat-polymerized; (AP) auto-polymerized; (NS) not stated.

Table 2. Quality assessment and risk of bias considering the aspects reported in the Materials and Methods sections.

Ref.	Sample Size Calculation	Sample Randomization	Control Group	Stating Clear Testing Method	Statistical Analyses Carried Out	Reliable Analytical Methods	Blinding of Evaluators	Risk of Bias
[19]	Yes	No	Yes	Yes	Yes	Yes	No	Medium
[20]	Yes	No	Yes	Yes	Yes	Yes	No	Medium
[21]	No	No	Yes	Yes	Yes	Yes	No	Medium
[22]	Yes	Yes	Yes	Yes	Yes	Yes	No	Low
[23]	Yes	No	Yes	Yes	Yes	No	No	Medium
[24]	Yes	Yes	Yes	Yes	Yes	Yes	No	Low
[25]	No	No	Yes	Yes	Yes	Yes	No	Medium

3. Results

Out of 479 studies identified, 7 [19–25] met the inclusion criteria and examined the effect of adding ZnO nanostructures to denture base resins on microbial adhesion, as detailed in Table 1. All studies used PMMA acrylic denture bases and employed various polymerization methods [19–25]. Specifically, six studies used heat-polymerized PMMA [19–25], while one study used auto-polymerized PMMA [24]. Six studies incorporated ZnONPs, and one study used ZnO nanorods [23]. Nanoparticle sizes ranged from 25 to 60 nm [19–23,25] and were not specified in one study [24].

In all included studies, ZnONP concentrations were represented as percentages [19–22,24], except for one study that used parts per million (ppm) [23]. All studies used unmodified material as a control [19–25]. Two studies applied a concentration of 0.2% by weight (wt) [19,20]. For experimental groups, Cierech et al. used 2.5, 5, and 7.5% by wt. [21], while Kamonkhantikul et al. tested concentrations of 1.25, 2.5, and 5% by wt. [22]. Apip et al. used 250, 500, and 1000 ppm [23], and Anwander et al. tested 0.1, 0.2, 0.4, and 0.8% by wt. [24]. Raj et al. used various concentrations: 1, 2, 5, 10, and 15% by wt. [25].

Treatment of the ZnONPs varied across studies, thus potentially impacting the effects on the denture base. In studies by Cierech et al., ZnONPs were treated through sedimentation—washing the material three times with deionized water, and then centrifuging and freeze-drying it [19–21]. Kamonkhantikul et al. applied salinization to the ZnONPs [22], while Apip et al. used ZnO nanowires [23]. Two studies did not mention ZnONP treatment methods [24–28].

In six studies, ZnONPs were added to the liquid monomer [18–22,24], while in Anwander et al., they were added to the powder polymer [24]. Specimen sizes also varied: Cierech et al. used 10 × 10 × 2 mm specimens [19,20], and four studies used different sizes [21–25]. For storage conditions, three studies stored samples in deionized water at different temperatures [20,23,24], while the remaining studies did not specify storage environments [19,20,22,25].

For microbial specimens, all included studies used a reference strain of *C. albicans* [19–25]. Three studies used *C. albicans* 14053 [12–14], three used *C. albicans* ATCC 10231 [23–25], and one used *C. albicans* ATCC 90028 [22].

All included studies evaluated four key factors: antifungal properties, surface roughness, density, and morphological changes of the microbes [19,20,22–25]. To examine nanopowder morphology, a scanning electron microscope (SEM) (Ultra Plus; Carl Zeiss Meditec AG, Jena, Germany) and a sputter coater (SCD 005/CEA 035, BAL-TEC, Switzerland) were used, with an InLens detector for imaging. The InLens detector allowed for the identification of surface contaminants. Density measurements were performed using a helium pycnometer (AccuPyc II 1340, Micromeritics, USA) following an in-house protocol. Surface roughness was measured using a Dektak XT stylus profiler. One study, however, focused on cytotoxicity release rather than roughness or density [21].

Various methods were used for microbial count assessment, including colony-forming units (CFU) [19,20,22], transmission electron microscopy (TEM) [23], energy-dispersive X-ray (EDX) microscopy [23], and light microscopy [25]. All studies concluded that ZnONPs enhanced antimicrobial properties against *C. albicans* [19–23,25], except one, which could not statistically demonstrate the impact of commercially available ZnO material on reducing biofilm adhesion [24].

Quality assessment

This review consists of two articles with a low risk of bias and five with a medium risk, as illustrated in Table 2 and Figure 2. None of the studies used blinded evaluators, but all stated their methods clearly.

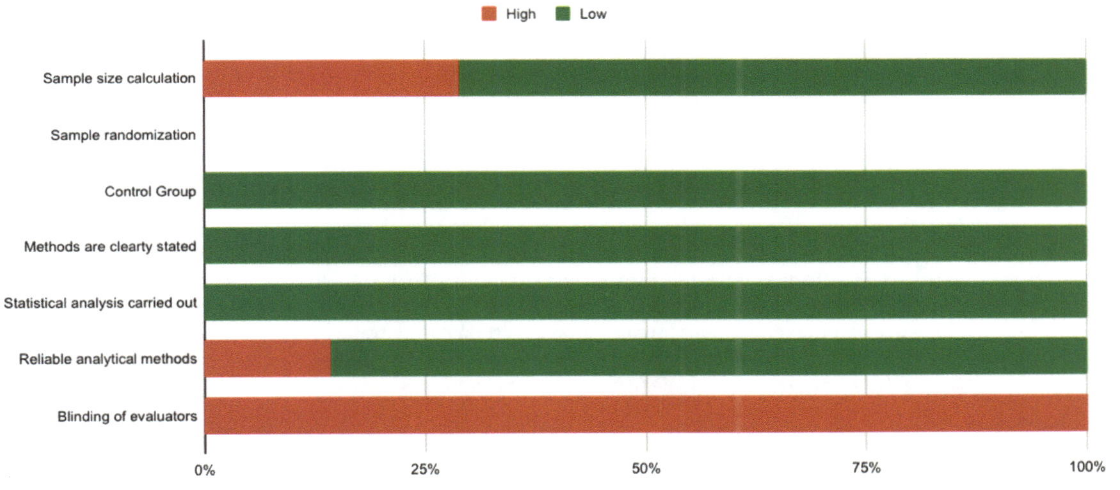

Figure 2. Risk of bias tool.

Different supportive findings, such as SEM and TEM [19–21,23,25], were used to analyze existing colonies on the specimen surface. Some showed colonies with different morphologies and a fully mature biofilm with a multilayered network of microbes and hyphae, while others showed damage to cells [23,25].

4. Discussion

ZnONPs are widely recognized for their multifunctionality, transparency, and stability, making them effective inorganic fillers [19–21]. When incorporated into a polymer, ZnONPs significantly enhance mechanical properties, such as tensile and impact strengths, as well as optical properties, contributing to a more esthetically favorable material [21]. Additionally, ZnONPs exhibit potent antimicrobial activity effective against a broad spectrum of microorganisms, including both Gram-negative and Gram-positive bacteria [22]. Owing to these attributes, ZnONPs have been selected as a reinforcing agent for denture base materials. Their incorporation results in improved mechanical strength, abrasion resistance, and antimicrobial properties, particularly anticandidal effects, while maintaining the biocompatibility of the denture base. These properties highlight the significant promise of ZnONPs in the development of advanced dental materials [25].

To counter *C. albicans* adhesion to PMMA surfaces, various approaches can be employed. These include the addition of bioactive glass, which decreases the *C. albicans* count at a concentration of 5% [26]. Other studies have investigated different nanoparticles and their antimicrobial effects, such as silver nanoparticles (AgNPs), titanium dioxide nanoparticles (TiONPs), and zirconium dioxide nanoparticles (ZrO_2NPs) [27,28]. Additionally, the application of topical cleaning agents or antifungal medications can help reduce biofilm formation. Various oral antifungal agents, including fluconazole, nystatin, amphotericin B, miconazole, ketoconazole, itraconazole, and clotrimazole, are recommended for treating denture stomatitis [26].

The findings of this review confirm the anti-adhesive activity of ZnO against *C. albicans* at a minimum inhibitory concentration of 0.75 mg/mL [19]. There was no difference in the morphology of HeLa cells treated with ZnONPs reported in a study by Cierech et al. and the structure of the cell monolayer between the control group and the cells treated with lower ZnONP concentrations (1–30 mg/L). However, in higher concentrations (50 mg), morphological changes (polygonal, flat cells turned spherical) were observed, but the structure of the monolayer remained unaffected (no gaps between adjacent cells, cells adhering to each other). Adherence was varied in cells treated with 100 mg/L of ZnONPs [21]. The use of ZnONPs reduces *C. albicans* adhesion, which in turn decreases the incidence of denture stomatitis [19–25]. In another study of a mono-species biofilm consisting of *C. albicans*, significantly fewer adherent cells (measured by relative absorbance values) were identified after 44 h compared to 20 h of biofilm formation in the denture [24]. Less adherence indicates a lesser risk of *C. albicans* infection.

ZnONPs have been examined for their effects on *C. albicans* adhesion. The studies included here conclude that there is an action of these nanoparticles on cell adherence to the acrylic surface. Specifically, the papers focus on the effect of different concentrations of ZnONPs incorporated into PMMA and how this influences microbial biofilm integrity and pathogen adherence [21–25]. Their results indicate that lower concentrations (30–50 mg/L) do not affect cell adhesion, while higher concentrations (100 mg/L) do [21]. This indicates that ZnONPs can enhance the surface properties of acrylic resins by inhibiting bacterial adherence and preventing biofilm formation at certain concentrations [21].

The antimicrobial effects of the materials can be affected by different factors, including nanoparticle concentration, shape (e.g., nanoparticles versus nanowires), and the method of introduction to the resin [24]. PMMA loaded with nanoparticle concentrations ranging from 100 to 250 ppm suggests an inverse relationship between microbial adhesion and nanoparticle concentration, as confirmed through SEM. This imaging method provides visual confirmation of the reduction of adherence, showing fewer fungal cells adhered to the surfaces of the ZnO-modified PMMA compared to controls. Furthermore, ZnO nanowires (ZnO-NWs) at high concentrations (up to 500–1000 ppm) show a role in impairing fungal adherence. Reports suggest a dose-dependent relationship, with greater reductions in adherence associated with higher concentrations of ZnO [23].

The anti-fungal effect of nanoparticles can be mediated through several mechanisms of action [29]. However, studies investigating the mechanisms of action of ZnONPs in particular are limited. However, the main reported mechanism of action excreted by ZnO on *C. albicans* is the release of metal ions and the formation of reactive oxidative species (ROS). The interaction of these two substances with the cell membrane leads to the inhibition of cell wall synthesis, cell signaling, enzyme activities, ribosome distribution, and DNA damage, the inactivation of protein synthesis, and molecular changes in cell proteins [15].

ZnONPs have the ability to reinforce the mechanical properties of denture bases [30]. Augmenting any material brings the advantages of the added material. However, adverse effects can also be associated with augmentation [30]. This can include brittleness in the case of high concentrations of ZnONPs. In addition, the color stability of denture bases can be influenced by nanoparticle incorporation as a result of the interaction between ZnONPs and the polymer, thus decreasing PMMA translucency [31]. The clinical use of ZnONPs in denture bases requires further study to ensure homogeneity, microbiological efficacy, mechanical strength, and biocompatibility [19]. Assessing ZnO's potential for mucosal irritation and its long-term effects is essential [20,30,31]. Despite ZnO's promise in preventing denture stomatitis, clinical trials are required to validate its safety and effectiveness [19,20,23,24].

The effect of ZnONPs on denture base wettability is often measured based on the contact angle. This angle affects how easily microbes can adhere to and spread across a surface. When ZnONPs are added to a PMMA, they often increase the surface roughness and alter the material by increasing the hydrophobicity through a higher contact angle. A higher contact angle therefore means a reduction in microbial adhesion and a surface less prone to retaining the moisture that helps plaque accumulation. Higher surface roughness tends to trap air between the surface and moisture, which increases the contact angle and reduces the area in contact with water (and, by extension, microorganisms). ZnO nanorods embedded in materials can achieve superhydrophobicity, with contact angles exceeding 160°, leading to reduced microbial adhesion and overgrowth [19,24,32].

The number of studies addressing this topic is limited, making it hard to write a high-quality comprehensive review. The quality of the review was also limited by two of the included publications having a high risk of bias. There is a lack of data regarding the long-term impact of ZnONPs on pathogens and human health. No clinical studies have been conducted to give more reliable information. This study focused only on the addition of ZnONPs to PMMA; all included studies were in vitro, with no clinical trials. Further studies are therefore needed to investigate ZnONP addition in different dental materials and the long-term related effects, and in vivo studies are recommended.

5. Conclusions

This review suggests that adding ZnONPs to denture base resins can help reduce denture stomatitis. The concentration and even distribution of ZnONPs are essential for effectiveness due to their impact on antimicrobial properties. However, all of the reviewed studies were conducted in vitro, with no clinical trials available. Future research, especially using clinical trials, is needed to confirm the effectiveness and safety of ZnONPs in real-world use with different dental materials.

Author Contributions: Conceptualization, N.M.M. and M.S.A.Q.; methodology, N.S.A. and M.M.G.; software, M.M.G.; validation, N.M.M., M.S.A.Q. and K.Z.A.; formal analysis, M.M.G.; investigation, N.S.A. and M.M.G.; resources, N.M.M. and M.S.A.Q.; data curation, N.M.M., M.S.A.Q. and N.S.A.; writing—original draft preparation, N.M.M., M.S.A.Q., N.S.A., K.Z.A. and M.M.G.; writing—review and editing, N.M.M., M.S.A.Q., N.S.A., K.Z.A. and M.M.G.; visualization, N.M.M. and M.S.A.Q.; supervision, M.M.G.; project administration, N.S.A. and M.M.G.; funding acquisition, N.M.M., K.Z.A. and M.S.A.Q. All authors have read and agreed to the published version of the manuscript.

Funding: This research received no external funding.

Institutional Review Board Statement: Not applicable.

Informed Consent Statement: Not applicable.

Data Availability Statement: The original contributions presented in the study are included in the article; further inquiries can be directed to the corresponding authors.

Conflicts of Interest: The authors declare no conflicts of interest.

References

1. Alqutaibi, A.Y.; Baik, A.; Almuzaini, S.A.; Farghal, A.E.; Alnazzawi, A.A.; Borzangy, S.; Aboalrejal, A.N.; AbdElaziz, M.H.; Mahmoud, I.I.; Zafar, M.S. Polymeric denture base materials: A review. *Polymers* **2023**, *15*, 3258. [CrossRef]
2. Zafar, M.S. Prosthodontic Applications of Polymethyl Methacrylate (PMMA): An Update. *Polymers* **2020**, *12*, 2299. [CrossRef] [PubMed]
3. Thu, K.M.; Yeung, A.W.K.; Samaranayake, L.; Lam, W.Y.H. Denture plaque biofilm visual assessment methods: A systematic review. *Int. Dent. J.* **2024**, *74*, 1–14. [CrossRef] [PubMed] [PubMed Central]
4. Fotovat, F.; Abbasi, S.; Nikanjam, S.; Alafchi, B.; Baghiat, M. Effects of various disinfectants on surface roughness and color stability of thermoset and 3D-printed acrylic resin. *Eur. J. Transl. Myol.* **2024**, *34*, 11701. [CrossRef] [PubMed] [PubMed Central]
5. Lopez, J.F.; Smith, R.L.; Gomez, M.J. Influence of surface roughness on bacterial adhesion to denture base materials. *J. Prosthet. Dent.* **2022**, *127*, 236–243.
6. Budtz-Jorgensen, E. Clinical aspects of Candida infection in denture wearers. *J. Am. Dent. Assoc.* **1978**, *96*, 474–479. [CrossRef]

7. McReynolds, D.E.; Moorthy, A.; Moneley, J.O.; Jabra-Rizk, M.A.; Sultan, A.S. Denture stomatitis—An interdisciplinary clinical review. *J. Prosthodont.* **2023**, *32*, 560–570. [CrossRef] [PubMed]
8. Silva, M.D.D.D.; Nunes, T.S.B.S.; Viotto, H.E.D.C.; Coelho, S.R.G.; Souza, R.F.; Pero, A.C. Microbial adhesion and biofilm formation by *Candida albicans* on 3D-printed denture base resins. *PLoS ONE* **2023**, *18*, e0292430. [CrossRef]
9. Barbeau, J.; Séguin, J.; Goulet, J.P.; de Koninck, L.; Avon, S.L.; Lalonde, B. Reassessing the presence of Candida albicans in denture-related stomatitis. *Oral Surg. Oral Med. Oral Pathol. Oral Radiol. Endod.* **2003**, *95*, 51–59. [CrossRef]
10. AL-Dwairi, Z.N.; Al-Quran, F.A.; Al-Omoush, S.A.; Baba, N.Z. Surface properties of denture base materials and their impact on the quality of life in complete denture wearers. *J. Prosthodont. Res.* **2020**, *64*, 31–37. [CrossRef]
11. Gendreau, L.; Loewy, Z.G. Epidemiology and etiology of denture stomatitis. *J. Prosthodont.* **2011**, *20*, 251–260. [CrossRef] [PubMed]
12. Nevzatoğlu, E.U.; Ozcan, M.; Kulak-Ozkan, Y.; Kadir, T. Adherence of Candida albicans to denture base acrylics and silicone-based resilient liner materials with different surface finishes. *Clin. Oral. Investig.* **2007**, *11*, 231–236. [CrossRef] [PubMed]
13. Saini, R.S.; Bavabeedu, S.S.; Quadri, S.A.; Gurumurthy, V.; Kanji, M.A.; Okshah, A.; Binduhayyim, R.I.H.; Alarcón-Sánchez, M.A.; Mosaddad, S.A.; Heboyan, A. Mapping the research landscape of nanoparticles and their use in denture base resins: A bibliometric analysis. *Discov. Nano* **2024**, *19*, 95. [CrossRef] [PubMed]
14. Karunakaran, H.; Krithikadatta, J.; Doble, M. Local and systemic adverse effects of nanoparticles incorporated in dental materials: A critical review. *Saudi Dent. J.* **2024**, *36*, 158–167. [CrossRef] [PubMed] [PubMed Central]
15. Djearamane, S.; Xiu, L.J.; Wong, L.S.; Rajamani, R.; Bharathi, D.; Kayarohanam, S.; De Cruz, A.E.; Tey, L.H.; Janakiraman, A.K.; Aminuzzaman, M.; et al. Antifungal properties of zinc oxide nanoparticles on Candida albicans. *Coatings* **2022**, *12*, 1864. [CrossRef]
16. Reddy, G.K.K.; Padmavathi, A.R.; Nancharaiah, Y.V. Fungal infections: Pathogenesis, antifungals and alternate treatment approaches. *Curr. Res. Microb. Sci.* **2022**, *3*, 100137. [CrossRef]
17. Ghannam, H.; Chahboun, A.; Turmine, M. Wettability of zinc oxide nanorod surfaces. *RSC Adv.* **2019**, *9*, 38289–38297. [CrossRef] [PubMed] [PubMed Central]
18. Page, M.J.; McKenzie, J.E.; Bossuyt, P.M.; Boutron, I.; Hoffmann, T.C.; Mulrow, C.D.; Shamseer, L.; Tetzlaff, J.M.; Akl, E.A.; Brennan, S.E.; et al. The PRISMA 2020 statement: An updated guideline for reporting systematic reviews. *BMJ* **2021**, *372*, 71. [CrossRef]
19. Cierech, M.; Wojnarowicz, J.; Szmigiel, D.; Baczkowski, B.; Grudniak, A.M.; Wolska, K.I.; Lojkowski, W.; Mierzwinska-Nastalska, E. Preparation and characterization of ZnO-PMMA resin nano-composites for denture bases. *Acta Bioeng. Biomech.* **2014**, *18*, 2. [CrossRef]
20. Cierech, M.; Klimek, L.; Trojanowska, I. Significance of polymethylmethacrylate (PMMA) modification by zinc oxide nanoparticles for fungal biofilm formation. *Int. J. Pharm.* **2016**, *510*, 323–335. [CrossRef]
21. Cierech, M.; Wojnarowicz, J.; Kolenda, A.; Krawczyk-Balska, A.; Prochwicz, E.; Wozniak, B.; Lojkowski, W.; Mierzwinska-Nastalska, E. Zinc oxide nanoparticles cytotoxicity and release from newly formed PMMA–ZnO nanocomposites designed for denture bases. *Nanomaterials* **2019**, *9*, 1318. [CrossRef] [PubMed]
22. Kamonkhantikul, K.; Arksornnukit, M.; Takahashi, H. Antifungal, optical, and mechanical properties of polymethylmethacrylate material incorporated with silanized zinc oxide nanoparticles. *Int. J. Nanomedicine* **2017**, *12*, 2353–2360. [CrossRef] [PubMed] [PubMed Central]
23. Apip, C.; Wong, L.J.; Alhabeeb, A.; Houbraken, J.; Janakiraman, A.K.; Jayapalan, J. An in vitro study on the inhibition and ultrastructural alterations of Candida albicans biofilm by zinc oxide nanowires in a PMMA matrix. *Saudi Dent. J.* **2021**, *33*, 944–953. [CrossRef] [PubMed]
24. Anwander, M.; Rosentritt, M.; Schneider-Feyrer, S.; Hahnel, S. Biofilm formation on denture base resin including ZnO, CaO, and TiO_2 nanoparticles. *J. Adv. Prosthodont.* **2017**, *9*, 482–485. [CrossRef]
25. Raj, I.; Mozetic, M.; Jayachandran, V.P.; Jose, J.; Thomas, S.; Kalarikkal, N. Fracture resistant, antibiofilm adherent, self-assembled PMMA/ZnO nanoformulations for biomedical applications: Physico-chemical and biological perspectives of nanoreinforcement. *Nanotechnology* **2018**, *29*, 305704. [CrossRef]
26. Gad, M.M.; Abu-Rashid, K.; Alkhaldi, A.; Alshehri, O.; Khan, S.Q. Evaluation of the effectiveness of bioactive glass fillers against Candida albicans adhesion to PMMA denture base materials: An in vitro study. *Saudi Dent. J.* **2022**, *34*, 730–737. [CrossRef]
27. Gad, M.M.; Fouda, S.M. Current perspectives and the future of Candida albicans-associated denture stomatitis treatment. *Dent. Med. Probl.* **2020**, *57*, 95–102. [CrossRef]
28. Nam, K.Y.; Lee, C.H.; Lee, C.J. Antifungal and physical characteristics of modified denture base acrylic incorporated with silver nanoparticles. *Gerodontology* **2012**, *29*, e413–e419. [CrossRef]
29. Sharma, N.; Jandaik, S.; Singh, T.G.; Kumar, S. Chapter 14—Nanoparticles: Boon to mankind and bane to pathogens. In *Nanobiomaterials in Antimicrobial Therapy*; Grumezescu, A.M., Ed.; William Andrew Publishing: Norwich, NY, USA, 2016; pp. 483–509, ISBN 9780323428644.
30. Akay, C.; Avukat, E. Effect of nanoparticle addition on polymethylmethacrylate resins. *Acta Sci. Dent. Sci.* **2019**, *3*, 91–97.

31. Szerszeń, M.; Cierech, M.; Wojnarowicz, J.; Górski, B.; Mierzwińska-Nastalska, E. Color stability of zinc oxide poly(methyl methacrylate) nanocomposite—A new biomaterial for denture bases. *Polymers* **2022**, *14*, 4982. [CrossRef]
32. Gad, M.M.; Abualsaud, R.; Khan, S.Q. Hydrophobicity of Denture Base Resins: A Systematic Review and Meta-analysis. *J. Int. Soc. Prev. Community Dent.* **2022**, *12*, 139–159. [CrossRef] [PubMed]

prosthesis

MDPI

Systematic Review

Clinical Effects of Interproximal Contact Loss between Teeth and Implant-Supported Prostheses: Systematic Review and Meta-Analysis

James Carlos Nery [1,*], Patrícia Manarte-Monteiro [1], Leonardo Aragão [2], Lígia Pereira da Silva [1], Gabriel Silveira Pinto Brandão [1] and Bernardo Ferreira Lemos [1]

[1] FP-I3ID, Faculty of Health Sciences, University Fernando Pessoa, 4200-150 Porto, Portugal; patmon@ufp.edu.pt (P.M.-M.); ligia@ufp.edu.pt (L.P.d.S.); 2022100042@ufp.edu.pt (G.S.P.B.); blemos@ufp.edu.pt (B.F.L.)

[2] Department of Physics and Astronomy "Augusto Righi", Università di Bologna, Via Irnerio 46, 40126 Bologna, Italy; leonardo.aragao@unibo.it

* Correspondence: 2022100344@ufp.edu.pt

Abstract: Dental rehabilitation with implants is a clinical reality in clinical practice. The Interproximal Contact Loss (ICL) between implant-supported prostheses adjacent to natural teeth is a relatively common occurrence. This systematic review and meta-analysis aims to evaluate the possible clinical effects of the periodontium regarding the ICL between teeth and implanted-supported prostheses. We also identified the main ICL assessment tools described in the literature. This study was registered on the PROSPERO (CRD42023446235), was based on the PICO strategy, and followed the PRISMA guidelines. An electronic search was carried out in the PubMed, B-on, Google Scholar, and Web of Science databases without setting a time limit for publications. Only systematic reviews and comparative clinical trials were included and analyzed. Nineteen publications were eligible for meta-analysis, with thirteen retrospective and six prospective clinical trials. A total of 2047 patients and 7319 prostheses in function were evaluated, and ICL was found in 51% with a confidence interval of 0.40 to 0.61. As ICL assessment tools, dental floss was used in 65%, matrices were used in 30%, and X-ray images were used in 5% of cases. The clinical follow-up ranged from 1 to 21 years, with 50% between 1 and 3 years, 25% between 3 and 10 years, and 25% between 10 and 21 years. ICL was found to occur more frequently in the mandible. No statistically significant difference existed between the anterior (55%) and posterior (47%) oral regions. On the mesial surface, ICL ranged from 13% to 81.4%, possibly due to the different follow-up periods and the diversity of methods used in the assessment. No differences were found for ICL between single or multiple implanted-supported prostheses. Food impaction was the most common effect of ICL and was more prevalent on the implant-supported prosthesis's mesial surface in the mandible's posterior region. There was evidence of peri-implant mucositis but without progression to peri-implantitis, and the form of retention or the number of elements was not relevant.

Keywords: dental implant; implant-supported prosthesis; loss of proximal contact; loss of interproximal contact; open contact; adjacent to natural tooth; food impaction

1. Introduction

Natural teeth show physiological tooth migration and movements that occur from tooth eruption, including functional inclination in the alveolus and the possibility of adapting to physiological and/or functional needs, whether horizontally, vertically, or rotationally. Occlusion is an important factor in maintaining interproximal contact surfaces and the physiological homeostasis of the periodontium [1,2].

The periodontal ligament keeps the tooth in union with the bone, allowing physiological movement. In contrast to teeth, implants have direct contact with the bone and are not subject to the mobility experienced by teeth. In implant rehabilitation, whether single or multiple, when adjacent to a natural tooth, this difference in behavior between the mobility of the tooth and the immobility or ankylosis of the implant is evident [3–5]. Figure 1a shows the relationship between an implant located in position 35 and perfect harmony in interproximal contact between the implant prosthesis and the adjacent natural teeth, with a healthy peri-implant and periodontal region.

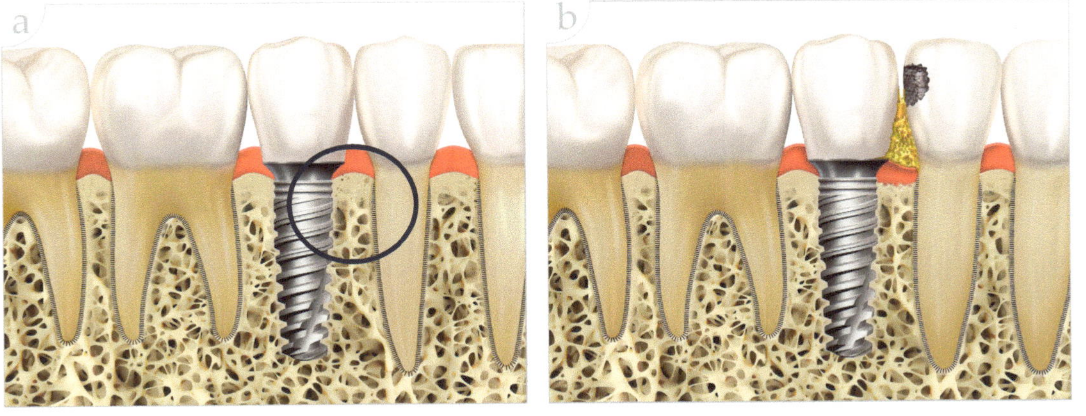

Figure 1. (**a**) Relationship between implant in position 35 and contact between implant prosthesis and adjacent natural teeth. The healthy peri-implant and periodontal regions are shown in the black circle. (**b**) Implant in position 35, interproximal contact loss (ICL) between prosthesis on implant and natural tooth with food impaction, caries in 34, and loss of gingival volume are visible.

The space opening induced by the implant-supported prosthesis adjacent to the natural tooth, also known as Interproximal Contact Loss (ICL), represents a typical clinical situation and has been reported in the literature, especially in systematic reviews, retrospective clinical studies, prospective publications, clinical case reports, and some classifications have been proposed [6,7]. Some ICL complications, such as Food Impaction (FI), caries lesions in the adjacent tooth, gingival inflammation, periodontal changes, and, in some studies, loss of bone support, have been reported. Various methods have been proposed to assess the measurement of the interproximal contact or ICL [8,9].

In order to prevent the installation and/or progression of those alterations, it is necessary to perform the monitoring of the interproximal space over time and adjust the alterations as soon as possible [10,11]. It is important to highlight the need for the evaluation of diagnostic and treatment methods in order to avoid ICL triggering and prevent its occurrence. This is shown in Figure 1b: Due to the lack of a contact point on the mesial face of the implant prosthesis adjacent to the natural tooth, Food Impaction (FI) can occur, causing damage to the gingival tissue and promoting demineralization of the tooth with the deposition of decayed tissue and the appearance of changes in the gingival mucosa that can evolve and generate bone loss in the damaged area [11–15].

For those purposes, this study aims to evaluate the possible clinical effects of the periodontium on the ICL between teeth and implanted-supported prostheses and also to identify the main ICL assessment tools described in the literature.

2. Materials and Methods

This systematic study formulated the research question, "Does the interproximal contact loss influence the outcomes for periodontium and other clinical effects?", based on the PICO model: Population—patients who received single or multiple implanted-supported prothesis adjacent to natural teeth, Intervention—assessment tools to evaluate and compare interproximal spaces between natural teeth and prostheses on implants, Comparison or

Control—interproximal spaces between natural teeth, and Outcome—clinical effects for the peri-implant and periodontal tissues of the probable loss interproximal contact between the natural tooth when adjacent to implanted-supported prosthesis over time. The following null hypothesis was formulated: the loss of proximal contact between the natural tooth and the prosthesis on implants has no clinical effects on the periodontium. This systematic review was registered on the International Prospective Register of Systematic Reviews (PROSPERO, CRD42023446235) and followed the PRISMA guidelines as described in Figure 2 [16].

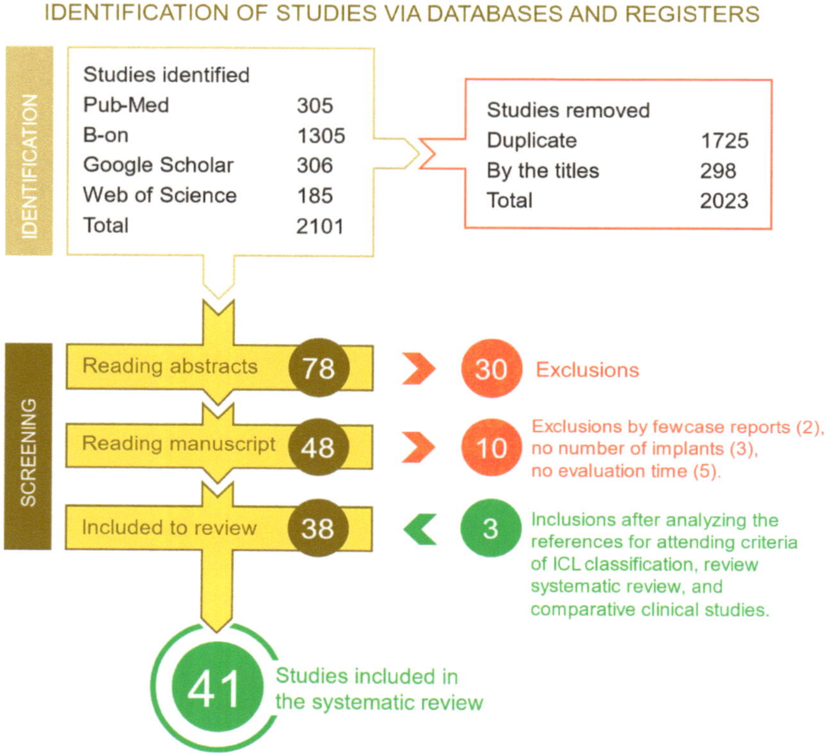

Figure 2. PRISMA flowchart of the strategy used in the literature search.

2.1. Inclusion Criteria, Exclusion Criteria and Eligibility

Prospective and retrospective observational clinical studies in humans with a follow-up of at least 1 year, with the presence of implant-supported dental crowns adjacent to natural teeth, were considered, which included assessment tools for the determination of interproximal space/contact, measurement of the contact gap, location of the gap in the oral cavity (maxilla, mandible, anterior, and/or posterior oral region), type of rehabilitation retention and number of prostheses (single or multiple), and clinical effects of the ICL. We excluded experimental laboratory studies, animal studies, case reports, and studies that did not include clinical data with follow-up for at least one year.

2.2. Search Strategies

Searches were carried out using the electronic databases PubMed/MEDLINE, B-on, Google Scholar, and Web of Science. The search strategy was conducted by two examiners (JCN and GB), and data collection was based on keywords and the Boolean operators "AND" and "OR": (dental implant or dental implants or implant-supported prosthesis) and "(loss of proximal contact or loss of interproximal contact or open contact or adjacent to natural tooth)" and "(food impaction)".

2.3. Selection and Evaluation of Studies

The Rayyan online platform was used to manage the references [17]. After selection, the articles were screened, and duplicates were excluded by the two examiners. The titles were then read, and the exclusions were made. The abstracts were read by the two examiners (JCN and GB), and when there were discrepancies, a third examiner (BL) was consulted. Potentially eligible texts were read by both researchers (JCN and GB), and at all stages, the Rayyan software was blinded (https://rayyan-ios.soft112.com accessed on 16 May 2024). The JCN researcher was responsible for deciding whether to include or exclude articles, and the second researcher (GB) was responsible for reviewing the information. Using the snowball technique, new articles were included after reading the bibliographical references of the articles selected for analysis. After this stage, the selected articles were exported to the Mendeley desktop 2.93.0 version.

Two thousand and twenty-one publications were collected, which, after excluding duplicates, resulted in 376 eligible papers. After reading the title, 78 studies were selected for evaluation of the abstract, and then a total of 30 publications were excluded. A total of 48 articles were identified for full-text reading. In addition, 10 publications were excluded (Table 1) because 2 were case reports with no follow-up period, 3 did not mention the number of implants placed, and 5 studies were laboratory comparisons using finite elements and superimposed models with no clinical evidence (Figure 2).

Table 1. Exclusion criteria and number of studies excluded after reading the manuscripts.

Exclusion Criteria	Number of Studies Excluded
Clinical data < 1 year follow up	2
Case reports without number of implants	3
Experimental laboratory studies	5
Animal studies	0

With the final reading of the articles by the two researchers (JCN and GB), 3 additional articles were included in the review as a result of reading the bibliographical references in the publications analyzed (ICL classification, systematic review, comparative clinical study). Thus, 41 studies were included in this systematic review, according to the PRISMA flowchart illustrated in Figure 2. The recommended methodology was used for the meta-analysis [16]. The CONSORT checklist, in Table 2, was used for the qualitative assessment of the studies [17].

The quality of the articles published by CONSORT criteria was assessed as Poor when they had a score of less than 10.50, Average when they had a score between 10.50 and 21, Good when they had a score between 22 and 31.50, and Excellent when they scored between 31.50 and 43 of the items analyzed [18]. Of the articles selected, 75% met the criteria between 16 and 21 (15 articles) and were therefore classified as average quality, while 25% scored between 22 and 31.50 (five articles) considered as good quality. It is important to point out that the fulfillment of the constant items showed that the risk of bias of the articles met the requirements for the strategy to be between average and good quality.

3. Results

The characteristics of the thirteen retrospective and six prospective studies evaluated are described in Table 3. A total of 2047 patients were described, except for the study [18] which did not provide information on the number of participants, with a total of 7319 restorations evaluated.

The meta-analysis results highlighted an effect size of 51% (confidence interval between 0.40 and 0.61) and heterogeneity: $I^2 = 99\%$, $\tau^2 = 0.84$, $p < 0.01$. However, two studies [1,18] showed extreme values of 13 and 85% (Figure 3).

Two studies did not report whether the implants were placed in the maxilla or mandible [8,19], and in another two studies, they were placed only in the mandible [20,21]. Another 15 studies reported the placement of implants in the maxilla or mandible in the same patient [1,7,22–34].

Table 2. Checklist for analyzing the methodology and risk of bias of the studies analyzed based on CONSORT (Consolidated Standards of Reporting Trial). The cross-marked studies (×) show compliance with the parameter assessed; blank cells represent non-compliance with the item analyzed. The omitted parameters 1b, 2b, 4a, 4b, 5, 12a, 14a, 14b, 19, 20, 21, and 22 shows compliance in all studies analyzed.

Study/Criteria	1a	2a	3a	3b	6a	6b	7a	7b	8a	8b	9	10	11a	11b	12b	13a	13b	15	16	17a	17b	18	23	24	25
[1] Yen et al. (2020)		×														×			×				×		
[7] Byun et al. (2015)		×					×								×	×			×				×		×
[8] Bombolaki et al. (2020)		×					×								×	×			×				×		
[14] Varthis et al. (2015)		×					×								×	×			×				×		
[15] Saber et al. (2020)		×					×								×	×			×				×		
[19] French et al. (2019)		×					×								×	×			×				×		
[20] Latimer et al. (2020)	×	×	×		×		×	×					×		×	×			×				×	×	
[22] Chanthassan et al. (2020)		×	×		×		×								×	×			×				×	×	
[23] Wong et al. (2015)		×			×		×								×	×			×				×		
[24] Liang et al. (2020)		×	×		×		×								×	×			×				×	×	
[25] Shi et al. (2019)		×	×		×		×	×					×	×	×	×			×				×	×	×
[26] Wei et al. (2008)					×		×								×				×						
[27] Koori et al. (2010)		×			×		×								×	×			×				×		×
[28] Pang et al. (2017)		×	×		×		×								×	×			×				×	×	×
[29] Wolfart et al. (2021)	×	×	×		×		×								×	×			×				×	×	×
[30] Manicone et al. (2021)		×	×		×		×								×	×			×						
[31] Jeong and Chang (2015)		×					×								×	×			×				×		×
[32] Abduo et al. (2021)	×	×	×		×		×								×	×	×		×				×	×	×
[33] Ren et al. (2016)		×	×		×		×								×	×			×				×		
[34] Kandathilparambil et al. (2020)		×			×		×								×	×			×				×	×	

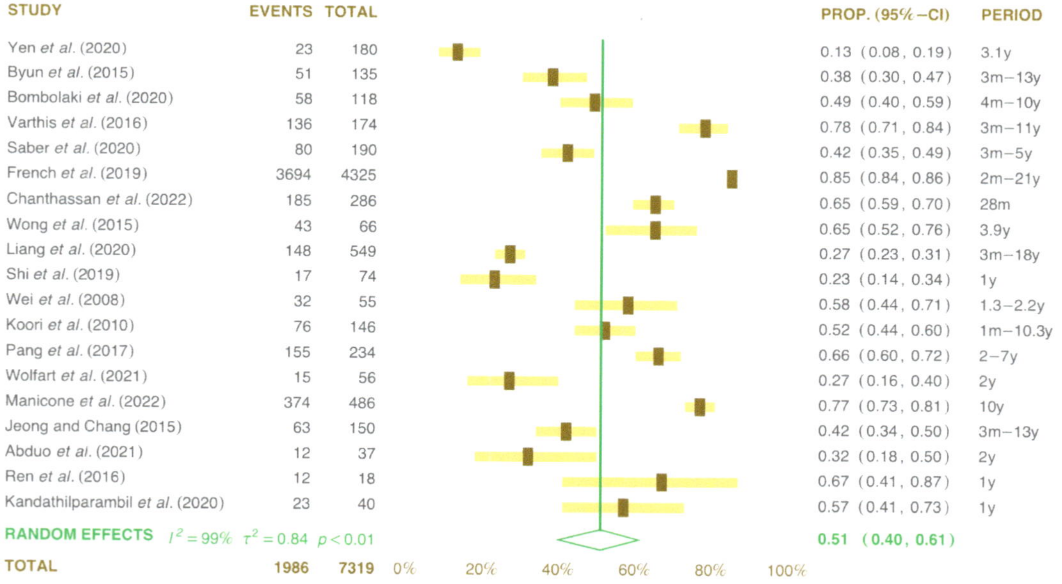

Figure 3. Forest plots: total cumulative results of events and number of prostheses installed from 13 retrospective and 6 prospective clinical studies evaluated by meta-analysis. The green diamond and solid line represent the results of all studies together [1,7,8,14,15,19,22–34].

Table 3. A description of the studies included in the systematic review and meta-analysis.

	Study	Type	TP	Mx Md	PN	A/P	M	D	Diagnostic	Prot	Ret	FI	Bio Alt	Evaluation Period
[1]	Yen et al. (2020)	Re	147	Mx/Md	180	AP	13.0%	2.3%	Rx 50 mm	S-M	CR	OK	NA	3.1 years
[7]	Byun et al. (2015)	Re	94	Mx/Md	135	AP	38.1%	24.6%	Dental floss	S-M	CR	OK	SA	3 months at 13 years
[8]	Bombolaki et al. (2020)	Re	83	33 Mx 50 Md	118	P	48.8%	26.7%	12 mm strips	S	SR	OK	SA	4 months at 10 years
[14]	Varthis et al. (2015)	Re	128	57.9% Ma 49.0% Md	174	AP	78.2%	21.8%	70 µm floss + Rx	S	CR	OK	NA	3 months at 11 years
[15]	Saber et al. (2020)	Re	83	34.1% Ma 31.5% Md	190	AP	42.1%	13.5%	70 µm floss + Rx	S	CR	OK	Mucosite	3 months at 5 years
[19]	French et al. (2019)	Re	NR	58.0% Mx 42.0% Md	4325	AP	85.4%	11.6%	50 µm floss + Rx	S-M	CR	OK	Mucosite	2 months at 21 years
[22]	Chanthassan et al. (2020)	Re	178	Mx Md	286	P	64.8%	35.2%	Floss	S	CR	OK	Mucosite	28 months
[23]	Wong et al. (2015)	Re	45	18 Mx 48 Md	66	P	65.0%		Matrix 38 µm	S-M	CR	OK	NA	3.9 years
[24]	Liang et al. (2020)	Re	317	Mx Md	549	P	27.0%	5.0%	Dental floss	S-M	CR	OK	NA	3 months at 18 years
[25]	Shi et al. (2019)	Pr	74	34 Ma 40 Md	74	P	23.0%	25.7%	Dental floss	S	CR	OK	NA	1 year
[26]	Wei et al. (2008)	Re	28	66 Mx 54 Md	55	AP	58.0%		Matrix 50 µm	NA	NA	OK	NA	1.3 to 2.2 years
[27]	Koori et al. (2010)	Re	105		146	AP	51.8%	15.6%	Matrix 50 µm	S-M	NA	OK	NA	1 month at 10.3 years
[28]	Pang et al. (2017)	Pr	150	122 Ma 177 Md	234	P	66.2%	36.9%	Matrix 50 to 100 µm	S-M	CR	OK	SA	2 at 7 years
[29]	Wolfart et al. (2021)	Pr	41	Mx Md	56	P	26.4%	14.6%	Matrix 50 µm	S	SR	OK	SA	2 years
[30]	Manicone et al. (2021)	Re	320	Mx Md	486	AP	77.0%	23.0%	Floss + Rx	S-M	CR	OK	Mucosite	10 years
[31]	Jeong and Chang (2015)	Re	100	Mx/Md	150	AP	42.0%		Dental floss	S-M	CR	OK	Bio Alt	3 months
[32]	Abduo et al. (2021)	Pr	35	NR	37	P	63.2%	36.2%	Dental floss	S	SR	NA	SA	2 years
[33]	Ren et al. (2016)	Pr	18	NR	18	P	64.8%		Matrix 50 µm	S	CR	OK	NA	1 year
[34]	Kandathilparambil et al. (2020)	Pr	40	Md	40	P	57.9%	38.9%	Matrix 50 µm + dig	S	CR	OK	NA	1 year

Legends: [Re] Restrospective; [Pr] Prospective; [TP] Total participants; [Mx] Maxilla; [Md] Mandible; [PN] Prostheses Number; [AP] Anterior-Posterior; [P] Posterior; [M] Mesial; [D] Distal; [S] Single; [S-M] Single-Multiple; [CR] Cement-retained; [SR] Screw-retained; [FI] Food Impaction; [Bio Alt] Biological alterations; [SA] No changes; [NA] Not evaluated. [OK] Evaluated.

The implants supported single and/or multiple prostheses. Ten studies with single prostheses [8,18,19,22,25,28–32] and the other nine did not distinguish whether the implants supported were single/multiple crowns [1,7,18,25,26,33,34]. In the meta-analysis, a result of 50% was obtained for single prostheses (confidence interval 0.38 to 0.62) and heterogeneity: $I^2 = 91\%$, $\tau^2 = 0.52$, $p < 0.01$. The analysis of single-multiple prostheses showed a result of 52% (confidence interval 0.35 to 0.69) and heterogeneity: $I^2 = 99\%$, $\tau^2 = 1.15$, $p < 0.01$ (Figure 4).

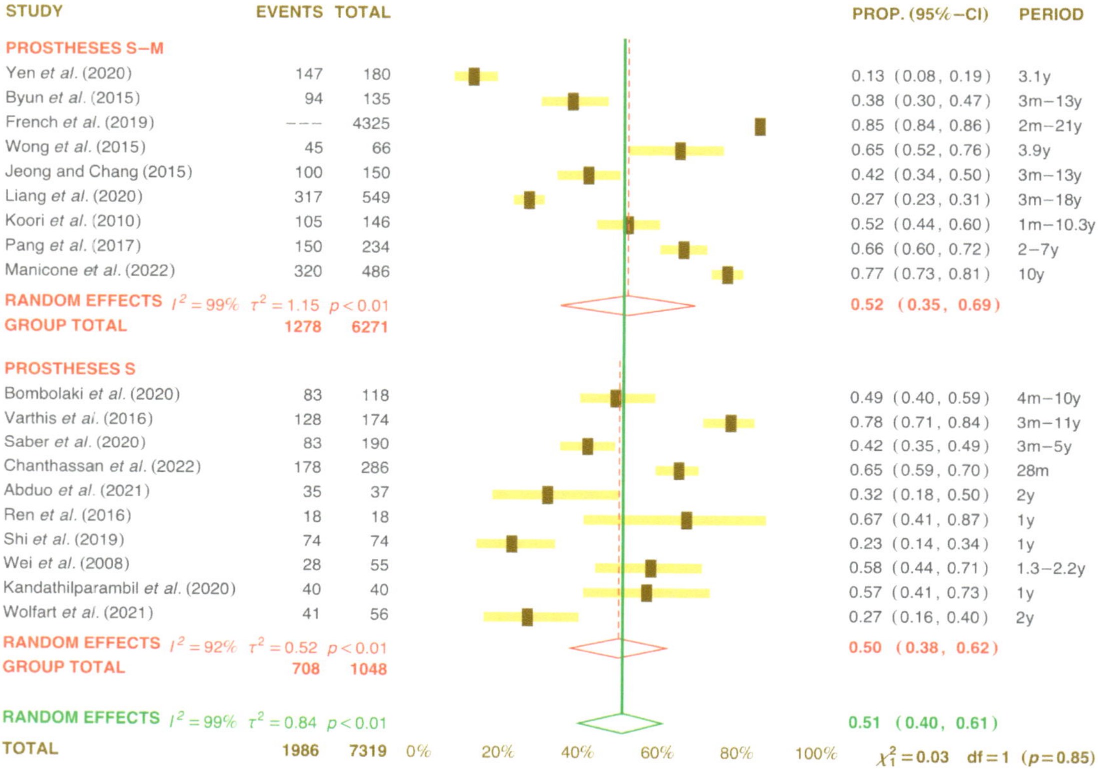

Figure 4. Forest plots: Analysis of loss of contact point according to the number of prostheses per implant. This is demonstrated by the graph in sub-division by Single/multiple (S-M) in 9 studies and Single (S) Prostheses in another 10 studies. The green diamond and solid line represent the results of all studies together, while the red diamonds and dotted lines represent each group's results [1,7,8,14,15,19,22–34].

As for implant location, in the anterior (A) or posterior (P) intra-oral region, nine studies reported placement in A-P [1,7,14,15,19,24,26,33,34], while ten studies were identified for the P region [8,22,25,27–32]. The results for region P were 47% (confidence interval 0.36 to 0.59) and heterogeneity: $I^2 = 95\%$, $\tau^2 = 0.51$, $p < 0.01$. In the publications that did not distinguish between A and P, the result was 55% (confidence interval 0.37 to 0.71) and heterogeneity: $I^2 = 99\%$, $\tau^2 = 1.12$, $p < 0.01$ (Figure 5).

For the type of retention, whether screw-retained (SR) or cement-retained (CR), three studies were found with screw-retained prostheses [8,28,30]. However, thirteen did not detail whether they were cement-retained or screw-retained [1,15,18,19,22,25,26,29,31,32], and another three did not report the type of retention used [7,33,34]. In the meta-analysis, for SR prostheses, we found a result of 37% (confidence interval 0.26 to 0.50) and heterogeneity: $I^2 = 77\%$, $\tau^2 = 0.13$, $p < 0.01$. In the CR analysis, we found a result of 55% (confidence interval 0.40 to 0.68) and heterogeneity: $I^2 = 99\%$, $\tau^2 = 1.06$, $p < 0.01$. Meta-analysis was also carried out on publications that did not report the type of retention and found a result of 48% (confidence interval 0.39 to 0.58) and heterogeneity: $I^2 = 77\%$, $\tau^2 = 0.08$, $p < 0.01$, results in Figure 6.

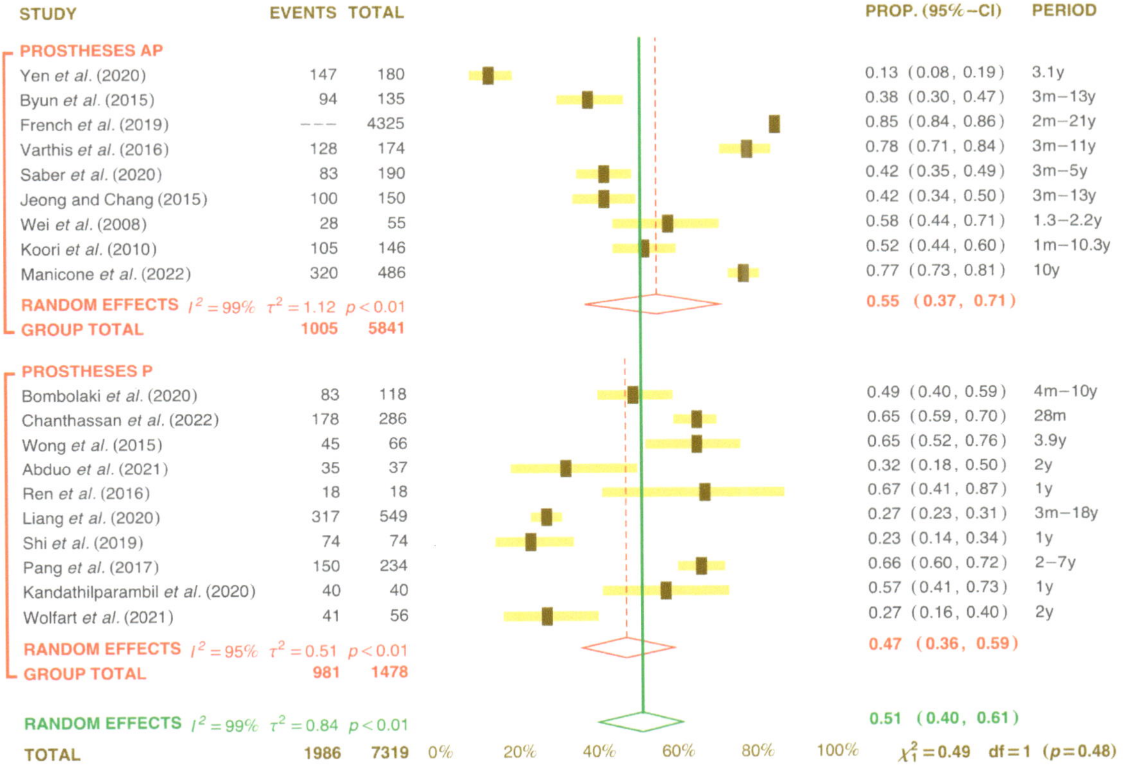

Figure 5. Forest plots: Analysis of loss of contact point according to the location of the prostheses. Demonstrated by graphs in subgroups by Antero-Posterior (AP) location in 9 studies and Posterior (P) in another 10 studies. The green diamond and solid line represent the results of all studies together, while the red diamonds and dotted lines represent each group's results [1,7,8,14,15,19,22–34].

To measure the interproximal space/contact, several methods as assessment tools were registered in 11 publications. Dental floss was the most validated [7,8,14,18,19,22,26,28,31,32,34] in 8 studies used metal matrix with different thicknesses [20,25,29,33,34]. In addition, 3 apical radiography images were taken [1,18,30], and 1 study used superimposition comparisons of digitalized models over time [35]. There was also a combination of more than one assessment tool [15,18,20,36], and in all of the studies, radiographs were used for follow-up at the control returns after the interproximal spaces had been closed. Since there were several ways to assess the interproximal space (the opening of the interproximal contact point), the meta-analysis of this variable was not carried out.

Various studies also recorded the time elapsed between the placement of the dental prosthesis and the interproximal opening space between the implant-supported prothesis and the adjacent tooth, which ranged from three months [32,33] to more than five years [7,24] or even ten years [19,27,35] or more, after installation [8,14,19]. A progressive increase in this opened space was observed over the time between installation and detection of the alteration [26,34].

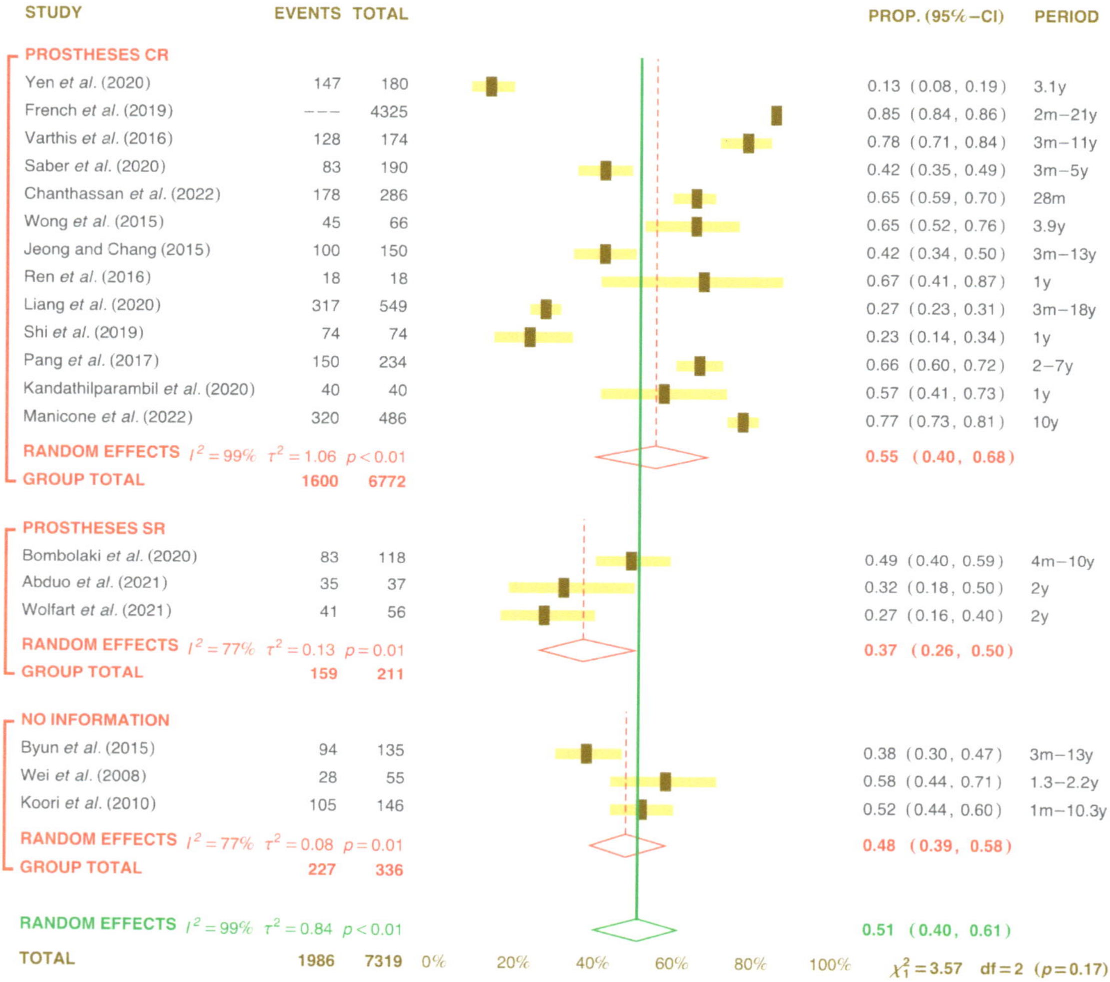

Figure 6. Forest plots: Analysis of loss of contact point according to the type of retention of the prostheses. Demonstrated by the graph in subgroups of cement-retained prostheses (CR), observed in 13 studies; retained by screws (SR) in 3 studies; no information regarding the type of retention in 3 studies. The green diamond and solid line represent the results of all studies together, while the red diamonds and dotted lines represent each group's results [1,7,8,14,15,19,22–34].

Food Impaction (FI) is known as a condition that influences and changes the gingival tissues. No gingival modifications were found in 12 studies [1,8,15,23,31–34]. Of the seven studies in which gingival conditions varied, six found gingival alterations [14,19,22,25,36], and one study found bone alterations [7]. In the meta-analysis, periodontal and peri-implant conditions were shown in 7 publications as alterations ranging from mucositis to loss of bone insertion, with biological variations of 60% (confidence interval 0.45 to 0.73) and heterogeneity: $I^2 = 99\%$, $\tau^2 = 0.63$, $p < 0.01$. In the other 12 publications, which reported no biological variations, a value of 45% was obtained (confidence interval 0.33 to 0.59) and heterogeneity: $I^2 = 96\%$, $\tau^2 = 0.83$, $p < 0.01$ (Figure 7).

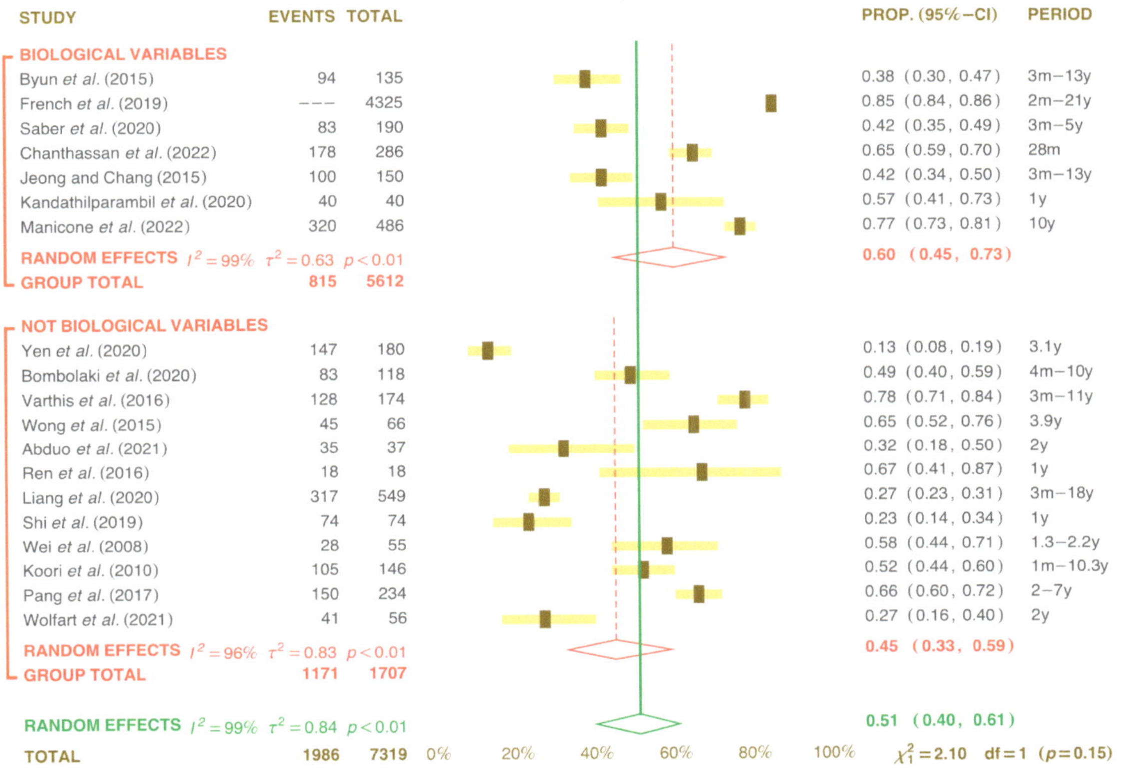

Figure 7. Forest plots: Analysis of subgroups in terms of loss of contact points and the occurrence of biological changes or no biological changes, subdividing the studies into two subgroups and demonstrating the proximity of both to the graph's central diamond. The green diamond and solid line represent the results of all studies together, while the red diamonds and dotted lines represent each group's results [1,7,8,14,15,19,22–34].

The clinical follow-up time varied between 1 and 21 years, and six trials recorded a follow-up of up to 3 years [23,28,29,32–34], seven studies between 3 and 10 years [1,8,19,23, 27,28,30] and another five publications [7,15,18,24,31] between 10 and 21 years of recall.

4. Discussion

Evaluating the periodontal and peri-implant consequences of ICL is important once it can be highly correlated with multiple aspects, such as the location of the prosthesis, type of retention, number of elements, time elapsed, and opening of interproximal space on periodontal health area.

In the present analysis, for the occurrence of ICL regarding location and type of retention, no changes were found for clinical effects in the region and for trigger changes in the periodontal or peri-implant area or even for the patient's general comfort [14,19,31].

Figure 3 highlighted the studies [1,7,24,25,29] that were outside the confidence interval of this analysis. Also, other studies [15,18,30] revealed a deviation from the confidence interval between 0.4 and 0.6.

4.1. Evaluation of Interproximal Contact Loss (ICL)

The assessment of ICL was carried out using different assessment tools, which prevented the authors from carrying out the meta-analysis. The methods varied from dental flossing, X-ray image evaluation, matrix strips of different dimensions, and clinical evaluation by routine examinations [3] or according to patients' reports [20,21]. Large discrepancies in the diagnosis (between 13% and 66%) were found in some studies [1,28], which can be explained by the lack of protocol for the analysis of resistance to the passage of dental floss of different thicknesses (50 to 70 μm) as used as the main tool for ICL detection [10,18,20,22,28,37]. Other studies qualified passage with minimal or no resistance to passage [7,8]; other studies scored this dental floss resistance with numbers 0, 1, 2, 3 [24,30,33] or with the letters a, b, and c [31,32].

The ICL was also measured using metallic strips [23,26,27,34,38] with a thickness variation of 0.1 to 0.5 mm as a tool to better measure the interproximal mesial–distal (M-D) space distance; an apical–coronal [20] metallic wire of different thicknesses was also used [18]. Dental floss was used to detect ICL and, at the same time, used to overlay the scanned models. This methodology was important to assess changes that occurred not only in the mesiodistal direction but also in other directions [3].

In the X-ray image evaluation, the measurement was carried out by checking the contact between the dental crowns and also in the follow-up after corrections. This tool assessed the bone measurement, which was taken from the implant shoulder (platform) to the bone crest closest to the implant, without detailing whether the implants were Bone Level or Tissue Level [1,14,15,19,37,39,40]. It is important to highlight the distortions that radiographic image evaluations can generate due to the difficulty of calibration and measuring the same position when repeating the technique.

Visual clinical assessment was carried out in all analyzed publications. The wide variation in ICL assessment methods may explain the discrepancies found in 13% [1], 23% [41], 65% [23], or even 78% in one reported trial [18]. The importance of establishing a standardized methodology for evaluating ICL is highlighted so that the results are more faithful and, possibly, have a smaller margin of distortion. It is also essential to perform some recommendations to the patient on the possibility of ICL occurrence, the clinical signs, and the need to return to the dentist for evaluation, which must be part of the guidelines and good practices in the prosthetic rehabilitations and delivery protocol.

4.2. Food Impaction (FI)

FI was found in all studies where ICL was evaluated, with confirmation of FI reported by the patients between 32% and 56% [20] of the studied populations [14,18,22,37]. However, in the studies in which there was no FI reported by the patients [7], gingival changes were still observed, such as mucositis and peri-implant changes [8,15,19,22,42]. The importance of monitoring these changes is reiterated, and if the change occurs, it must be monitored according to the loss of supporting tissue; observation may be necessary [11] in case of limited tissue loss or clinical return for evaluation and correction of cases of peri-implant disease [11,38,43] (Figure 1b).

If changes occur without loss of supporting tissue, this should only be monitored [11]. The patient must be warned of the possibility of FI occurrence and the need for periodic clinical monitoring in order to evaluate and adjust where cavities and peri-implant disease may be present [10,11,38,43]. It is still well observed that AI was the main cause in the development of cavities on the distal surfaces of the teeth when they were adjacent to the implant and highlights the need for information about this fact to the patient to justify strategies to prevent the development of carious lesions, such as fluorides, adequate brushing, flossing and the use of silver diamine fluoride [44].

4.3. Factors for Interproximal Contact Loss (ICL)

The contact surfaces must be correctly distributed in the buccal-palatal/lingual and apical–coronal directions for the correct performance of gingival protection functions, maintenance of the interproximal space [2,4] and assistance in the correct transmission of occlusal force to the bone region [10,26]. Incorrect interproximal surface contact or anatomical contouring were reported as predictive for ICL [1,6,7,9,10,15,22–24,29,37,41].

In a finite element study, it was detected that the correct anatomical contour of the interproximal surface contact contributed to better load distribution on the adjacent tooth and the implanted-supported dental crown [4]. The mesial region of the prosthesis over the implant was the most affected by the ICL [7,8,14,15,18,20,22,24,27,28,30,34]; however, a high variation was detected, from 10% [34] to 78% [14].

Regarding the location of the implant in the dental arch, ICL occurred in the maxilla from 14% [19] to 30% [8], but it was more frequently detected in the posterior region of the mandible, which ranged from 15% [1] to 54% [20]. It should be noted that in some publications, only the posterior region (P) of the mandible was analyzed [33,34]. In other studies, there was no distinction regarding the intra-oral region (A or P). Some studies that analyzed [1,14,30] the AP subgroups and those that [22,24,25,29] were carried out in the P subgroup were outside the diamond confidence limit and stood out.

Occlusion seems to be fundamental for the stability of the interproximal surface contact [1,3,5,9,10,14,18], highlighting the anterior force component [10,19,27] and the occlusion distribution with the antagonist arch [7,10,24,26]. An important finding was that the ICL between single or multiple prostheses, when adjacent to the natural tooth, varied between 16% and 75% [15,20]. The studies highlighted the use of single retention [18,32] and those that did not distinguish whether the prostheses were single or multiple [1,18,24,30] but were outside the confidence interval of 0.35 to 0.69.

The time elapsed between the placement of the prosthesis and the detection of the interproximal space was very variable, being the earliest at 3 months [18,22] and may increase over time [7,8]. On the other hand, it can be activated later, 8 or more years after the prosthesis connection [19]. No differences were found for the type of retention, as prostheses retained by cement or by screws [29], and the occurrence of ICL. However, it was also reported that for those retained by screws, they were easier to maintain; it was easier to reestablish the correct interproximal contact, and they could be restored even with the prosthesis or the adjacent tooth [18,28,37].

Patient age was another topic analyzed in this systematic review [36,38] and highlighted as a determining factor for the establishment of ICL, except in [23], which relates the loss of attachment to be aggravated with advancing age and increased tooth mobility. Some studies analyzed the influence of gender and found that males had a higher prevalence of ICL [1,5,6,15].

4.4. Clinical Effects of Interproximal Contact Loss (ICL)

The main effect reported in ICL was FI; some publications directly stated a relationship. It was detected prior to the evaluation by some patients due to discomfort, and in some cases by filling a questionnaire relating to pain symptoms affecting quality of life [15,19,37]. However, other trial results [20,22,23] showed little perception of FI noted by the patient.

The consequences of FI on the gingival tissues near the tooth or implant-supported prosthesis are not very clear in the literature [1,8,14,26,32,37]. However, inflammation of the periodontal and peri-implant regions was detected by probing the periodontal groove and peri-implant [7,18,19,22,24,26,30,34]. Patients who presented bone loss between the natural tooth and the implanted-supported prosthesis may have an increased risk for ICL [42]. In this meta-analysis, the group with biological changes resulted in 60% and a range from 45% to 73% with a randomized effect of $I^2 = 99\%$, $\tau^2 = 0.63$, $p < 0.01$. However, it should be highlighted that the study [14] was outside the edge of the diamond on the chart. For the group where no biological variations occurred, the meta-analysis resulted in 45% and a range of 33% to 59% with a randomized effect of $I^2 = 99\%$, $\tau^2 = 0.84$, $p < 0.01$, but it should be noted that the study [1] was left out of the diamond graph in Figure 7.

It is important to educate the patient regarding the need for correct hygiene with dental floss [33]. Some publications have shown the development of carious lesions in the natural tooth adjacent to the prosthesis over the implant [2,6,8,10,14,19,22,40,44]. Although ICL was more prevalent in the mesial region of the implant prosthesis than in the distal region, there was an important variation between the literature analyzed as the divergence of results was relevant in the mesial region from 4.1% [39] to 85.4% [15] and distally with a variation of 2.3% [1] to 38.9% [34].

The diversity of interproximal space/contact assessment tools and methods may explain some discrepancies regarding the type of prosthesis retention (screwed or cemented) and as not being related to the ICL occurrence [1,5,10,15,20,23]. There has not yet been confirmation of the influence of the type of retention on the occurrence of ICL [14,18,19, 22,25,26,29,31,32,35,36,41]. The diversity in contact point assessment methods and the discrepancy regarding the type of prosthesis retention may explain the lack of correlation between the types of retention and the ICL, according to the meta-analysis result of 55%, confidence interval from 0.40 to 0.68 ($p < 0.01$); in contrast to only three studies that mention screw-retained prostheses [8,28,38] with a range of 37% (0.26 to 0.50) $p < 0.01$. Three publications stand out in that they do not mention the type of retention applied [7,26,27].

Important arguments were made in [33] so that the retention is screwed, aiming to facilitate the ICL reconstruction, either with light-cured resins [11] or with ceramic additions to the prostheses [10] or, even, through the reconstruction of the adjacent tooth with resin or with inlays/onlays [9]. Correct reconstruction of the interproximal surface contact is essential for reestablishing periodontium homeostasis, for patient comfort, and for the rehabilitation performance over time (Figure 8).

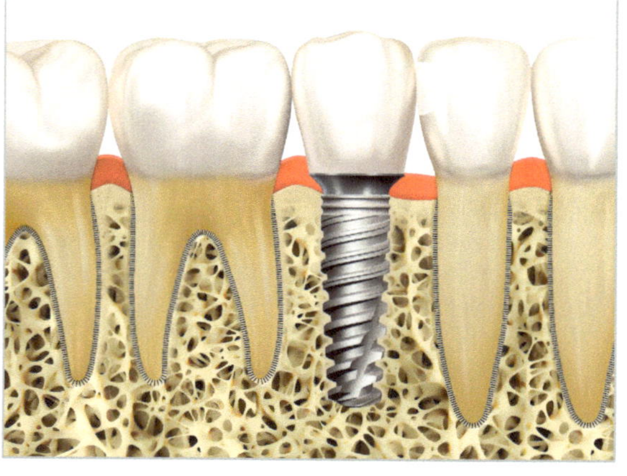

Figure 8. Implant in position 35, removal of caries, and reconstruction of interproximal contact loss (ICL) between prosthesis on implant adjacent to natural tooth. Note the deficient periodontal space.

An important finding reported [34] the impact of using an occlusal splint (2 mm thick) on the incidence of ICL. The study revealed that the group that did not utilize a containment drip had the highest incidence of ICL at 30%, whereas the group that used a containment drip saw a lower incidence of 15% during the study period. These results suggest promising avenues for future research in this area.

The occurrence of ICL in various locations in the oral cavity was analyzed, both in the maxilla and mandible, being mostly found in the mandible in the posterior region in the mesial portion of the prosthesis over the implant [35,36,38]. However, in the meta-analysis, the P region with 47% (0.36 to 0.59) with $p < 0.01$ demonstrates three studies [31,32,38] that escaped the confidence interval of the diamond chart (0.36 to 0.59). Furthermore, 55% (0.37 to 0.71) of the publications did not distinguish between Anterior and Posterior locations with $p < 0.01$, and four studies in this region [1,14,15,30] served outside the confidence interval (0.37 to 0.73) of the diamond chart.

Several hypotheses have been discussed in the literature regarding the ICL occurrence being the most accepted, the passive eruption of the teeth [3,7,10], the continuous growth of the face [10,19], and the anterior force component [1,10,26], as well as the excess/direction of the load that can cause mesialization and/or movement of the tooth [31]. Those considerations should be observed when monitoring each of the cases. To reduce the possibility of ICL, reconstruction and correct adjustment of contact points are essential for adequate occlusal load distribution to the tooth and bone tissue, as well as to the implant and underlying region, promoting homeostasis in those fields [2,6,24,37] and minimizing the occurrence of ICL [4,5], a greater adjustment or tightening of the surface contact is also recommended [33]. Tooth wear surfaces due to passive eruption must be observed, and the neutralization of the anterior load component should be monitorized [10].

Due to the high prevalence of ICL, it is always recommended that implant prostheses be retained with screws to facilitate the reconstruction of contact in the region, whether by adding them to the tooth or to the prosthesis. Some authors [27] also suggested the inclusion in recommendations to patients that ICL is a predictable event that could occur at any time, with the need for monitoring and/or reconstruction.

5. Conclusions

This review assessed the clinical effects of Interproximal Contact Loss (ICL), the impact of the type of prosthesis retention, the number of prosthetic elements, and the anterior or posterior positioning of the rehabilitation. Additionally, the primary assessment tools described in clinical trials for detecting ICL and measuring interproximal space/contact over time were examined.

ICL occurrence was found to be a complex phenomenon influenced by multiple factors. Changes were most frequently observed on the mesial surface of the implant prosthesis, particularly in the posterior mandibular region. In the anterior region, the earliest instances were detected at 3 months. Subsequent ICL frequency tendency increased over time. ICL played a role in food impaction, leading to tissue changes (mucositis) that could progress to peri-implantitis, making it a risk factor for gingival and periodontal conditions.

Factors such as continuous tooth movement, anterior load distribution, and occlusal contact were identified as significant contributors to ICL, particularly on the mesial surface of implant-supported prostheses. While the use of an occlusal splint may not prevent ICL, it can help mitigate some of the clinical effects when ICL occurs.

Due to the multiplicity of assessment tools and methods, new calibration and protocols are required to develop and standardize clinical and research evaluation of ICL, allowing reliable and early detection of interproximal space.

Author Contributions: Conceptualization: J.C.N. and B.F.L.; Methodology: J.C.N. and G.S.P.B.; Software: L.A.; Writing—original draft preparation: J.C.N.; writing—review and editing: B.F.L., L.P.d.S., L.A. and P.M.-M.; visualization: J.C.N., B.F.L., L.P.d.S. and L.A.; supervision: P.M.-M. and B.F.L. All authors have read and agreed to the published version of the manuscript.

Funding: This research received no external funding.

Acknowledgments: The authors are grateful to the University Fernando Pessoa, Porto, Portugal, and to the participants for their cooperation.

Conflicts of Interest: The authors declare no conflicts of interest.

References

1. Yen, J.Y.; Kang, L.; Chou, I.C.; Lai, Y.L.; Lee, S.Y. Risk assessment of interproximal contact loss between implant-supported fixed prostheses and adjacent teeth. A retrospective radiographic study. *J. Prosthet. Dent.* **2022**, *127*, 86–92. [CrossRef] [PubMed]
2. Khairnar, M. Classification of food impaction-revisited and its management. *Indian J. Dent. Adv.* **2013**, *5*, 113–119. [CrossRef]
3. Jo, D.W.; Kwon, M.J.; Kim, J.H.; Kim, Y.K.; Yi, Y.J. Evaluation of adjacent tooth displacement in the posterior implant restoration with proximal contact loss by superimposition of digital models. *J. Adv. Prosthodont.* **2019**, *11*, 88–94. [CrossRef] [PubMed]
4. Dai, W.; Lu, X. The influence of contact area between implant and its adjacent teeth on finite element analysis. *Vibroeng. PROCEDIA* **2019**, *22*, 182–187. [CrossRef]

5. Gasser, T.J.W.; Papageorgiou, S.N.; Eliades, T.; Hämmerle, C.H.F.; Thoma, D.S. Interproximal contact loss at implant sites: A retrospective clinical study with a 10-year follow-up. *Clin. Oral Implant. Res.* **2022**, *33*, 482–491. [CrossRef] [PubMed]
6. Chopra, A.; Sivaraman, K.; Narayan, A.I.; Balakrishnan, D. Etiology and classification of food impaction around implants and implant-retained prosthesis. *Clin. Implant. Dent. Relat. Res.* **2019**, *21*, 391–397. [CrossRef] [PubMed]
7. Byun, S.J.; Heo, S.M.; Ahn, S.G.; Chang, M. Analysis of proximal contact loss between implant-supported fixed dental prostheses and adjacent teeth in relation to influential factors and effects. A cross-sectional study. *Clin. Oral Implant. Res.* **2015**, *26*, 709–714. [CrossRef]
8. Bompolaki, D.; Edmondson, S.A.; Katancik, J.A. Interproximal contact loss between implant-supported restorations and adjacent natural teeth: A retrospective cross-sectional study of 83 restorations with an up to 10-year follow-up. *J. Prosthet. Dent.* **2022**, *127*, 418–424. [CrossRef]
9. Li, Q.L.; Ying Cao, C.; Xu, Q.J.; Xu, X.H.; Yin, J.L. Atraumatic Restoration of Vertical Food Impaction with an Open Contact Using Flowable Composite Resin Aided by Cerclage Wire under Tension. *Scientifica* **2016**, *2016*, 4127472. [CrossRef] [PubMed]
10. Greenstein, G.; Carpentieri, J.; Cavallaro, J. Open contacts adjacent to dental implant restorations: Etiology, incidence, consequences, and correction. *J. Am. Dent. Assoc.* **2016**, *147*, 28–34. [CrossRef]
11. Sfondouris, T.; Prestipino, V. Chairside management of an open proximal contact on an implant-supported ceramic crown using direct composite resin. *J. Prosthet. Dent.* **2019**, *122*, 1–4. [CrossRef] [PubMed]
12. Fathi, A.; Mosharraf, R.; Ebadian, B.; Javan, M.; Isler, S.; Dezaki, S.N. Prevalence of Proximal Contact Loss between Implant-Supported Prostheses and Adjacent Natural Teeth: An Umbrella Review. *Eur. J. Dent.* **2022**, *16*, 742–748. [CrossRef] [PubMed]
13. Manicone, P.F.; De Angelis, P.; Rella, E.; Papetti, L.; D'Addona, A. Proximal Contact Loss in Implant-Supported Restorations: A Systematic Review and Meta-Analysis of Prevalence. *J. Prosthodont.* **2022**, *31*, 201–209. [CrossRef] [PubMed]
14. Varthis, S.; Randi, A.; Tarnow, D.P. Prevalence of Interproximal Open Contacts Between Single-Implant Restorations and Adjacent Teeth. *Int. J. Oral Maxillofac. Implant.* **2016**, *31*, 1089–1092. [CrossRef] [PubMed]
15. Saber, A.; Chakar, C.; Mokbel, N.; Nohra, J. Prevalence of Interproximal Contact Loss Between Implant-Supported Fixed Prostheses and Adjacent Teeth and Its impact on Marginal Bone Loss: A Retrospective Study. *Int. J. Oral Maxillofac. Implant.* **2020**, *35*, 625–630. [CrossRef] [PubMed]
16. Schwarzer, G.; Carpenter, J.R.; Rücker, G. *Meta-Analysis with R, Use R!* Springer: Cham, Switzerland, 2015. [CrossRef]
17. Schulz, K.F.; Altman, D.G.; Moher, D. CONSORT 2010 statement: Updated guidelines for reporting parallel group randomised trials. *BMJ* **2010**, *340*, c332. [CrossRef]
18. Ouzzani, M.; Hammady, H.; Fedorowicz, Z.; Elmagarmid, A. Rayyan—A web and mobile app for systematic reviews. *Syst. Rev.* **2016**, *5*, 210. [CrossRef]
19. French, D.; Naito, M.; Linke, B. Interproximal contact loss in a retrospective cross-sectional study of 4325 implants: Distribution and incidence and the effect on bone loss and peri-implant soft tissue. *J. Prosthet. Dent.* **2019**, *122*, 108–114. [CrossRef] [PubMed]
20. Latimer, J.M.; Gharpure, A.S.; Kahng, H.J.; Aljofi, F.E.; Daubert, D.M. Interproximal open contacts between implant restorations and adjacent natural teeth as a risk-indicator for peri-implant disease—A cross-sectional study. *Clin. Oral Implant. Res.* **2021**, *32*, 598–607. [CrossRef]
21. Mehanna, S.; Habre-Hallage, P. Proximal contact alterations between implant-supported restorations and adjacent teeth in the posterior region: A 3-month prospective study. *J. Clin. Exp. Dent.* **2021**, *13*, 479–486. [CrossRef]
22. Chanthasan, S.; Mattheos, N.; Pisarnturakit, P.P.; Pimkhaokham, A.; Subbalekha, K. Influence of interproximal peri-implant tissue and prosthesis contours on food impaction, tissue health and patients quality of life. *Clin. Oral Implant. Res.* **2022**, *33*, 768–781. [CrossRef] [PubMed]
23. Wong, A.T.; Way, P.Y.; Pow, E.H.; Leung, K.C. Proximal contact loss between implant-supported prostheses and adjacent natural teeth: A retrospective study. *Clin. Oral Implant. Res.* **2015**, *26*, 68–71. [CrossRef] [PubMed]
24. Liang, C.H.; Nien, C.Y.; Chen, Y.L.; Hsu, K.W. The prevalence and associated factors of proximal contact loss between implant restoration and adjacent tooth after function: A retrospective study. *Clin. Implant. Dent. Relat. Res.* **2020**, *22*, 351–358. [CrossRef] [PubMed]
25. Shi, J.Y.; Zhu, Y.; Gu, Y.X.; Lai, H.C. Proximal Contact Alterations between Implant-Supported Restorations and Adjacent Natural Teeth in the Posterior Region: A 1-Year Preliminary Study. *Int. J. Oral Maxillofac. Implant.* **2019**, *34*, 165–168. [CrossRef] [PubMed]
26. Wei, H.; Tomotake, Y.; Nagao, K.; Ichikawa, T. Implant prostheses and adjacent tooth migration: Preliminary retrospective survey using 3-dimensional occlusal analysis. *Int. J. Prosthodont.* **2008**, *21*, 302–304. [PubMed]
27. Koori, H.; Morimoto, K.; Tsukiyama, Y.; Koyano, K. Statistical analysis of the diachronic loss of interproximal contact between fixed implant prostheses and adjacent teeth. *Int. J. Prosthodont.* **2010**, *23*, 535–540. [PubMed]
28. Pang, N.S.; Suh, C.S.; Kim, K.D.; Park, W.; Jung, B.Y. Prevalence of proximal contact loss between implant-supported fixed prostheses and adjacent natural teeth and its associated factors: A 7-year prospective study. *Clin. Oral Implant. Res.* **2017**, *28*, 1501–1508. [CrossRef] [PubMed]
29. Wolfart, S.; Rittich, A.; Gro, B.K.; Hartkamp, O.; von der Stück, A.; Raith, S.; Reich, S. Cemented versus screw-retained posterior implant-supported single crowns: A 24-month randomized controlled clinical trial. *Clin. Oral Implant. Res.* **2021**, *32*, 1484–1495. [CrossRef]

30. Manicone, P.F.; De Angelis, P.; Papetti, L.; Rella, E.; De Angelis, S.; D'Addona, A. Analysis of Proximal Contact Loss Between Implant Restorations and Adjacent Teeth: A 10-Year Retrospective Study. *Int. J. Periodont. Restor. Dent.* **2022**, *42*, 113–119. [CrossRef]
31. Jeong, J.S.; Chang, M. Food Impaction and Periodontal/Peri-Implant Tissue Conditions in Relation to the Embrasure Dimensions Between Implant-Supported Fixed Dental Prostheses and Adjacent Teeth: A Cross-Sectional Study. *J. Periodontol.* **2015**, *86*, 1314–1320. [CrossRef]
32. Abduo, J.; Lee, C.L.; Sarfarazi, G.; Xue, B.; Judge, R.; Darby, I. Encode Protocol Versus Conventional Protocol for Single-Implant Restoration: A Prospective 2-Year Follow-Up Randomized Controlled Trial. *J. Oral Implant. Implantol.* **2021**, *47*, 36–43. [CrossRef] [PubMed]
33. Ren, S.; Lin, Y.; Hu, X.; Wang, Y. Changes in proximal contact tightness between fixed implant prostheses and adjacent teeth: A 1-year prospective study. *J. Prosthet. Dent.* **2016**, *115*, 437–440. [CrossRef] [PubMed]
34. Kandathilparambil, M.R.; Nelluri, V.V.; Vayadadi, B.C.; Gajjam, N.K. Evaluation of biological changes at the proximal contacts between single-tooth implant-supported prosthesis and the adjacent natural teeth—An in vivo study. *J. Indian Prosthodont. Soc.* **2020**, *20*, 378–386. [CrossRef] [PubMed]
35. Abduo, J.; Lau, D. Proximal contact loss between implant prostheses and adjacent natural teeth: A qualitative systematic review of prevalence, influencing factors and implications. *Heliyon* **2022**, *8*, e10064. [CrossRef]
36. Bento, V.A.A.; Gomes, J.M.L.; Lemos, C.A.A.; Limirio, J.P.J.O.; Rosa, C.D.D.R.D.; Pellizzer, E.P. Prevalence of proximal contact loss between implant-supported prostheses and adjacent natural teeth: A systematic review and meta-analysis. *J. Prosthet. Dent.* **2023**, *129*, 404–412. [CrossRef] [PubMed]
37. Liu, X.; Liu, J.; Zhou, J.; Tan, J. Closing open contacts adjacent to an implant-supported restoration. *J. Dent. Sci.* **2019**, *14*, 216–218. [CrossRef] [PubMed]
38. Oh, W.S.; Oh, J.; Valcanaia, A.J. Open Proximal Contact with Implant-Supported Fixed Prostheses Compared with Tooth-Supported Fixed Prostheses: A Systematic Review and Meta-analysis. *Int. J. Oral Maxillofac. Implant.* **2020**, *30*, 99–108. [CrossRef] [PubMed]
39. Ghasemi, S.; Oveisi-Oskouei, L.; Torab, A.; Salehi-Pourmehr, H.; Babaloo, A.; Vahed, N.; Abolhasanpour, N.; Taghilou, S.; Ghasemi, A. Prevalence of proximal contact loss between implant-supported fixed prosthesis and adjacent teeth and associated factors: A systematic review and meta-analysis. *J. Adv. Periodont. Implant. Dent.* **2022**, *14*, 119–133. [CrossRef] [PubMed]
40. Murugan Jeyasree, R.; Muthuraj, T. An Insight into Classification, Diagnosis and Comprehensive Management of Food Impaction. In *Periodontology—New Insights*; IntechOpen: London, UK, 2022. [CrossRef]
41. Kim, J.Y.; Lim, Y.J.; Heo, Y.K. Modification of framework design for an implant-retained fixed restoration helps when proximal contact loss occurs. *J. Dent. Sci.* **2019**, *14*, 213–215. [CrossRef] [PubMed]
42. Papalexopoulos, D.; Samartzi, T.K.; Tsirogiannis, P.; Sykaras, N.; Sarafianou, A.; Kourtis, S.; Mikeli, A. Impact of maxillofacial growth on implants placed in adults: A narrative review. *J. Esthet. Restor. Dent.* **2023**, *35*, 467–478. [CrossRef]
43. Manouchehri, E.; Alirezaei, S.; Roudsari, R.L. Compliance of Published Randomized Controlled Trials on the Effect of Physical Activity on Primary Dysmenorrhea with the Consortium's Integrated Report on Clinical Trials Statement: A Critical Appraisal of the Literature. *Iran. J. Nurs. Midwifery Res.* **2020**, *25*, 445–454. [CrossRef] [PubMed]
44. French, D.; French, M.-C.; Linke, B.; Lizotte, J. Two-center observational case series describing Decay Adjacent to Fixed Implant Restations (DATFIR) and evaluation of case parameters. *J. Oral Sci.* **2018**, *4*, 32–41. Available online: https://www.researchgate.net/publication/325021520_Two-Center_Observational_Case_Series_Describing_Decay_Adjacent_to_Fixed_Implant_Restorations_DATFIR_and_Evaluation_of_Case_Parameters (accessed on 10 June 2024).

prosthesis

MDPI

Technical Note

Innovating Prosthodontic Rehabilitation: A Streamlined Two-Step Technique for Mobile Denture Fabrication

Luca Fiorillo [1,2,3,*], Cesare D'Amico [1,4], Francesca Gorassini [5], Marta Varrà [6], Emanuele Parbonetti [6], Salvatore Varrà [6], Vincenzo Ronsivalle [7] and Gabriele Cervino [1]

1 School of Dentistry, Department of Biomedical and Dental Sciences and Morphofunctional Imaging, University of Messina, Via Consolare Valeria, 1, 98125 Messina, Italy; cdamico@unime.it (C.D.); gcervino@unime.it (G.C.)
2 Multidisciplinary Department of Medical-Surgical and Odontostomatological Specialties, University of Campania "Luigi Vanvitelli", 80121 Naples, Italy
3 Department of Dental Cell Research, Dr D.Y. Patil Dental College and Hospital, Dr D.Y. Patil Vidyapeeth, Pimpri, Pune 411018, India
4 School of Dentistry, Aldent University, 1001 Tirana, Albania
5 FiDent, Centro Medico Odontoiatrico, 89121 Reggio di Calabria, Italy
6 Private Practice, 89025 Reggio di Calabria, Italy
7 Department of General Surgery and Surgical-Medical Specialties, School of Dentistry, University of Catania, Via S. Sofia 78, 95124 Catania, Italy; vincenzo.ronsivalle@hotmail.it
* Correspondence: lfiorillo@unime.it

Abstract: This manuscript introduces a novel two-step technique for fabricating mobile dentures post-extraction to streamline prosthodontic rehabilitation. The study utilizes various materials, including dental polymers, metals, ceramics, and composite materials, each chosen for their unique properties that contribute to the final prosthesis's functionality, durability, and esthetics. The detailed procedure involves an initial occlusal registration immediately following tooth extraction, capturing precise occlusal relationships and a comprehensive dental impression. This approach reduces clinical visits and leverages optimal alveolar ridge morphology. The expected results highlight the efficiency of the technique, reducing treatment time without compromising quality and potentially improving patient satisfaction and prosthodontic outcomes. This innovative method conclusively promises rapid, efficient, and patient-centered dental rehabilitation, emphasizing the need for future research to validate its effectiveness and explore long-term outcomes.

Keywords: dentures; complete; dental impression materials; dental prosthesis design; patient care planning; prosthodontics; dental materials

1. Introduction

1.1. Background

The fabrication of mobile dentures following dental extractions is a cornerstone in prosthodontic rehabilitation, aiming to restore masticatory function, esthetics, and overall quality of life [1]. Removable prosthodontics are a treatment option for patients who have lost one or more teeth. Fixed prosthodontics, such as bridges and crowns, are cemented or screwed to the patient's existing teeth or dental implants, offering a non-removable solution that closely replicates the natural appearance and function genuine teeth. This approach ensures excellent stability and efficiency in chewing, contributing positively to oral health. However, fixed prosthodontics often require the alteration or preparation of adjacent healthy teeth, which can lead to potential long-term dental issues and necessitates a complex and rigorous maintenance regimen to prevent problems such as decay under the prosthetic or periodontal disease at the abutment sites. On the other hand, removable prosthodontics, like complete and partial dentures, provide a versatile and less invasive alternative that is generally more cost-effective and easier for patients to manage in daily cleaning and maintenance. Despite these benefits, they might lack the stability and comfort of fixed alternatives, as they can move during speech or eating, potentially leading to sore spots and diminished patient satisfaction. Removable options also typically require more frequent adjustments to accommodate changes in the oral cavity, adding to their long-term upkeep. Deciding between fixed and removable prosthodontics depends on several patient-specific factors, including the condition of remaining teeth, preference for permanence versus flexibility, esthetic considerations, and financial constraints. Each option presents unique advantages and challenges that should be carefully considered regarding the patient's lifestyle, expectations, and overall oral health goals [2]. Removable prosthodontics are made of a base material, such as acrylic resin, that supports artificial teeth. Removable prosthodontics can be either complete or partial. Complete dentures replace all teeth in an arch, while partial dentures replace only some of the teeth. Manufacturing removable prosthodontics is a complex process involving several steps [3]. Traditional protocols for denture fabrication typically involve multiple clinical and laboratory steps, often extending over several weeks or even months. This time-consuming process can be a significant source of distress for patients, prolonging the adjustment and adaptation to the new prosthesis. In recent years, there has been a paradigm shift towards minimizing the number of clinical visits and reducing the overall treatment time without compromising the quality of the prosthodontic outcome [4,5].

During the realization of dentures, the pre-clinician diagnostic phases are essential. Initially, a comprehensive oral examination is conducted to assess the patient's systemic and oral health, including the condition of any remaining teeth, the health of the oral mucosa, and the anatomical features of the jaw and oral cavity. This could lead to instrumental examinations, like radiography, to evaluate the periodontal status of residual teeth. Carefully analyzing the soft and hard tissues to highlight anomalies, such as exostoses or mandibular tori, is essential. Subsequently, we can proceed with the joint evaluation of the patient, highlighting the presence of temporomandibular anomalies. To evaluate the peri-oral soft tissues, we must consider the high esthetic value of a removable prosthetic rehabilitation. The importance of a thorough diagnosis and analysis of temporomandibular disease (TMD) in patients before the realization of removable dentures cannot be overstated. Accurate assessment of the temporomandibular joint (TMJ) and its function is crucial because TMD can significantly impact the overall success and functionality of the dentures. Patients with undiagnosed or untreated TMD may experience increased discomfort and complications with their prosthesis due to the added strain on their compromised joint structures [6]. This can lead to improper fitting of the denture, increased wear and damage, and decreased patient satisfaction and quality of life. Only after assessing the patient's psychological state and being sure of their acceptance of a removable rehabilitation will it be possible to proceed to the subsequent phases [6,7].

This is followed by detailed dental impressions, which serve as the primary models for creating the dentures. Precise measurements of the maxillary and mandibular arches are taken to establish the spatial relationship between the two, ensuring that the dentures will align correctly during mastication. Additionally, esthetic considerations such as the color, size, and shape of the prosthetic teeth are decided in consultation with the patient, aiming to mimic natural teeth as closely as possible. This meticulous preparation phase is vital for crafting a functional, comfortable, and esthetically pleasing dental prosthesis that meets the specific needs and expectations of the patient.

This technique focuses on possibly rehabilitating an edentulous patient in the shortest possible time to avoid functional and esthetic discomfort. Studies such as those by Jogezai et al. have explored immediate loading protocols, emphasizing the importance of rapid rehabilitation in enhancing patient satisfaction and treatment acceptance [8]. However, the need for a concise yet comprehensive protocol for mobile denture fabrication, especially following dental extractions, remains a subject of clinical interest and scientific inquiry [9,10].

1.2. Materials Used in Removable Prosthesis Fabrication

Fabricating removable dental prostheses involves various materials, each selected for specific properties contributing to the final product's functionality, durability, and esthetics. The primary materials include dental polymers, metals, ceramics, and composite materials [11–15].

1.2.1. Dental Polymers

Polymethylmethacrylate (PMMA) is a widely used material for denture bases. It is chosen for its ease of manipulation, acceptable color stability, and compatibility with the oral environment. PMMA's properties include good esthetic qualities, such as translucency and color options that mimic natural gum tissue, and adequate mechanical strength. However, it can be prone to fracture under high impact and may undergo dimensional changes over time due to water absorption [16,17].

1.2.2. Metals

Metals used in removable prostheses include cobalt–chrome alloys, stainless steel, and sometimes gold. These materials are used mainly in the framework of partial dentures or for clasps and attachments. Metals offer high strength, rigidity, and resistance to wear and deformation, which are crucial for the structural integrity of partial dentures. Cobalt–chrome alloys are particularly valued for their favorable mechanical properties, biocompatibility, and minimal allergic potential [18].

1.2.3. Ceramics

Ceramic materials, such as porcelain, are sometimes used for artificial teeth in dentures due to their exceptional esthetic qualities, including color stability and resistance to wear. Porcelain teeth provide a high degree of natural appearance due to their translucency and color-matching capabilities. However, their brittleness and the potential for abrasion to opposing natural teeth limit their use [19,20].

1.2.4. Composite Materials

Dental composites are increasingly being used for artificial teeth and modifications or repairs to dentures. These materials offer an excellent balance between esthetics and mechanical properties, with improved wear resistance compared to acrylic teeth and better impact strength than porcelain. Composites can be easily adjusted and polished, making them ideal for intraoral modifications [21–23].

1.2.5. Soft Liners

Soft liner materials, including silicone-based materials and soft acrylics, are used to improve the comfort of dentures, especially in cases with bony undercuts or sensitive tissues. These materials can absorb masticatory forces, reducing pressure points and improving the distribution of occlusal loads [24].

1.3. Articulators

Articulators in prosthodontics are mechanical devices used for simulating the temporomandibular joint and jaw movements, aiding in the accurate fabrication of dental prostheses, including mobile dentures. There are several types, ranging from simple to highly complex:

1. Non-adjustable or simple articulators can replicate basic opening and closing movements. They are used for simple restorations that do not require extensive occlusal detailing.

2. Semi-adjustable articulators can simulate more natural jaw movements, including lateral and protrusive movements. They are used for more complex restorations, balancing functionality and cost.

3. Fully adjustable articulators: These replicate the full range of mandibular movements and can be customized to match the patient's specific jaw movements, which are

recorded via face-bow transfers and other measurements. They are used for the most complex and precise dental restorations, including high-end mobile dentures, where occlusal harmony and function are crucial.

4. Virtual or digital articulators: with the advancement of digital dentistry, virtual articulation software simulates jaw movements digitally, allowing for precise design and fabrication of dentures using CAD/CAM technologies [25–27].

Each type of articulator has its specific use, depending on the complexity of the dental prosthesis being fabricated and the level of occlusal detail required.

1.4. Aim

The present manuscript introduces a novel two-step technique aimed at streamlining the process of mobile denture fabrication post-extraction, potentially redefining the conventional timelines and procedures of removable prosthodontics. Its primary aim is to reduce the number of clinical visits required, thus enhancing patient satisfaction and potentially improving prosthodontic outcomes. By integrating occlusal registration and immediate impression techniques immediately post-extraction, the technique captures optimal alveolar ridge morphology, which is crucial for the functional and esthetic success of the dentures. This approach promises to maintain high standards of care and streamline the overall treatment timeline, offering a more efficient and patient-centered approach to prosthodontic rehabilitation.

2. Materials and Equipment

In standard procedures, the realization of a removable prosthodontic appliance is a meticulous multi-step process, combining clinical assessments and dental laboratory techniques to create a functional and esthetically pleasing dental prosthesis. A detailed patient evaluation and preliminary impressions are initially obtained to develop diagnostic models. These models help in the accurate planning of the prosthesis. Following this, definitive impressions are made using specialized materials to capture an exact negative of the oral structures used to create master casts. Subsequent steps involve recording the jaw's relations to establish the spatial relationship between the maxillary and mandibular arches, which is crucial for the correct alignment of prosthetic teeth. Artificial teeth are then selected and arranged based on esthetic and functional criteria, ensuring that they mimic natural teeth as closely as possible in size, shape, and color. A trial insertion follows, where a wax model of the denture set is tested within the patient's mouth, allowing for necessary adjustments before the final processing. This phase transitions into the processing and finishing of the denture, where the wax model is converted into the actual prosthesis using materials like acrylic resin. After thorough polishing and refinements, the prosthesis is ready for delivery. The final steps include post-insertion adjustments to optimize comfort and functionality, ensuring the integration of the prosthesis into the patient's daily life. This comprehensive approach strives to restore the patient's masticatory function and esthetics. It emphasizes minimizing clinic visits and enhancing overall treatment efficiency, underscoring the evolution of modern prosthodontics toward patient-centered care [5,28].

The proposed technique for mobile denture fabrication encompasses two pivotal steps that are strategically designed to optimize clinical efficiency while ensuring prosthetic precision and patient comfort using standard materials (Table 1) (Figures 1–5).

Table 1. The first column shows the different phases. The second column shows the detailed procedures. The third column shows the main author of each procedure.

<table>
<tr><td rowspan="2">Phase 1: Occlusal registration and initial impressions</td><td>Occlusal registration immediately post-extraction (Figure 1b)</td><td rowspan="2">In-office phase</td></tr>
<tr><td>Alginate dental impression following tooth extraction (Figure 1c,d)</td></tr>
</table>

Table 1. *Cont.*

Phase 2: Prosthesis assembly and adjustment	Mounting and first assembly of the prosthesis (Figure 2)	Dental technician phase
	Further adjustments and final assembly (Figure 2)	In-office phase
	Final impressions and prosthesis completion (Figures 3 and 4)	In-office phase and delivery to dental technician for finalization
	Delivery and post-delivery adjustments (Figure 5)	In-office phase

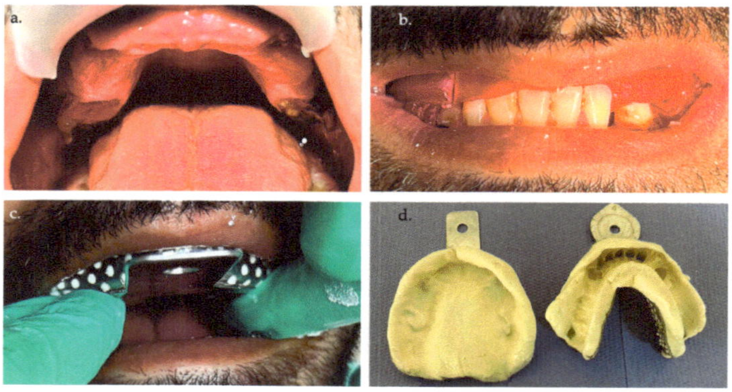

Figure 1. (**a**) Intraoral photo of the patient at the first appointment in the pre-dental extraction phase. (**b**) Centric registration with red wax in the pre-extraction phase. (**c**) The patient undergoes an alginate impression of the dental arches. (**d**) Cleaned and disinfected impressions are ready to be sent to the dental laboratory.

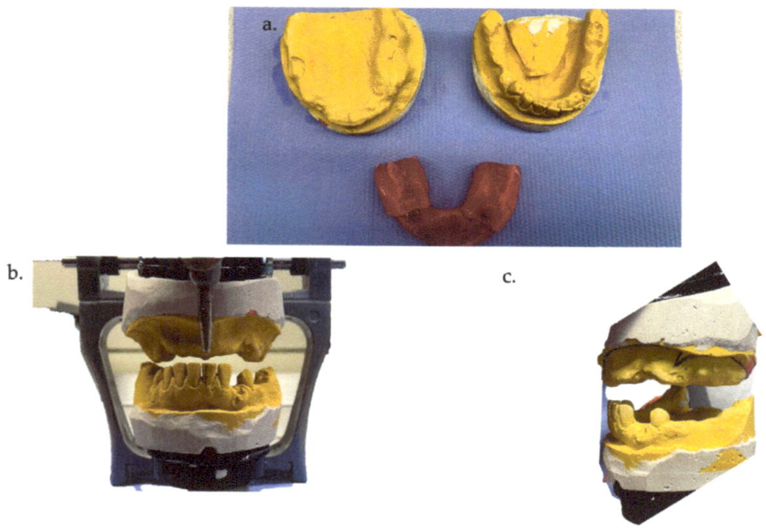

Figure 2. (**a**) Dental models in plaster and red wax with first centric registration. (**b**) Dental models mounted in articulator front view. (**c**) Dental models mounted in articulator lateral view.

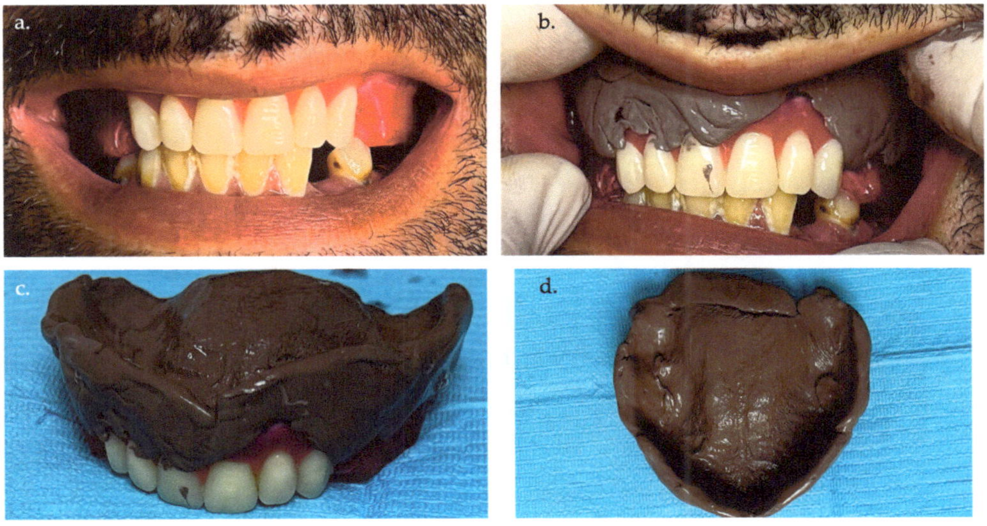

Figure 3. (**a**) Test teeth with resin base in position; the wax rims in the lateral posterior sector are evident. The functional and esthetic test is satisfactory, considering the single step up to this point. (**b**) Relining the resin base with polysulfide material is performed with trial assembly and simultaneous occlusal and mucosal functionalization (heated wax rims). (**c**) Resin base with polymerized material, front view. (**d**) Resin base with polymerized material, palatal view.

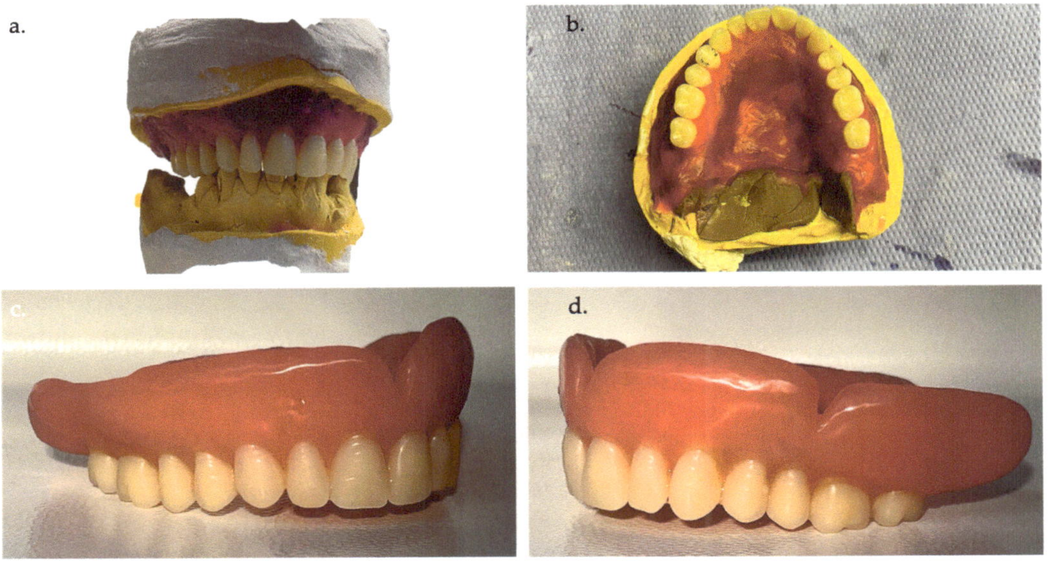

Figure 4. (**a**) The assembly of the new plaster model obtained in the definitive articulator shows that the dental technician has already assembled the teeth in the lateral posterior sectors. (**b**) Occlusal view of the resin base with relining material (polysulphide) during the creation of the definitive model. (**c**) Completed removable total prosthesis, right side view. (**d**) Completed removable total prosthesis, left lateral view.

a.

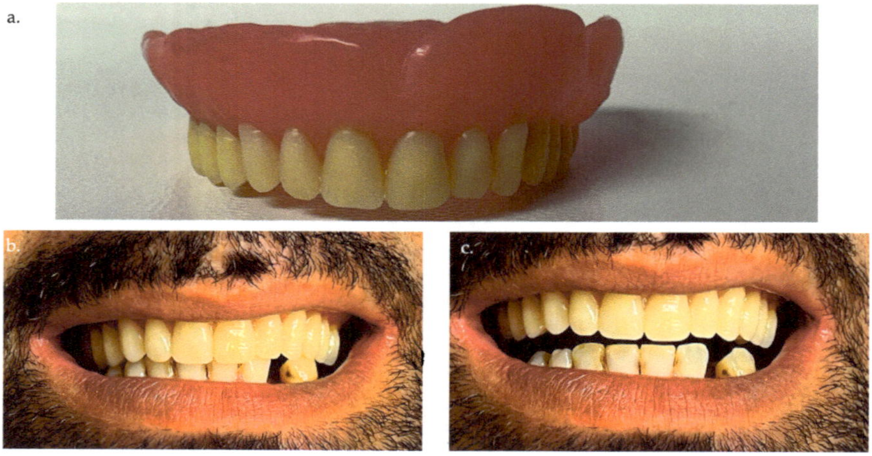

Figure 5. (**a**) Completed removable total prosthesis, front view. (**b**,**c**) Delivered denture.

3. Detailed Procedures

The patient comes to our observation requesting a removable rehabilitation of the upper arch. Patient ASA2 has no contraindications to the treatment and does not complain of painful symptoms in the oral cavity. He does not report any anomalies, noises, or pain during chewing attributed to TMJ pathologies. Upon intraoral physical examination, the patient presents with several root residues in the upper arch. The lower arch presents a condition of oligodontia and does not require treatment. In the first step, an occlusal registration is performed immediately following the extraction of the designated teeth. This step is crucial for capturing the precise occlusal relationships and vertical dimensions of occlusion, which are fundamental for the functional success of the prosthesis. The occlusal registration is conducted during the same appointment as the extraction of the teeth, thereby reducing the number of clinical visits. If the patient is already edentulous or his vertical height is significantly altered, following this protocol will be much more difficult. This is followed by an alginate dental impression immediately after extracting the remaining teeth. This approach not only streamlines the clinical workflow but also capitalizes on the residual alveolar ridge morphology before significant resorption occurs, a concept supported by studies emphasizing the importance of immediate impressions in capturing the optimal ridge anatomy [29].

It is necessary to report as much information as possible on the first wax block, such as the midline and the smile line. At this point, the dental technician can mount the plaster models obtained from our dental impressions in the articulator. This assembly will follow the occlusal registration obtained through the first wax mark; this will allow the dental technician to carry out a first assembly of the teeth on the resin base of the prosthesis, using the front sector only, respecting the color taken during the first session.

The dental technician will create a wax base on the plaster model obtained to assemble the dental elements blocked in wax. However, a wax rim will remain in the posterior and lateral posterior sectors, which will be used for occlusal registration during the second step [30].

Having obtained this rigid resin base with the teeth of the frontal sector mounted, it will be possible to proceed with carrying out all the esthetic and phonetic tests, evaluating the function of the canine and incisor guides and, above all, the median and canine line from an esthetic point of view. A precise and reliable centric position will be obtained by heating the wax rims of the posterior sector and bringing the patient into the centric position. If the teeth in the frontal sector are not congruent with the occlusion, they will be adapted, or if this is not possible, the protocol will start again with a new wax rim positioned on the resin base once the teeth have been removed.

Having ensured the absence of any pain reported by the patient during the tests and a stable position of the resin base on the palate, it will be possible to proceed with the precision impression that will be functionalized with polysulphide-based pastes [31–34]. Polysulfides are hydrophobic elastomers of natural origin with a relatively low cost, excellent tear resistance, and viscoelastic characteristics. They also have a long polymerization time, which allows the detection of secondary mucodynamic impressions in totally or partially edentulous patients. It is a material often criticized for its difficult processing and organoleptic characteristics. The poor dimensional stability of the material requires the technician to pour the impressions within 30 min of removal from the oral cavity [35].

The impression material is then kneaded and placed on the resin base, closing the patient in a centric position and functionalizing the soft tissues (mucosa of the lips and the genials). The resin base with the polymerized impression material can be removed, and the patient can proceed to the last delivery step.

The prosthesis can thus be delivered following an occlusal and esthetic check. The post-extraction mucous membranes and alveoli will not yet have reached complete maturation and healing, but this allows the patient to have a full set of teeth and restore function and esthetics within a few days.

4. Expected Results

4.1. Clinical Implications

The choice of material for a removable prosthesis has significant implications for clinical outcomes. Dentists must consider the patient's specific needs, including the condition of the oral tissues, masticatory efficiency, esthetic preferences, and any allergies or sensitivities. The longevity and success of a removable prosthesis also depend on proper maintenance by the patient, including regular cleaning and avoiding practices that could damage the prosthesis, such as using it to open bottles or chewing hard foods.

Creating a removable prosthodontic with standard protocols involves meticulously detailed procedures that span several critical steps designed to ensure that the prosthesis meets both functional and esthetic requirements for the patient. This process typically encompasses the following:

1. Patient evaluation and preliminary impressions: initial consultation and assessment of the patient's oral health status are conducted, followed by taking preliminary impressions of the arches to create diagnostic models;
2. Definitive impressions and master cast fabrication: this entails using more precise impression materials to capture detailed oral structures to create master casts;
3. Jaw relation records: this entails determining and recording the spatial relationship between the upper and lower jaws to guide the proper alignment of the denture teeth;
4. Selection and arrangement of denture teeth: the appropriate size, shape, and color of artificial teeth are chosen based on esthetic and occlusal requirements, and this is followed by their precise arrangement;
5. Trial insertion: this is a test fitting of the denture with the arranged teeth set in wax, allowing for adjustments in esthetics and occlusion before final processing;
6. Processing and finishing: this is the conversion of the wax denture into its final form by replacing the wax with a denture base material, usually acrylic resin, which is followed by polishing and finishing touches;
7. Delivery and post-insertion adjustments: the finished denture is handed to the patient, followed by making necessary adjustments to enhance comfort and function, ensuring satisfactory integration into the oral cavity [36–38].

Each step requires detailed attention to technical and clinical details, ensuring that the final prosthesis meets the patient's function, comfort, and appearance needs. From a clinical perspective, this approach significantly reduces the number of patient visits required, directly enhancing patient convenience and satisfaction. Financially, by optimizing materials and reducing chair time, the technique potentially lowers the overall cost of denture fabrication for patients. For clinicians, the technique offers a more efficient workflow, allowing for rapid prosthodontic rehabilitation while maintaining high standards of care. This innovative method redefines conventional timelines and procedures in removable prosthodontics, emphasizing efficiency and patient-centered care.

4.2. Literature Discussions

The innovative two-step technique for mobile denture fabrication presents a significant advancement in removable prosthodontics, potentially setting a new standard for efficiency and patient-centered care. One of the most notable advantages of this technique is the reduction in the number of clinical visits, which directly correlates with patient convenience and satisfaction. Additionally, by performing the impression immediately post-extraction, the technique harnesses the most favorable ridge anatomy, potentially enhancing the stability and retention of the final denture. The strategic timing of the occlusal registration also ensures that the vertical dimension of the occlusion is accurately captured, a crucial determinant of the functional and esthetic success of the prosthesis. However, it is imperative to consider this technique's potential limitations and challenges. The condensed timeline necessitates meticulous clinical execution and may leave limited room for error correction. The immediate post-extraction phase is also characterized by tissue inflammation and healing dynamics, which may introduce variables that could impact the final denture fit and comfort. Comparative studies, such as those by Heartwell et al. [39], have emphasized the importance of considering the biological and healing factors in immediate denture protocols. Therefore, while the proposed two-step protocol offers substantial benefits, it requires a high level of clinical expertise and a thorough understanding of the biological processes involved in post-extraction healing.

Cherkashin BF et al. [40] developed a protocol for setting artificial teeth in completely removable prosthodontics without bite ridge landmarks based on comprehensive dentistry literature and clinical experience, leading to more efficient prosthesis fabrication for patients with complete secondary dementia. Von Stein-Lausnitz M et al. [41] conducted a double-blinded randomized controlled trial to assess the impact of face-bow registration on occlusal parameters in complete denture remounting, revealing no substantial differences between face-bow and mean setting methods in changing the vertical dimension [42]. Steinmassl PA et al. [43] evaluated various CAD/CAM denture systems, highlighting these technologies' potential to reduce patient visits and improve denture adaptation, with each system offering unique advantages based on the dentist's expertise and patient needs. Alhallak KR et al. [44] review the use of 3D printing technologies for manufacturing removable prosthodontics, focusing on advantages like time efficiency, satisfying clinical outcomes, and securing patient records while noting challenges such as material strength, esthetics, and biocompatibility, urging further studies. Suzuki Y et al. [45] analyze the laboratory efficiency of additive manufacturing for removable prosthodontic frameworks, comparing surface roughness, fitness accuracy, and retentive forces with conventional methods, finding areas where additive manufacturing falls short or matches traditional techniques. Davda K et al. [46] investigated the trueness and precision of copy denture templates made via conventional and 3D printing methods, finding that 3D-printed templates significantly improved trueness and accuracy over traditional methods, highlighting the potential of 3D printing in dental prosthetics. Bilgin MS et al. [47] review CAD/CAM and rapid prototyping technologies for removable prosthodontic fabrication, covering technological advancements, techniques, and the transition towards digital planning and manufacturing while noting current limitations and the need for technical expertise in traditional methods. In their clinical study, Lo Russo et al. assessed the accuracy of a two-step scanning strategy for intraoral scans of edentulous mandibular arches, comparing these with conventional polysulfide impressions. The results showed no significant distortion or differences in 3D deviations when comparing the scans to the impressions, indicating the reliability of the two-step scanning method. However, the analysis revealed significant regional variations, although these did not differ significantly between corresponding regions on both arch sides [48].

Accurate diagnosis and thorough analysis of temporomandibular disorders (TMDs) are essential before the fabrication of removable dentures. TMDs can significantly affect dentures' masticatory function, comfort, and overall success. The temporomandibular joint (TMJ) acts as a crucial interface in the biomechanics of mastication, which is directly influenced by the occlusal dynamics introduced by dentures [49]. In a systematic review, Nimonkar et al. [50] critically examine five peer-reviewed articles selected based on temporomandibular disease and removable dentures. These articles suggest that complete dentures, particularly when carefully designed to restore the vertical dimension and proper occlusal relationships, can significantly alleviate TMD symptoms in edentulous patients. Despite the inherent limitations of the data due to study heterogeneity, the findings advocate for the therapeutic role of complete dentures in potentially reversing TMD-related manifestations, enhancing masticatory function, and improving overall quality of life for edentulous individuals. The discussions within the review emphasize the necessity for well-fabricated dentures and highlight that while current studies support the benefits of dentures in TMD management, further high-quality randomized clinical trials are required to solidify these findings and optimize treatment protocols. This review provides a foundational perspective on the intersection of prosthodontics and temporomandibular health, pointing towards a nuanced understanding of the biomechanical and rehabilitative interactions facilitated by complete dentures in managing TMDs [49–51].

The two-step technique described herein for fabricating mobile dentures immediately post-extraction is designed to address current challenges in prosthodontic rehabilitation, particularly the need for rapid restoration of function and esthetics with minimal clinic visits. This technique leverages advanced material selection and streamlined clinical procedures to optimize treatment outcomes and patient satisfaction. In daily dental practice, the introduction of this technique has the potential to significantly influence the approach to denture fabrication. By reducing the number of patient visits and treatment duration, dental practitioners can enhance the efficiency of their practice and patient throughput. The materials used, including dental polymers, metals, ceramics, and composites, are selected for their durability, esthetic qualities, and biocompatibility, ensuring that the final prosthesis meets both functional and esthetic requirements [16,48].

It is crucial to understand that the success of this innovative technique hinges not only on the materials and technologies used but also on the meticulous execution of clinical and laboratory steps. The educational takeaway is the importance of a thorough understanding of material properties and the appropriate application of clinical techniques to optimize the outcome of the prosthesis. The two-step technique offers a rapid, efficient solution compared to traditional methods. However, it requires precise execution and coordination between the dental team and laboratory technicians. The primary limitation is the dependency on the clinician's skill and the patient's oral condition at the time of extraction, which can vary widely and affect the fit and function of the immediate denture.

In contrast, digital techniques, such as CAD/CAM and 3D printing, provide high precision and reproducibility without the same level of operator dependency. These methods allow for the creation of dentures with consistent quality and fit utilizing digital impressions that can be archived and reused for future prostheses. However, digital methods can be cost-prohibitive and require significant investments in technology and training. The aspect of sustainability in prosthodontic practices, particularly through the selection of materials and techniques, is increasingly important [13,14,29,36,51]. Moreover, the durability of the selected materials reduces the frequency of prosthesis replacement, thereby decreasing waste and promoting sustainability in dental practice.

4.3. Limitations

The technique presented here has different limitations, some absolute and some relative to the case. Those relating to the case are linked to the fact that the patient does not have intact dental elements but only root residues. Therefore, this could make occlusal registration more difficult. Furthermore, the patient is partially edentulous in the posterior sectors of the lower arch, which could facilitate delivery. Regarding the absolute limits, we must consider the need for a specific learning curve to carry out this technique. Materials such as polysulfide-based pastes are complex in their use and functionalization. The wax block must always be created first and used at the first meeting by the clinician. Suppose there are many extractions to be performed. In that case, it is necessary to record the occlusion in such a way as to obtain contact with both the mucosa and the teeth, then record the centricity after several extractions. The first dental assembly is highly operator-dependent and may not be suitable for the second step. The resin base with the dental test remains inside the resin, which is positioned in the flask.

5. Conclusions

The novel two-step technique for mobile denture fabrication post-extraction marks a significant leap forward in removable prosthodontics. By consolidating the clinical steps and harnessing the immediate post-extraction phase for impression and occlusal registration, this technique offers a promising avenue for rapid, efficient, and patient-centric prosthodontic rehabilitation. The reduction in clinical visits and the strategic use of immediate post-extraction tissue morphology underscore the patient-centered approach of this technique. However, the success of this innovative method hinges on meticulous clinical execution and a profound understanding of tissue dynamics post-extraction. Future research and clinical trials are warranted to validate this technique further, explore its long-term outcomes, and establish its position in the repertoire of prosthodontic treatment options. The potential for integrating advanced materials and technologies such as CAD and 3D printing into this technique could also be investigated, which would further enhance the precision and customization of dentures, particularly in terms of occlusal registration. This technique's possibilities to redefine the standards of care in removable prosthodontics are immense, promising a new era of efficiency and patient satisfaction in dental rehabilitation.

Author Contributions: Conceptualization, L.F. and G.C.; methodology, L.F., F.G., M.V., E.P. and S.V.; software, not applicable; validation, L.F., C.D. and G.C.; formal analysis, L.F. and V.R.; investigation, L.F.; resources, L.F.; data curation, L.F.; writing—original draft preparation, L.F.; writing—review and editing, L.F. and G.C.; visualization, L.F.; supervision, G.C.; project administration, L.F.; funding acquisition, G.C. All authors have read and agreed to the published version of the manuscript.

Funding: This research received no external funding.

Institutional Review Board Statement: Ethical review and approval were waived for this study because the implementation of this technique did not harm or cause harm to any human being; in any case, the materials used are already in use in dentistry.

Informed Consent Statement: Informed consent was obtained from all subjects involved in the study.

Data Availability Statement: The full data are available in the manuscript.

Conflicts of Interest: The authors declare no conflicts of interest.

References

1. Tsunoda, A.; Kanazawa, H.; Ishige, T.; Kishimoto, S. A missing denture. *Lancet* **2004**, *364*, 1884. [CrossRef] [PubMed]
2. Fiorillo, L.; Cervino, G.; de Stefanq, R.; Iannellq, G.; Cicciu, M. Socioeconomic behaviors on dental professions: A Google Trends investigation in Italy. *Minerva Stomatol.* **2020**, *69*, 309–316. [CrossRef] [PubMed]
3. Dai, M.; Song, Q.; Lin, T.; Huang, X.; Xie, Y.; Wang, X.; Zheng, L.; Yue, J. Corrigendum: Tooth loss, denture use, and all-cause and cause-specific mortality in older adults: A community cohort study. *Front. Public Health* **2023**, *11*, 1360927. [CrossRef] [PubMed]
4. Palomer, T.; Ramírez, V.; Ortuño, D. Relationship between oral health and depression: Data from the National Health Survey 2016–2017. *BMC Oral Health* **2024**, *24*, 188. [CrossRef] [PubMed]
5. Campbell, S.D.; Cooper, L.; Craddock, H.; Hyde, T.P.; Nattress, B.; Pavitt, S.H.; Seymour, D.W. Removable partial dentures: The clinical need for innovation. *J. Prosthet. Dent.* **2017**, *118*, 273–280. [CrossRef] [PubMed]

6. Furquim, B.D.; Flamengui, L.M.; Conti, P.C. TMD and chronic pain: A current view. *Dent. Press J. Orthod.* **2015**, *20*, 127–133. [CrossRef]
7. Galletti, C.; Lombardo, C.; La Barbiera, C.; Boronat-Català, M.; Almiñana-Pastor, P.J.; Sala Fernàndez, C.; Ramírez-Sebastià, A.; Muscatello, M.R.; Bruno, A.; Mento, C. Dental anxiety, Quality of Life and body image: Gender differences in Italian and Spanish population. *Minerva Dent. Oral Sci.* **2024**, *73*, 14–19. [CrossRef]
8. Jogezai, U.; Laverty, D.; Walmsley, A. Immediate dentures part 1: Assessment and treatment planning. *Dent. Update* **2018**, *45*, 617–624. [CrossRef]
9. Wang, C.; Gao, M.; Yu, Y.; Zhang, W.; Peng, X. Clinical analysis of denture rehabilitation after mandibular fibula free-flap reconstruction. *Beijing Da Xue Xue Bao Yi Xue Ban* **2024**, *56*, 66–73.
10. Sokolowski, A.; Horak, D.; Behlau, A.; Madreiter-Sokolowski, C.; Lorenzoni, M.; Sokolowski, A. Evaluation of two printing techniques for maxillary removable partial denture frameworks. *J. Prosthet. Dent.* **2024**, *131*, 707.e1–707.e8. [CrossRef]
11. Kim, M.C.; Byeon, D.J.; Jeong, E.J.; Go, H.B.; Yang, S.Y. Color stability, surface, and physicochemical properties of three-dimensional printed denture base resin reinforced with different nanofillers. *Sci. Rep.* **2024**, *14*, 1842. [CrossRef] [PubMed]
12. Alhotan, A.; Raszewski, Z.; Chojnacka, K.; Mikulewicz, M.; Kulbacka, J.; Alaqeely, R.; Mirdad, A.; Haider, J. Evaluating the Translucency, Surface Roughness, and Cytotoxicity of a PMMA Acrylic Denture Base Reinforced with Bioactive Glasses. *J. Funct. Biomater.* **2023**, *15*, 16. [CrossRef] [PubMed]
13. Yan, S.; Zhou, J.L.; Zhang, R.J.; Tan, F.B. Evaluation of the influence of different build angles on the surface characteristics, accuracy, and dimensional stability of the complete denture base printed by digital light processing. *Heliyon* **2024**, *10*, e24095. [CrossRef] [PubMed]
14. Papathanasiou, I.; Kamposiora, P.; Papavasiliou, G.; Ferrari, M. The use of PEEK in digital prosthodontics: A narrative review. *BMC Oral Health* **2020**, *20*, 217. [CrossRef] [PubMed]
15. Cervino, G.; Cicciù, M.; Herford, A.S.; Germanà, A.; Fiorillo, L. Biological and chemo-physical features of denture resins. *Materials* **2020**, *13*, 3350. [CrossRef] [PubMed]
16. Yerliyurt, K.; Tasdelen, T.B.; Egri, O.; Egri, S. Flexural Properties of Heat-Polymerized PMMA Denture Base Resins Reinforced with Fibers with Different Characteristics. *Polymers* **2023**, *15*, 3211. [CrossRef]
17. Garcia, A.; Sugio, C.Y.C.; de Azevedo-Silva, L.J.; Gomes, A.C.G.; Batista, A.U.D.; Porto, V.C.; Soares, S.; Neppelenbroek, K.H. Nanoparticle-modified PMMA to prevent denture stomatitis: A systematic review. *Arch. Microbiol.* **2021**, *204*, 75. [CrossRef] [PubMed]
18. Ananya; Rani, P.; Sinha, T.; Prakash, J. Maxillary Cast Partial Denture and Mandibular Implant-Supported Metal-Ceramic Prosthesis with a Split Framework to Compensate for Mandibular Flexure: A Case Report. *Cureus* **2023**, *15*, e49071. [CrossRef]
19. Kancyper, S.; Sierraalta, M.; Razzoog, M.E. All-ceramic surveyed crowns for removable partial denture abutments. *J. Prosthet. Dent.* **2000**, *84*, 400–402. [CrossRef]
20. Yoon, T.H.; Madden, J.C.; Chang, W.G. A technique to restore worn denture teeth on a partial removable dental prosthesis by using ceramic onlays with CAD/CAM technology. *J. Prosthet. Dent.* **2013**, *110*, 331–332. [CrossRef]
21. Pavarina, A.C.; Machado, A.L.; Vergani, C.E.; Giampaolo, E.T. Preparation of composite retentive areas for removable partial denture retainers. *J. Prosthet. Dent.* **2002**, *88*, 218–220. [CrossRef] [PubMed]
22. Shepherd, K.A.; Janzen, J.; Brackett, W.W.; Romero, M.F. Prototyping of increased vertical dimension of occlusion using an existing removable partial denture and composite resin: A case report. *Gen. Dent.* **2023**, *71*, 72–76. [PubMed]
23. Mete, J.J.; Dange, S.P.; Khalikar, A.N.; Vaidya, S.P. Rehabilitation of anterior edentulous space by glass fiber reinforced composite removable partial denture during preadolescent period: A case report. *J. Indian Prosthodont. Soc.* **2011**, *11*, 195–198. [CrossRef] [PubMed]
24. Whitsitt, J.A.; Battle, L.W.; Jarosz, C.J. Enhanced retention for the distal extension-base removable partial denture using a heat-cured resilient soft liner. *J. Prosthet. Dent.* **1984**, *52*, 447–448. [CrossRef] [PubMed]
25. Aljohani, A.O.; Sghaireen, M.G.; Abbas, M.; Alzarea, B.K.; Srivastava, K.C.; Shrivastava, D.; Issrani, R.; Mathew, M.; Alsharari, A.H.L.; Alsharari, M.A.D.; et al. Comparative Evaluation of Condylar Guidance Angles Measured Using Arcon and Non-Arcon Articulators and Panoramic Radiographs—A Systematic Review and Meta-Analysis. *Life* **2023**, *13*, 1352. [CrossRef]
26. Palaskar, J.N.; Joshi, N.P.; Hindocha, A.D.; Gunjal, A.P.; Balsaraf, K.D. Evaluation of Sagittal Inclination of Occlusal Plane and Horizontal Condylar Guidance Using Various Anterior Reference Points on Arcon and Nonarcon Articulators. *Contemp. Clin. Dent.* **2022**, *13*, 255–260. [CrossRef]
27. Lepidi, L.; Galli, M.; Mastrangelo, F.; Venezia, P.; Joda, T.; Wang, H.L.; Li, J. Virtual Articulators and Virtual Mounting Procedures: Where Do We Stand? *J. Prosthodont.* **2021**, *30*, 24–35. [CrossRef]
28. Cervino, G.; Sambataro, S.; Stumpo, C.; Bocchieri, S.; Murabito, F.; Fiorillo, L.; Meto, A.; Zecca, P.A.; Caprioglio, A.; Cicciù, M. Determination of the vertical dimension and the position of the occlusal plane in a removable prosthesis using cephalometric analysis and golden proportion. *Appl. Sci.* **2021**, *11*, 6948. [CrossRef]
29. Baek, M.-W.; Lim, H.-C.; Noh, K.; Choi, S.-H.; Lee, D.-W. Dimensional Changes in Extraction Sockets: A Pilot Study Evaluating Differences between Digital and Conventional Impressions. *Appl. Sci.* **2022**, *12*, 7662. [CrossRef]
30. Beckett, L.S. Accurate occlusal relations in partial denture construction. *J. Prosthet. Dent.* **1954**, *4*, 487–495. [CrossRef]
31. Smith, P.K. The effect on the accuracy of polysulphide impression material after treating preparations with various agents. *Aust. Dent. J.* **1971**, *16*, 337–339. [CrossRef] [PubMed]

32. Tolley, L.G.; Craig, R.G. Viscoelastic properties of elastomeric impression materials: Polysulphide, silicone and polyether rubbers. *J. Oral Rehabil.* **1978**, *5*, 121–128. [CrossRef] [PubMed]
33. Darvell, B.W. Aspects of the chemistry of polysulphide impression material. *Aust. Dent. J.* **1987**, *32*, 357–367. [CrossRef] [PubMed]
34. Peroz, I. The dimensional stability of polyether, polysulphide and silicone impression materials as well as the hardness of casts after disinfection. *Dtsch. Zahnarztl. Z.* **1988**, *43*, 1066–1071. [PubMed]
35. Al-Ansari, A. Which final impression technique and material is best for complete and removable partial dentures? *Evid.-Based Dent.* **2019**, *20*, 70–71. [CrossRef] [PubMed]
36. Goncalves, T.; Forgerini, G.; Drummond, L.B.; Wanghon, Z.M.L.; Carmo Filho, L.C.D.; Philippi, A.G. Digital workflow to produce 3D-printed incisal facial indirect restoration with lingual rest seats to support a removable partial denture. *J. Prosthet. Dent.* **2023**, in press. [CrossRef] [PubMed]
37. Ciocca, L.; Maltauro, M.; Pierantozzi, E.; Breschi, L.; Montanari, A.; Anderlucci, L.; Meneghello, R. Evaluation of trueness and precision of removable partial denture metal frameworks manufactured with digital technology and different materials. *J. Adv. Prosthodont.* **2023**, *15*, 55–62. [CrossRef]
38. Benyahia, H.; El Benaissi, A.; Bahij, L.; Merzouk, N.; Regragui, A. Impression techniques in Removable partial denture: Epidemiological study. *Tunis Med.* **2023**, *101*, 41–46.
39. Heartwell, C.M.; Salisbury, F.W. Immediate complete dentures: An evaluation. *J. Prosthet. Dent.* **1965**, *15*, 615–624. [CrossRef]
40. Cherkashin, B.F.; Furtsev, T.V. Protocol for teeth set-up in full denture without the use of wax rims and its theoretical justification. *Stomatologiia* **2023**, *102*, 50–55. [CrossRef]
41. von Stein-Lausnitz, M.; Sterzenbach, G.; Helm, I.; Zorn, A.; Blankenstein, F.H.; Ruge, S.; Kordaß, B.; Beuer, F.; Peroz, I. Does a face-bow lead to better occlusion in complete dentures? A randomized controlled trial: Part I. *Clin. Oral Investig.* **2018**, *22*, 773–782. [CrossRef] [PubMed]
42. Risciotti, E.; Squadrito, N.; Montanari, D.; Iannello, G.; Macca, U.; Tallarico, M.; Cervino, G.; Fiorillo, L. Digital Protocol to Record Occlusal Analysis in Prosthodontics: A Pilot Study. *J. Clin. Med.* **2024**, *13*, 1370. [CrossRef] [PubMed]
43. Steinmassl, P.A.; Klaunzer, F.; Steinmassl, O.; Dumfahrt, H.; Grunert, I. Evaluation of Currently Available CAD/CAM Denture Systems. *Int. J. Prosthodont.* **2017**, *30*, 116–122. [CrossRef] [PubMed]
44. Alhallak, K.R.; Nankali, A. 3D Printing Technologies for Removable Dentures Manufacturing: A Review of Potentials and Challenges. *Eur. J. Prosthodont. Restor. Dent.* **2022**, *30*, 14–19. [CrossRef] [PubMed]
45. Suzuki, Y.; Shimizu, S.; Waki, T.; Shimpo, H.; Ohkubo, C. Laboratory efficiency of additive manufacturing for removable denture frameworks: A literature-based review. *Dent. Mater. J.* **2021**, *40*, 265–271. [CrossRef] [PubMed]
46. Davda, K.; Osnes, C.; Dillon, S.; Wu, J.; Hyde, P.; Keeling, A. An Investigation into the Trueness and Precision of Copy Denture Templates Produced by Rapid Prototyping and Conventional Means. *Eur. J. Prosthodont. Restor. Dent.* **2017**, *25*, 186–192. [CrossRef] [PubMed]
47. Bilgin, M.S.; Baytaroğlu, E.N.; Erdem, A.; Dilber, E. A review of computer-aided design/computer-aided manufacture techniques for removable denture fabrication. *Eur. J. Dent.* **2016**, *10*, 286–291. [CrossRef]
48. Lo Russo, L.; Sorrentino, R.; Esperouz, F.; Zarone, F.; Ercoli, C.; Guida, L. Assessment of distortion of intraoral scans of edentulous mandibular arch made with a 2-step scanning strategy: A clinical study. *J. Prosthet. Dent.* **2023**, in press. [CrossRef]
49. Minervini, G.; Fiorillo, L.; Russo, D.; Lanza, A.; D'Amico, C.; Cervino, G.; Meto, A.; Di Francesco, F. Prosthodontic Treatment in Patients with Temporomandibular Disorders and Orofacial Pain and/or Bruxism: A Review of the Literature. *Prosthesis* **2022**, *4*, 253–262. [CrossRef]
50. Nimonkar, S.; Godbole, S.; Belkhode, V.; Nimonkar, P.; Pisulkar, S. Effect of Rehabilitation of Completely Edentulous Patients with Complete Dentures on Temporomandibular Disorders: A Systematic Review. *Cureus* **2022**, *14*, e28012. [CrossRef]
51. Santos, M.; Azevedo, L.; Fonseca, P.; Viana, P.; Araújo, F.; Villarinho, E.; Fernandes, G.; Correia, A. The Success Rate of the Adhesive Partial Fixed Prosthesis after Five Years: A Systematic Review. *Prosthesis* **2023**, *5*, 282–294. [CrossRef]

prosthesis

MDPI

Case Report

Clinical Protocol for Implant-Assisted Partial Removable Dental Prostheses in Kennedy Class I: A Case Report

Irina Karakas-Stupar [1], Lucia K. Zaugg [1], Nicola U. Zitzmann [1,*], Tim Joda [1,2], Stefan Wolfart [3] and Taskin Tuna [3]

[1] Department of Reconstructive Dentistry, University Center for Dental Medicine Basel, University of Basel, 4058 Basel, Switzerland; irina.stupar@unibas.ch (I.K.-S.); lucia.zaugg@unibas.ch (L.K.Z.); tim.joda@zzm.uzh.ch (T.J.)

[2] Clinic of Reconstructive Dentistry, Center of Dental Medicine, University of Zurich, 8032 Zurich, Switzerland

[3] Department of Prosthodontics and Biomaterials, Center of Implantology, RWTH University Hospital Aachen, 52074 Aachen, Germany; swolfart@ukaachen.de (S.W.); ttuna@ukaachen.de (T.T.)

* Correspondence: n.zitzmann@unibas.ch; Tel.: +41-61-267-26-25

Abstract: Patients with Kennedy Class I are usually treated with clasp-retained removable partial dentures (RPDs) as the prosthetic gold standard. For additional stabilization of the RPD, clinicians are often confronted with the question of secondary implant placement, which requires the fabrication of new prostheses. This case report is part of an ongoing multi-center randomized controlled study (RCT) investigating conventional RPDs without and with supplementary implants. A design of the RPD framework, including matrix housings, is crucial to enable subsequent implant retention or support. Ultra-short implants (Straumann TL 4.1 × 4 mm) offer the opportunity for additional support and retention in the edentulous posterior region, where bone availability is often reduced. This future-oriented and minimally invasive approach with virtual treatment planning and guided implant surgery offers the possibility of simplified functional and cost-effective aftercare.

Keywords: prosthodontics; removable dental prosthesis; clasp-retained removable partial dentures; dental implant; computer-aided implant surgery (CAIS); patient-reported outcome measures (PROM); randomized controlled study (RCT)

1. Introduction

Population demographics show that overall edentulism is decreasing in the aging population due to improved prophylaxis and care, while the proportion of partially edentulous individuals is increasing as a result of longer life expectancy, increasing population aging, and the fact that more teeth are retained at older age due to improved oral health care and prevention [1–5].

Posterior molars are more likely to be lost than anterior teeth [1,5], resulting in remaining anterior dentition with bilateral edentulous posterior areas typically classified as Kennedy Class I [6]. The prosthetic gold standard for the treatment of cases with bilateral free-end situations are removable partial dentures (RPDs) with clasps, not at least due to the fact that fixed restorations with dental implants are often considered too costly, time-consuming, and invasive, particularly when additional augmentative measures are required [4]. In contrast, RPD treatment is minimally invasive, cost-effective, and patients can be provided with dentures in a timely manner [2–4].

RPDs with clasps in a Kennedy Class I situation are often associated with technical or biological complications due to the lack of posterior support [7] and frequent requirement of adjustments such as relining or fracture repair [8]. For instance, in a prospective study over 2 years, clasp-supported RPDs in Kennedy Class I situations showed partial non-occlusion in 35% of prostheses at 6 months, which increased to over 50% at 2 years in function [9]. In this context, various groups of authors reported that the reduction in residual ridge height is closely related to edentulism and denture use [10], and that non-denture wearers tend to have more residual edentulous ridge tissue compared to denture wearers. [11]. In a retrospective study of Kennedy Class II patients (unilateral free-end) using cone-beam computed tomography (CBCT) analyses, vertical and horizontal alveolar bone resorption in the edentulous sites was higher in RPD wearers than in patients without RPDs [12]. Thus, the alveolar ridge in the saddle region appears to be more susceptible to resorption due to the pressure load exerted by the prosthetic saddle. It can therefore be expected that improperly fitted RPDs will transfer even more unfavorable forces to the alveolar ridge, which may lead to further progression of resorption of the residual ridge [4].

A retrospective study reported that approximately 40% of partial dentures are no longer in use within 5 years [13] because of factors such as socio-demographics, pain, and esthetics [2]. Several reviews relate this to incorrect or inaccurate planning and execution in the fabrication of RPDs [2,4]. Considering that the number of partially edentulous patients will increase and that not every patient will receive fixed implant restorations for financial reasons or due to other factors such as a loss of several teeth and/or severe soft and hard tissue loss, it is necessary to give more importance to the topic of RPD [2,4]. Therefore, to optimize general RPD and minimize potentially damaging forces on abutment teeth and supporting tissues, at a minimum, the improved planning of denture design in conjunction with accurate assessment of tooth status, positions, and preparation, as well as patient education, consistent follow-up, and further development of new fabrication technologies and materials are required [2].

Provided that the RPD saddles are extended to rest at the retromolar pads, alveolar ridge resorption due to non-physiological loading is deemed to be minimized [14] Additional implants in the posterior region of RPDs, also known as implant-assisted removable partial dentures (IARPDs), have been described to improve function and stability [5,15]. According to a recent systematic review, conversion from conventional RPDs to IARPDs by subsequent implant placement and integration into the prosthesis, lead to an improvement in overall oral function (mastication, pronunciation, and esthetics) in partially edentulous Class I mandibular Kennedy patients. There was also a significant improvement in patient-reported outcome measures (PROMs), particularly in the field of physical pain and psychological impairment, and masticatory performance improved significantly in terms of maximum bite force, active occlusal contact area and mandibular jaw movement [16]. Another recent systematic review also reported that IARPDs showed favorable clinical outcomes and significantly higher patient satisfaction than before treatment and compared to conventional RPDs [17]. According to this review, stud attachments were most commonly used in IARPDs, and the different attachment systems did not affect implant survival or patient satisfaction [17]. However, both reviews addressed a lack of high-quality long-term RCT studies and a need for an IARPD treatment protocol based on well-structured, long-term clinical trials, in which implant location, type, and size, as well as type of attachment, metal framework design, and surgical and loading protocols are determined [16,17].

In light of these findings, a multicenter RCT was designed to evaluate the effect of additional implants in Kennedy Class I patients in terms of PROMs and cost–benefit analyses comparing maintenance care costs (initially and during long-term follow-up) for the treatment of partially edentulous patients with two additional posterior implants using two types of attachments that either support or retain RPDs. The case report presented here is part of this ongoing multicenter study and is intended to illustrate the clinical digital workflow and technical protocol for restoring a Kennedy Class I situation with a conventional RPD taking into account direct or later implant placement using ultra-short implants for additional support or retention in the posterior region.

2. Detailed Case Description

A 52-year-old male patient introduced himself to the University Center for Dental Medicine Basel (UZB) with the request for a prosthetic rehabilitation. He was in general good health, had never smoked, and related previous tooth loss to caries lesions. Initial periodontal and radiographic screening revealed healthy conditions of the remaining dentition with severe tooth wear and a collapsed bite (Figure 1).

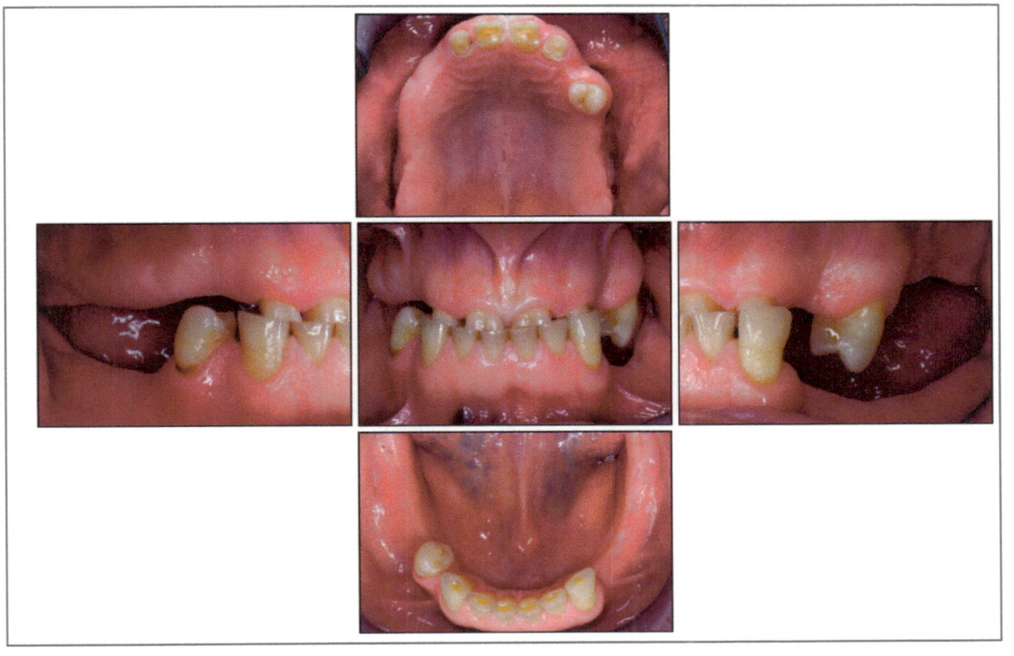

Figure 1. Clinical situation before treatment.

Due to the complex intermaxillary situation requiring an increase in the vertical dimension of occlusion (VDO), an overdenture prosthesis with anterior post copings and a clasp at the intact left premolar was planned in the maxilla, while a clasp-retained RPD was indicated in the mandible. The intact anterior dentition in the mandible determined the occlusal plane, and the increase in VDO was planned in the maxilla, which was restored first providing a diagnostic set-up in the mandible. For the RDP in the mandible, intraoral scans with and without the set-up were analyzed with a software program (3Shape Dental Manager, Copenhagen, Denmark), which enabled an exact planning of the prosthesis path of insertion and the position of the retentive clasp undercuts at the abutment teeth. This digital analysis replaced the traditional way with mounting casts in a surveyor to determine the path of insertion and undercuts at the abutment teeth (Figure 2a,b). Required occlusal rest cavities were prepared and a new intraoral scan was taken, which in turn was incorporated into the further planning. The RPD was then designed in the planning software and configured with 2 clasps at the most distal abutment teeth and an extension of the bilateral saddles to ensure support on the retromolar pads. In addition, the housings in the metal framework were scheduled for later implant healing abutments or retention elements in the area of the second molars (Figure 2c).

Figure 2. (**a**,**b**) Planning tooth preparations for three clasp retentions, of which only the two most distal abutment teeth were restored with clasps, (**c**) final removable partial denture with housings for future implant retention or support (arrows).

At the same time, two implants were planned using the coDiagnostiX® software (Version 10.7.0, Straumann AG, Basel, Switzerland). Therefore, the intraoral scan of the analog tooth set-up (Figure 3) or a virtual set-up of the previously planned denture design were imported for orientation. This scan of the set-up and the intraoral scan of the mandible without set-up were overlaid with the CBCT data (Figure 3). By switching between the prosthetic planning software (3Shape Dental Manager, version 2.21.2.2 (2021-1), Copenhagen, Denmark) and the implant planning software (coDiagnostiX®, Straumann AG, Basel, Switzerland), the implant axes and positions were aligned according to the path of insertion of the RPD, and the housings in the virtual RPD were planned accordingly. Subsequently, the surgical guide for navigated implant placement was fabricated by 3D printing (Rapidshape P-Series, Straumann AG; Basel, Switzerland).

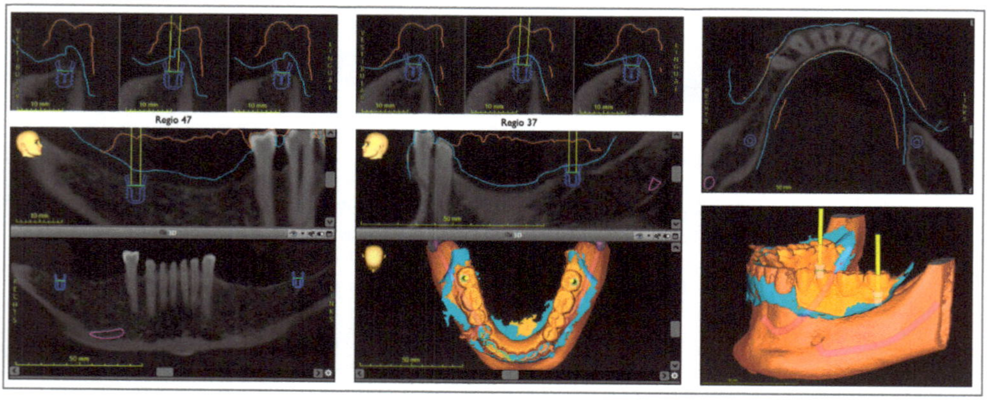

Figure 3. Digital planning of implants in the second molar region.

Implant placement was performed fully guided with the printed surgical guide (Figure 4a). Two ultra-short tissue-level implants with a diameter of 4.1 mm and 4 mm intraosseous length (Straumann AG, Basel) were inserted in the posterior molar region (Figure 4b) and left for submerged healing; in the case of an existing interims prosthesis, the saddles have to be relieved in the wound area (Figure 4c).

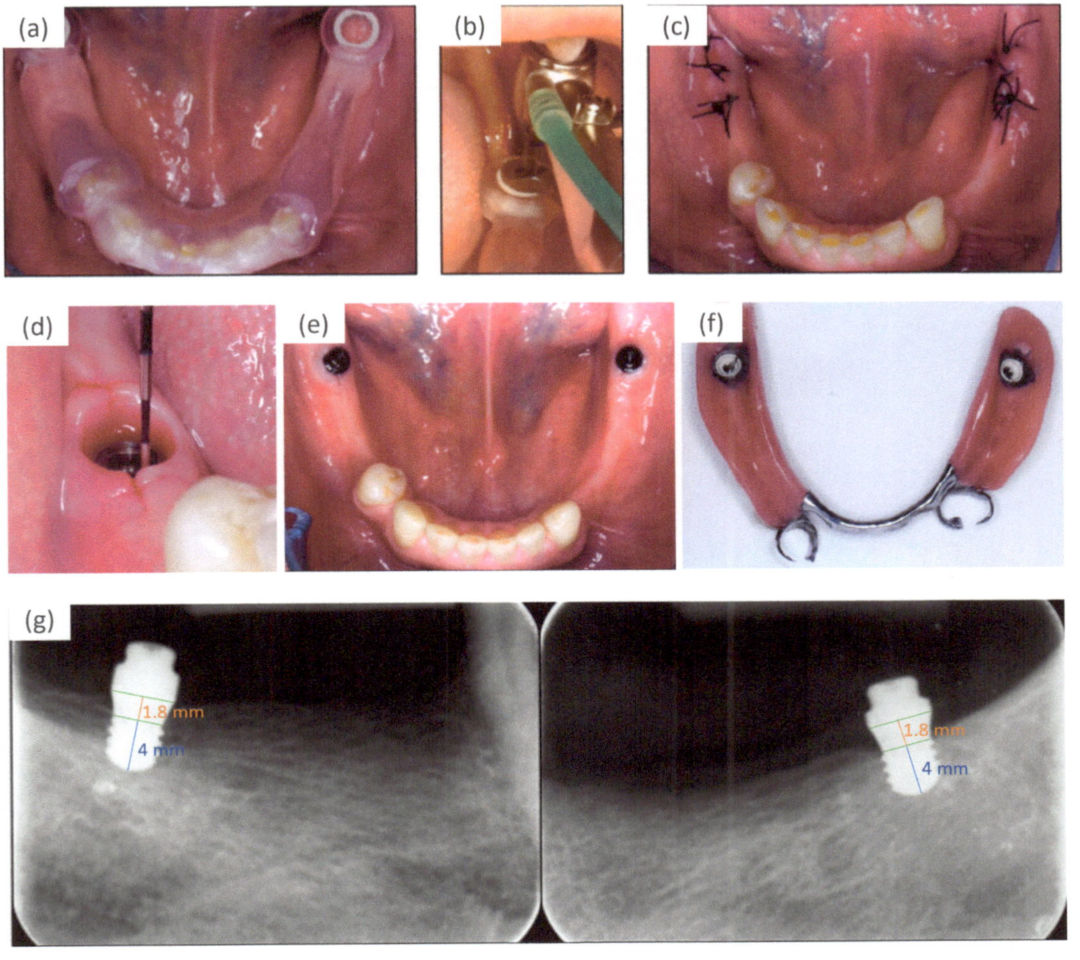

Figure 4. (**a**) Surgical guide in place, (**b**) guided preparation of implant site, (**c**) implant in place for submerged healing, (**d**) selection of height of retention elements according to mucosal height and vertical space, (**e**) retention elements connected, (**f**) view from basal with matrices fixed in the RPD, (**g**) radiographs after abutment connection; the markings on the implants indicate the 4 mm length of the intraosseous part and 1.8 mm of the polished part, which is placed at the level of the soft tissue margin.

When the fully navigated implant placement corresponded to the virtual implant planning, the virtually planned prosthesis was fabricated using a chromium-cobalt restoration by a milling process in a high precision milling machine (PFM 4024-5D, Primacon, Peissenberg, Germany) completed with denture teeth and the prostheses saddle. The RPD was incorporated after completion of wound healing approximately 3 weeks after surgery. In the event of significant manual corrections of the implant position, a new intraoral scan or an analog impression has to be taken to visualize the approximate implant position. The virtual RPD planning can be adjusted accordingly, and the prosthesis fabricated.

Following a 3-month submerged implant healing period, abutment connection was conducted, and retention elements were inserted (Novaloc, Straumann; Figure 4d–g). The height of the elements were selected taking into account the height of the mucosal peri-implant tissues and the vertical space (Figure 4d). Matrices were fixed in the RPD using self-curing methylmethacrylate resin (Unifast Trad, GC Europe AG, Luzern, Switzerland) (Figure 4f). The patient's perception of the RPD before and after connection to the implant was assessed with questionnaires on function, stability, and satisfaction [18]; the OHIP-G14 and masticatory performance with a color mixing ability test [19]. Based on these, the patient indicated a beneficial impact on his perception and function after prosthetic rehabilitation with implant retention. Figure 5 shows the patient's completed prosthetic restoration in both jaws (Figure 5) and Figure 6 provides an overview of the patient's clinical treatment schedule (Figure 6).

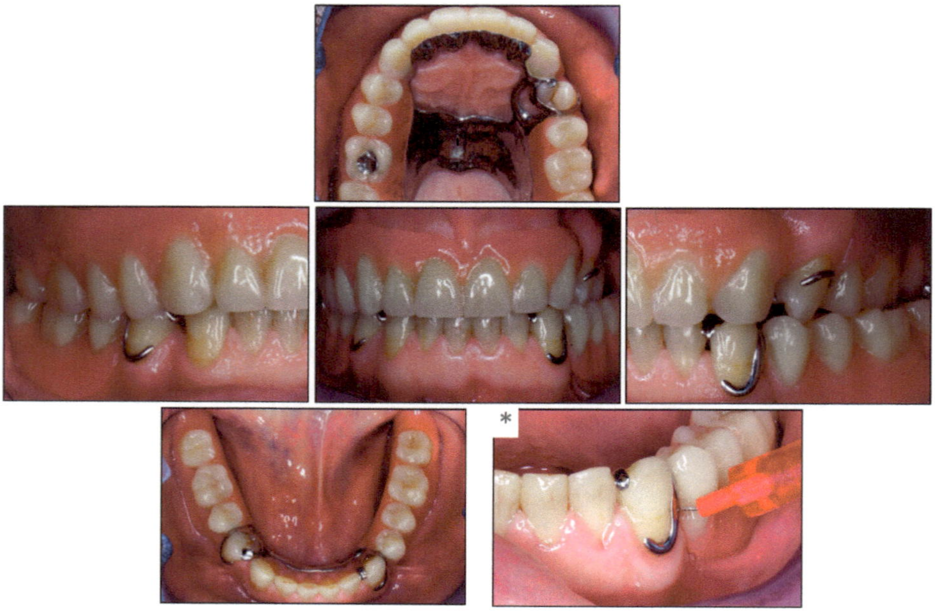

Figure 5. Clinical situation after prosthetic rehabilitation. (*) Illustration of the cleaning ability in the proximal space distal to the abutment tooth with an interdental brush.

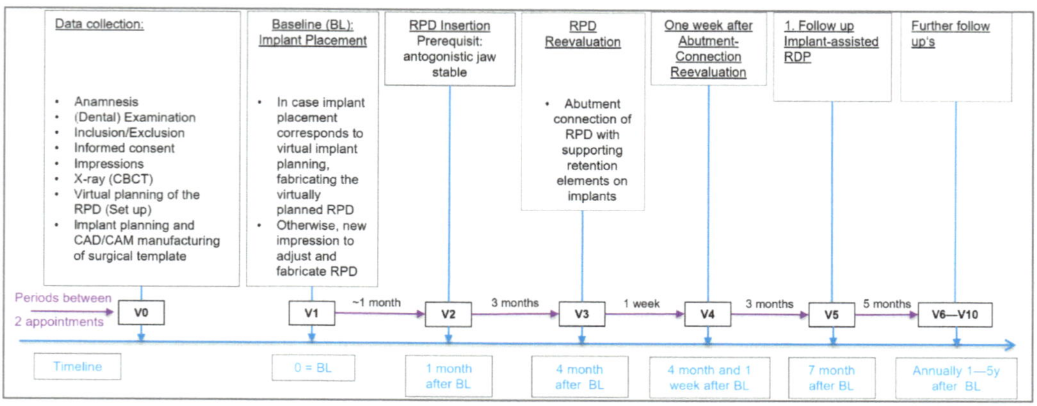

Figure 6. Overview of the patient's clinical treatment schedule (simplified from the original study protocol).

3. Discussion

The presented case report described a partially edentulous Class I Kennedy patient restored with an RPD and a fully guided surgical protocol using ultra-short implants in the posterior regions.

The advanced atrophy of the mandibular alveolar crest can be challenging in clinical practice due to lack of bone and proximity to anatomically vulnerable structures such as the nervus alveolaris inferior in the mid-posterior part of the mandible or the sinus in the posterior part of the maxilla. The use of ultrashort implants simplifies implant placement because invasive bone augmentation procedure is usually not required, and burden and costs are reduced, while still providing the benefit of support or retention of the RPD. Furthermore, in the case of technical or biological complications, the ultra-short implant design simplifies implant removal if required, keeping the intervention minimally invasive and enable a back-off strategy at any point of therapy [20,21]. The tissue-level implant design initially intended for an unsubmerged healing protocol seems to be advantageous due to the ease of maintenance, particularly in the posterior regions, which are difficult to access for personal oral hygiene [22]. The use of ultra-short 4 mm implants are primarily considered in severely reduced alveolar ridges. Certainly, the described concept of IARPD can also be implemented if according to implant planning longer implants are feasible without further bone augmentation.

The long-term performance of the ultra-short implants used in this study remains to be evaluated during follow-up. In a study using IARPDs in Kennedy Class I situations, in which 6-mm-long and 4.1-mm-wide implants were used together with longer and narrower implants, the mean marginal bone loss (MBL) after one year of function was reported 1.10+/−0.53 mm for the 6mm long implants [23], which is critical for ultra-short implants. In this study, free-end saddles were retained on 4 implants with locators, of which the posterior ones measured 6 mm length [23]. According to a finite element analysis study, implants in IARPDs are subjected to high stresses, which is why it was recommended to reduce stress by, among other things, using more than one implant to support a free end saddle [24]. No data from clinical studies with IARPDs are available reporting about the retention elements applied in the current ongoing study. Using healing abutments for RPD support, their loosening has been frequently reported as a common technical complication, which can be easily remedied [16]. The most important issue seems to be that the alveolar ridge remains more stable with the implant in situ. In case of implant loss, posterior ultra-short implants can be easily replaced at the same or an adjacent distal site. To enable implant replacement and avoiding subsequent prosthesis remake, an extended housing should be incorporated into the metal framework in the saddle area during initial IARPD fabrication. Even when the patient is against an IARPD treatment and prefers a simple RPD at the beginning, it still may be beneficial to incorporate a wide housing in the framework to enable implant placement at a later timepoint without weakening the prosthesis stability.

Treatment in the present case was performed according to the described protocol without any significant deviation, i.e., solely based on the intraoral scans, the virtual prosthesis and implant planning, with fully navigated implant insertion and subsequent prosthesis fabrication using a digitally milled chromium-cobalt restoration completed with denture teeth. The prosthesis could be inserted and subsequently "connected" with the retention elements selected intraorally according to the height of the healed peri-implant mucosal tissues and the available vertical space, and matrices inserted chairside.

Patient-centered outcomes were measured with the Oral Health Impact Profile (OHIP) and the assessment of chewing efficiency and evaluation of implant survival/success will be followed up over five years in function. As this study is still ongoing, the long-term results of all the included patients are to be awaited. Nonetheless, this specific case showed promising results in terms of patients' perception of function and oral health (OHQoL-G14) comparing conventional RDPs to implant-supported RDPs with retentive components in posterior sites of Kennedy Class I or, as in other examined individuals in this study, with implant healing abutments for prosthesis support. Personal oral hygiene was easily conducted with and without the prostheses in place (Figure 5,*) and the amount of maintenance care requirements were rated as low.

Author Contributions: Conceptualization, N.U.Z., T.J., S.W. and T.T.; methodology, N.U.Z., T.J., S.W. and T.T.; resources, I.K.-S. and L.K.Z.; clinical treatment and supervision, L.K.Z. and N.U.Z.; writing—original draft preparation, I.K.-S., L.K.Z. and T.T.; writing—review and editing, N.U.Z., T.J. and S.W.; project administration, N.U.Z., T.J. and S.W.; funding acquisition, N.U.Z., T.J. and S.W. All authors have read and agreed to the published version of the manuscript.

Funding: This study was supported by the International Team of Implantology, Switzerland (ITI Research grant No. ITI-1376_2019).

Institutional Review Board Statement: This study was conducted in accordance with the Declaration of Helsinki and approved by the Ethics Committee of Northwest and Central Switzerland (ID 2019-02167, approval date: 3 January 2020).

Informed Consent Statement: Written informed consent was obtained from the patient for publication of this case report and any accompanying images.

Data Availability Statement: Additional data can be provided by the corresponding author on request.

Acknowledgments: The contribution of Sebastian Kühl, Department of Oral Surgery, is acknowledged.

Conflicts of Interest: All authors declare that they have no conflict of interest.

References

1. Bergman, B.; Hugoson, A.; Olsson, C.O. A 25 year longitudinal study of patients treated with removable partial dentures. *J. Oral Rehabil.* **1995**, *22*, 595–599. [CrossRef]
2. Campbell, S.D.; Cooper, L.; Craddock, H.; Hyde, T.P.; Nattress, B.; Pavitt, S.H.; Seymour, D.W. Removable partial dentures: The clinical need for innovation. *J. Prosthet. Dent.* **2017**, *118*, 273–280. [CrossRef] [PubMed]
3. Friel, T.; Waia, S. Removable Partial Dentures for Older Adults. *Prim. Dent. J.* **2020**, *9*, 34–39. [CrossRef]
4. Kim, J.J. Revisiting the Removable Partial Denture. *Dent. Clin. N. Am.* **2019**, *63*, 263–278. [CrossRef]
5. Zitzmann, N.U. Point of Care: When restoring the teeth of partially edentulous patients with removable partial dentures, do you consider placing implants to enhance the retention and stability of the prosthesis? *J. Can. Dent. Assoc.* **2005**, *71*, 551–552.
6. Applegate, O. The Rationale of Partial Denture Choice. *J. Prosth. Dent.* **1960**, *10*, 891–907. [CrossRef]
7. Wagner, B.; Kern, M. Clinical evaluation of removable partial dentures 10 years after insertion: Success rates, hygienic problems, and technical failures. *Clin. Oral Investig.* **2000**, *4*, 74–80. [CrossRef] [PubMed]
8. Schneider, C.; Zemp, E.; Zitzmann, N.U. Oral Health Improvements in Switzerland over 20 Years. *Eur. J. Oral Sci.* **2017**, *125*, 55–62. [CrossRef] [PubMed]
9. Sassen, H. Functional parameters and occlusion of partial dentures as a function of the type of attachment. *Dtsch. Zahnarztl. Z.* **1990**, *45*, 576–578.
10. Canger, E.M.; Celenk, P. Radiographic evaluation of alveolar ridge heights of dentate and edentulous patients. *Gerodontology* **2012**, *29*, 17–23. [CrossRef]
11. Pietrokovski, J.; Harfin, J.; Levy, F. The influence of age and denture wear on the size of edentulous structures. *Gerodontology* **2003**, *20*, 100–105. [CrossRef] [PubMed]
12. Ozan, O.; Orhan, K.; Aksoy, S.; Icen, M.; Bilecenoglu, B.; Sakul, B.U. The effect of removable partial dentures on alveolar bone resorption: A retrospective study with cone-beam computed tomography. *J. Prosthodont. Implant. Esthet. Reconstr. Dent.* **2012**, *22*, 42–48. [CrossRef]
13. Koyama, S.; Sasaki, K.; Yokoyama, M.; Sasaki, T.; Hanawa, S. Evaluation of factors affecting the continuing use and patient satisfaction with Removable Partial Dentures over 5 years. *J. Prosthodont. Res.* **2010**, *54*, 97–101. [CrossRef] [PubMed]
14. Zitzmann, N.U.; Rohner, U.; Weiger, R.; Krastl, G. When to choose which retention element to use for removable dental prostheses. *Int. J. Prosthodont.* **2009**, *22*, 161–167. [PubMed]
15. Shahmiri, R.A.; Atieh, M.A. Mandibular Kennedy Class I implant-tooth-borne removable partial denture: A systematic review. *J. Oral Rehabil.* **2010**, *37*, 225–234. [CrossRef]
16. Park, J.A.-O.; Lee, J.A.-O.; Shin, S.A.-O.; Kim, H.A.-O.; Molinero-Mourelle, P.; Bischof, F.; Yilmaz, B.; Schimmel, M.; Abou-Ayash, S. Effect of conversion to implant-assisted removable partial denture in patients with mandibular Kennedy classification I: A systematic review and meta-analysis. *Clin. Oral Implant. Res.* **2020**, *31*, 360–373. [CrossRef]

17. Putra Wigianto, A.Y.; Goto, T.A.-O.; Iwawaki, Y.; Ishida, Y.; Watanabe, M.; Ichikawa, T. Treatment outcomes of implant-assisted removable partial denture with distal extension based on the Kennedy classification and attachment type: A systematic review. *Int. J. Implant. Dent.* **2021**, *7*, 111. [CrossRef]
18. Zitzmann, N.U.; Marinello, C.P. Treatment outcomes of fixed or removable implant-supported prostheses in the edentulous maxilla. Part I: Patients' assessments. *J. Prosthet. Dent.* **2000**, *83*, 424–433. [CrossRef]
19. Schimmel, M.; Christou, P.; Miyazaki, H.; Halazonetis, D.; Herrmann, F.R.; Müller, F. A novel colourimetric technique to assess chewing function using two-coloured specimens: Validation and application. *J. Dent.* **2015**, *43*, 955–964. [CrossRef]
20. Schimmel, M.; Janner, S.F.M.; Joda, T.; Wittneben, J.G.; McKenna, G.; Brägger, U. Mandibular implant-supported fixed complete dental prostheses on implants with ultrashort and standard length: A pilot treatment. *J. Prosthet. Dent.* **2021**, *126*, 137–143. [CrossRef]
21. Schimmel, M.; Müller, F.; Suter, V.; Buser, D. Implants for elderly patients. *Periodontology 2000* **2017**, *73*, 228–240. [CrossRef]
22. Derks, J.; Schaller, D.; Håkansson, J.; Wennström, J.L.; Tomasi, C.; Berglundh, T. Effectiveness of Implant Therapy Analyzed in a Swedish Population: Prevalence of Peri-implantitis. *J. Dent. Res.* **2016**, *95*, 43–49. [CrossRef] [PubMed]
23. Jensen, C.; Speksnijder, C.M.; Raghoebar, G.M.; Kerdijk, W.; Meijer, H.J.A.; Cune, M.S. Implant-supported mandibular removable partial dentures: Functional, clinical and radiographical parameters in relation to implant position. *Clin. Implant. Dent. Relat. Res.* **2017**, *19*, 432–439. [CrossRef] [PubMed]
24. Eom, J.W.; Lim, Y.J.; Kim, M.J.; Kwon, H.B. Three-dimensional finite element analysis of implant-assisted removable partial dentures. *J. Prosthet. Dent.* **2017**, *117*, 735–742. [CrossRef] [PubMed]

prosthesis

MDPI

Article

Influence of Simulated Skin Color on the Accuracy of Face Scans

Ido Brintouch, Aisha Ali, Georgios E. Romanos and Rafael A. Delgado-Ruiz *

School of Dental Medicine, Stony Brook University, Stony Brook, NY 11794, USA; brintouchido@gmail.com (I.B.); aisha.ali2@stonybrookmedicine.edu (A.A.); georgios.romanos@stonybrookmedicine.edu (G.E.R.)

* Correspondence: rafael.delgado-ruiz@stonybrookmedicine.edu; Tel.: +1-631-632-6913

Abstract: Aims: this study aims to investigate the impact of simulated skin color and the use of fiducial markers on the accuracy of 3D facial scans, comparing two types of structured light scanners under constant ambient lighting conditions. Materials and Methods: Three mannequins with different skin colors—black, white, and pink—were scanned using two light based hand-held scanners (infrared light and blue-light). Each mannequin was scanned with and without fiducial markers placed on defined anatomical landmarks. A total of one hundred thirty-two scans were performed and converted into standard tessellation language (STL) files. STL files from each scanner were compared to their respective control scans using point cloud comparison software. Accuracy was evaluated based on root mean square (RMS) values. Descriptive statistics summarized the data, and a *t*-test was performed to assess differences in RMS values between scans with and without fiducial markers for each scanner type. Results: The infrared light scanner showed the highest accuracy for the white mannequin, as evidenced by lower RMS values compared to the pink and black mannequins. Adding fiducial markers significantly enhanced scan accuracy for the pink and black mannequins. The blue-light scanner achieved accuracy for the white and pink mannequins comparable to that of the infrared scanner. However, it was unable to scan the black mannequin, even with the use of markers. Conclusions: Within the limitations of this study, simulated skin color significantly affects the accuracy of facial 3D scans. Scans of lighter (white) tones demonstrate higher accuracy compared to darker tones. Fiducial markers enhance the accuracy for an infrared scanner; however, a blue-light scanner is unable to capture dark simulated skin, even with the addition of fiducial markers.

Keywords: face scanning; accuracy; skin tone

1. Introduction

Surface scanning technology plays a vital role in fields such as dentistry, cosmetics, construction, manufacturing, and inspections, as well as in identity verification protocols [1,2]. Surface scanners have been shown to provide acceptable accuracy compared to other methods [3]. In dentistry, surface scanners (intraoral and facial) are widely used in prosthodontics, orthodontics, and oral surgery [4]. Specifically, 3D facial scanners provide precise reconstructions of oral and facial features and textures [1], enhance patient satisfaction, enable personalized outcomes, and support the manufacturing of highly accurate prostheses [5].

The accuracy of facial scanning depends on various parameters, including the type of scanner and scanning conditions [6–18]. Applied to face scanning, stationary scanners offer higher accuracy than handheld scanners [11], as the movement associated with the handheld scanner and the movement of the patient's face (micro-movements of facial muscles and whole head macro-movements) can lead to less reliable models [12]. In relation to the scanning conditions, the longer it takes for a person to be scanned, the higher the likelihood of information becoming conflicting [7,8].

Facial scanning can be achieved using various techniques, including photogrammetry, photographs, video, and light-based 3D scanning. Light-based scanners use different light sources, such as white, blue, or infrared light [16,17]. Different types of light-based scanners offer unique advantages. White light scanners are generally safer for the eyes and are effective at capturing structures with irregular surfaces. In contrast, blue light scanners provide higher precision and capture fine details more accurately but require eye protection due to potential risks, such as cataracts or macular degeneration, if used without proper safeguards [17,18]

Besides the type of light, the way light interacts with an object is also a major factor when addressing light conditions. Light works by carrying energy to a material and, once reached, the energy will be transferred to that material [14]. Once the energy has reached an object, absorption and reflection occur based on the type of material the light has interacted with. Black or darker colors are recognized to be great absorbers of light, while lighter colors are known to reflect light. For example, black objects are known to perfectly absorb light containing many different wavelengths [15]. When light is absorbed instead of reflected it contributes to the formation of far fewer shadows and less recognition of edges and corners. On the contrary, lighter backgrounds are well known to reflect light and create a well-defined edge or shadow. This concept of edges and shade matching being easily definable when scanning is vastly important to ensure an accurate final result [19]. A parameter that until now has not been investigated but which may influence a scan's accuracy is the scanned surface's color. This parameter is relevant because the color of the scanned object interacts with the light source either by reflecting or absorbing it, thus potentially affecting the accuracy of the scanner [20].

In a recent study, Varda et al., 2022, evaluated the influence of ambient light and object color on the 3D scanning process. In their study, the authors scanned two identical geometrical samples, one white and one black, using two three-dimensional scanners and four variations in the ambient light [21]. Their study showed that both scanners possessed similar accuracy, and white objects were more consistently scanned than black objects [21]. Building on these findings, this study aims to investigate the impact of simulated skin color and the use of fiducial markers on the accuracy of 3D facial scans, comparing two types of light-based scanners (infrared light vs. blue light) under constant ambient lighting conditions. The null hypothesis is that neither skin color, the presence of markers, nor the type of scanner has any effect on scanner accuracy.

2. Materials and Methods

2.1. Groups and Sample Size

Mannequin heads with three different simulated skin tones—dark (black mannequin), very light (white mannequin), and medium (pink mannequin)—were used in this in vitro study. The black mannequin model ref# HTC, the white mannequin ref# Style 3, and the pink mannequin ref# Meahus, were procured online from www.amazon.com. The black and white mannequins were made of styrofoam, while the pink mannequin was constructed from silicone rubber. The dimensions of the silicone mannequin were slightly larger than those of the styrofoam mannequins. Facial scanning was performed using two protocols: the first without fiducial markers (control group, C) and the second with fiducial markers (test group, T). For the scans with fiducial markers, nine markers were placed at anatomically significant landmarks: exocanthion (right and left), cheilion (right and left), pogonion, philtrum, helical crus of the ears (right and left), and the glabella (Figure 1).

The sample size was determined using the Raosoft sample size calculator http://www.raosoft.com/samplesize.html (accessed on 10 March 2024) to achieve 80% statistical power, with a 7% margin of error and an effect size of 0.5. This analysis indicated that 132 scans were needed (66 per scanner type). For each scanner, thirty-three scans were conducted with fiducial markers (test group) and thirty-three without markers (control group). Each skin tone (black, white, and pink) was represented equally with eleven scans per skin tone.

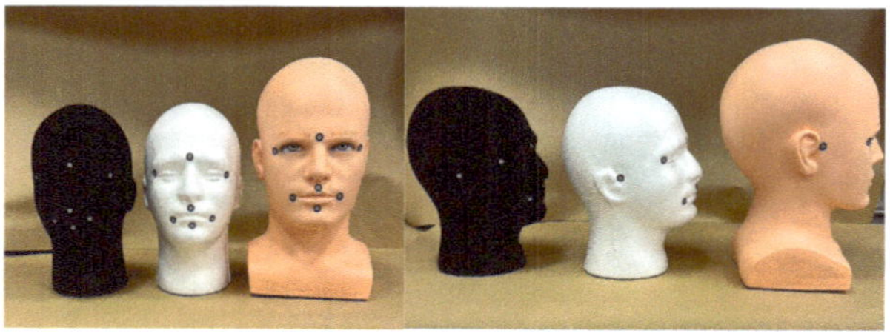

Figure 1. Frontal and a lateral view of the mannequin heads with fiducial markers placed. Nine locations were marked: left and right exocanthion, left and right cheilion, the pogonion, the philtrum, the right and left helical crus of the ears, and the glabella.

2.2. Scanning Procedure and Post-Processing

The mannequins were placed on a Revopoint Dual-Axis Turntable stand (Revopoint, Shenzhen, China). With a rotational speed/20 s per full rotation (Figure 2).

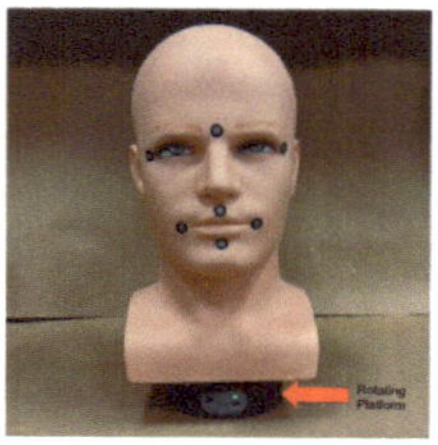

Figure 2. A red arrow pointing to the rotating base/platform placed below each mannequin.

Scanning was conducted using two types of light-based scanners. The first scanner, the POP-3 Plus® Portable 3D scanner (Revopoint, Shenzhen, China), utilizes dual-camera infrared structured light, providing an accuracy of 0.05 mm and a resolution of 0.08 mm. The second scanner, the Seal® 3D scanner (3D-Maker Pro, Shenzhen, China), is a portable 24-bit color scanner based on blue structured light, with an accuracy of 0.02 mm and a resolution of 0.07 mm (Figure 3).

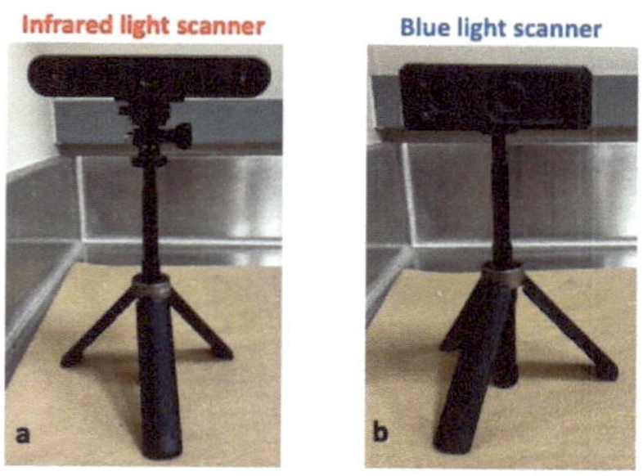

Figure 3. Scanners used in this experiment. The image shows the infrared light scanner (**a**) and the blue light scanner (**b**) mounted on a tripod.

Each scanner was mounted on a tripod for stability during the scanning process, while each mannequin head was placed on the rotational stand at a fixed distance of 22 inches (55.88 cm) from the scanner (Figure 4).

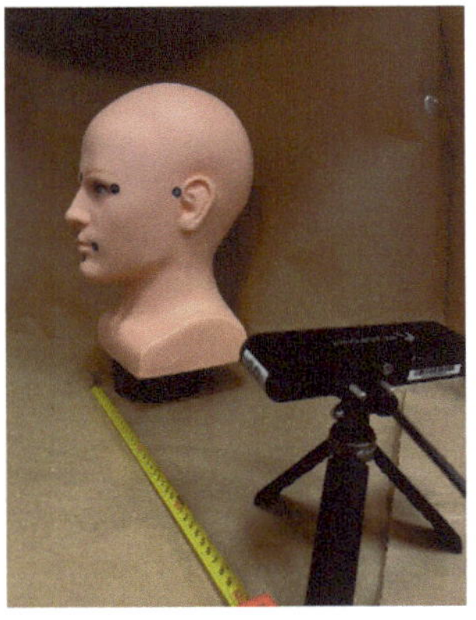

Figure 4. Setting scanning distance. The image shows one of the mannequin heads on the rotational base, a meter with the set distance, and the camera mounted in the tripod.

Calibration was conducted for each scanner by scanning each mannequin head 10 times, using various scanning options, such as unique shapes, dark object, face mode, and body mode, and high resolution for both scanners. This iterative process was used to optimize the final scanning parameters, which included high accuracy, feature tracking, and general object mode, with color scanning enabled. The duration of each scan was set for two complete rotations of the head, capturing approximately 900 to 1000 frames for each completed scan. Following the completion of scans, image post-processing was conducted using the designated software for each scanner: RevoScan 5 Version 4.3.2 (Revopoint, Shenzhen, China) for the POP-3 Plus® scanner (Revopoint, Shenzhen, China), and JMStudio-MAC-2.3.5 (3D-Maker Pro, Shenzhen, China) for the Seal® 3D scanner (3D-Maker Pro, Shenzhen, China). The post-processing included Digital Trimming, Fusion, Isolation, Overlap, Smooth, Simplify, Mesh, Fill Holes, Texture Mapping, and File export.

- Digital trimming: Reduced noise and redundant information.
- Fusion: Merged all captured data into a single unified model.
- Isolation: Removed unrelated/background data not connected to the main model.
- Overlap: Detected and eliminated overlapping data points.
- Smooth: Removed noise and duplicated data to create a cleaner model.
- Simplify: Compressed the data to reduce the overall file size.
- Mesh: Enhanced the model's quality by controlling point density and detail.
- Fill Holes: Repaired areas with missing data to ensure uniformity.
- Texture Mapping: Helped replicate surface textures.
- File export: Files were exported as standard tessellation language (STL) for comparative evaluation of point clouds.
- Eleven evaluations were completed without markers and eleven evaluations were completed with markers for each mannequin group for each scanner. For a total of one hundred and thirty-two evaluations.

2.3. Global Deviations

Global deviations between scans were assessed using the open-access surface-matching software CloudCompare Version 2.13.0 Kharkiv (https://www.cloudcompare.org/main.html, accessed on 1 June 2024). For each experimental group, a designated reference scan was compared to all other scans within the same group to evaluate deviations.

STL files (control scan and comparison scans) were imported into CloudCompare software. The models were aligned using the 'Best Alignment' tool, and three points were identified on each STL file to achieve finer registration. Cloud registration was employed with parameters set as follows: RMS difference of 1×10^{-5}, final overlap of 100%, and a maximum thread count of 7/8. Distances between the STL models were then calculated by selecting the reference, comparing models, and using the 'Compute Stat. Parameters' tool. This tool provided RMS (root mean square) values and generated merged STL files with a color-coded scalar field, visually highlighting areas of deviation between the reference and compared models. The RMS value expressed in millimeters was used to evaluate the accuracy of the scans.

Descriptive statistics including mean, standard deviations, and margin of error were presented for all the groups. Box plots were used to illustrate the findings of all groups. Paired *t*-tests were completed to determine intra-group differences between scanners completed with and without markers. To determine the statistical significance was determined when $p < 0.05$.

2.4. CRIS Guidelines

In this experimental in vitro study, the Checklist for Reporting In Vitro Studies (CRIS) guidelines were implemented to improve the quality of the reported data [22]. According to the CRIS guidelines, sample size calculation must be described, meaningful difference between groups must be presented, sample preparation and handling must be described, sample allocation and statistical analysis must be described.

3. Results

Cloud comparisons showed small deviation for scanners obtained on white mannequins compared to pink and black mannequins and improved accuracy for scans completed with fiducial markers compared to scans completed without fiducial markers (Figures 5–7).

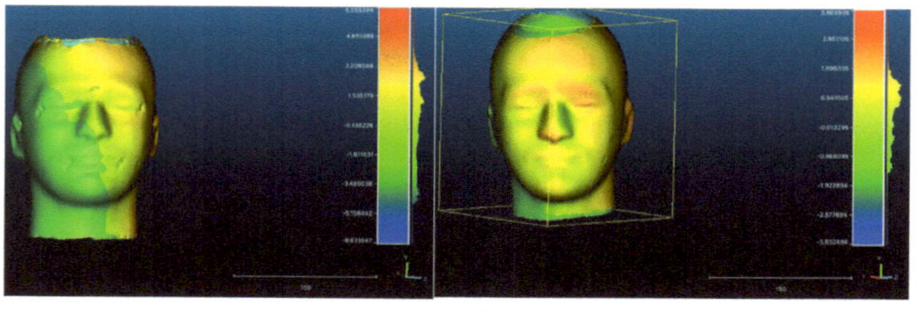

Figure 5. Illustrative images of white mannequin heads showing scans overlaid on the control scan. The left image includes fiducial markers, while the right image does not. The color scale represents the RMS values, indicating the accuracy of the scans. Best accuracy of all groups achieved with the white mannequins.

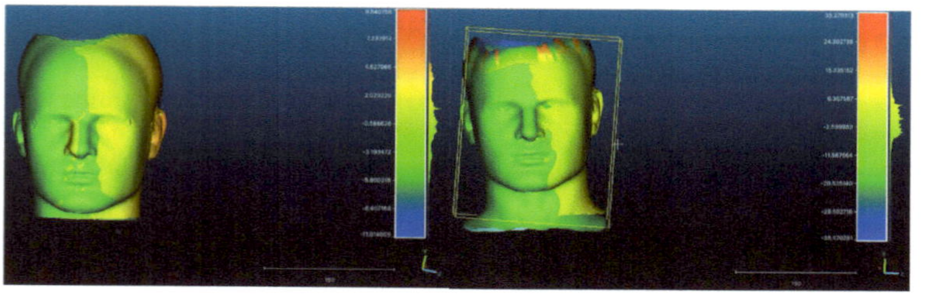

Figure 6. Illustrative images of pink mannequin heads with scans overlaid on the control scan. The left image includes fiducial markers, while the right image does not. The color scale represents the RMS values, indicating scan accuracy. Accuracy on pink mannequins was lower than on white mannequins but higher than on black mannequins.

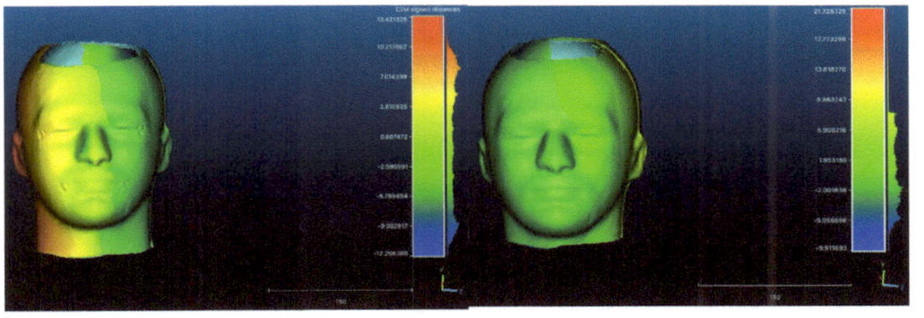

Figure 7. Illustrative images of black mannequin heads with scans overlaid on the control scan. The left image includes fiducial markers, while the right image does not. The color scale represents the RMS values, indicating scan accuracy. Accuracy on black mannequins was the lowest of all groups (white and pink).

The infrared scanner showed better accuracy for white and pink mannequins and lower accuracy for black mannequin scans. The accuracy for pink and black mannequins improved when markers were used (Figure 8).

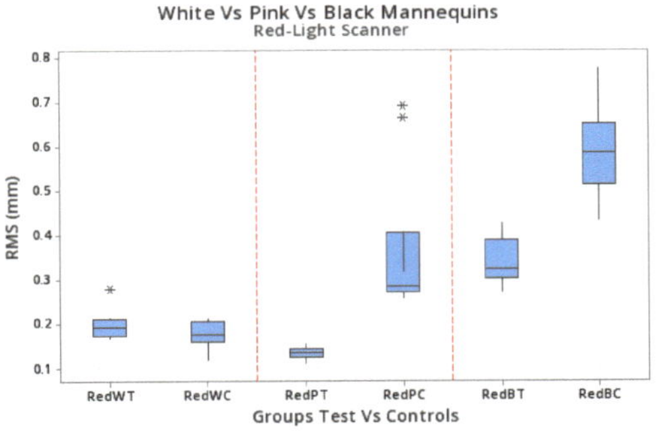

Figure 8. Infrared scanner. Box-plot comparison of RMS values for object color scans in white (W), pink (P), and black (B) mannequins with (T) and without markers (C). The asterisk (*) represents outlier.

The blue-light scanner demonstrated lower scanning accuracy. White and pink mannequins showed a behavior comparable to the infrared scanner, it was unable to capture scans of the black mannequins. Errors such as "object not detected", "out of sight", "too close", and blank measurements were consistently recorded during attempts to scan black mannequins, regardless of the presence of fiducial markers (Figure 9).

3.1. Global Deviations for the Infrared Scanner in Mannequins with and Without Markers

The white mannequin group with markers (RedWT) had lower RMS and lower standard deviation (0.17753 ± 0.02781) compared to the white group with markers (RedWC) (0.1977 ± 0.0317). The pink mannequin with markers (RedPT) (0.13542 ± 0.01310) had lower RMS and lower standard deviation compared to the pink mannequin without markers (RedPC) (0.3612 ± 0.1685). The black mannequin with markers (RedBT) showed lower RMS and lower standard deviations (0.3417 ± 0.0506) compared with the black mannequin without markers (RedBC) (0.5902 ± 0.1063) (Table 1).

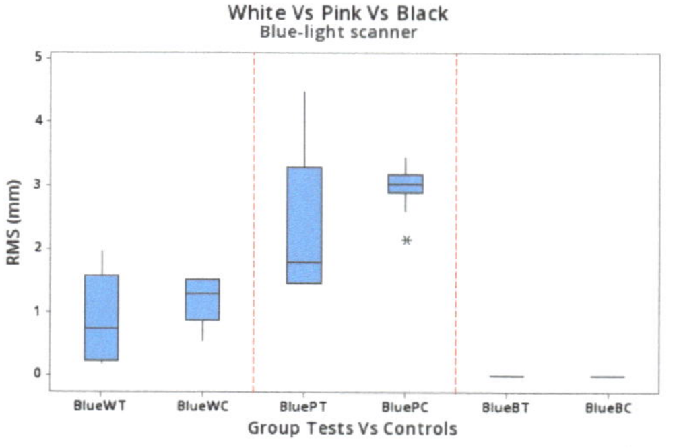

Figure 9. Blue-light Scanner. Box-plot comparison of RMS values for object color scans in white (W), pink (P), and black (B) mannequins with (T) and without markers (C). The asterisk (*) represents outlier.

Table 1. Descriptive statistics of RMS values for object color scans for infrared scanner on white (W), pink (P), and black (B) mannequins with (T) and without markers (C).

Group	N	Mean	StDev	95% CI
RedWT (Test)	10	0.1977	0.0317	(0.1779, 0.2175)
RedWC (Control)	10	0.17753	0.02781	(0.15771, 0.19734)
RedPT (Test)	10	0.13542	0.01310	(0.05602, 0.21482)
RedPC (Control)	10	0.3612	0.1685	(0.2818, 0.4406)
RedBT (Test)	10	0.3417	0.0506	(0.2864, 0.3970)
RedBC (Control)	10	0.5902	0.1063	(0.5349, 0.6455)

3.2. *Global Deviations for the Blue-Light Scanner in Mannequins with and Without Markers*

The white mannequin group with markers (BlueWT) showed lower RMS (0.8280 ± 0.45934) compared to the white group with markers (BlueWC) (1.19940 ± 0.330039). The pink mannequin with markers (BluePT) (2.26478 ± 1.0608) had lower RMS compared to the pink mannequin without markers (BluePC) (2.98472 ± 0.381317). Scanning of the black mannequin with or without markers was not possible (Table 2).

Table 2. Descriptive statistics of RMS values for object color scans for blue-light scanner on white (W), pink (P), and black (B) mannequins with (T) and without markers (C).

Group	N	Mean	StDev	95% CI
BlueWT (Test)	10	0.8280	0.45934	(0.1965, 1.9653)
BlueWC (Control)	10	1.19940	0.330039	(0.55231, 1.5223)
BluePT (Test)	10	2.26478	1.0608	(1.4749, 4.4803)
BluePC (Control)	10	2.98472	0.381317	(2.15074, 3.45292)
BlueBT (Test)	10	-	-	-
BlueBC (Control)	10	-	-	-

3.3. *Statistical Comparison of Infrared Scans with and Without Markers*

Paired t-test showed no differences in the RMS values of white mannequin with or without markers $p > 0.05$. Meanwhile, there were statistically significant differences in the pink mannequins and black mannequins with and without markers $p < 0.05$ (Table 3).

Table 3. T-value and P-value comparisons of RMS values for white (W), pink (P), and black (B) colored objects with (D) and without markers (C).

Samples	T-Value	Adjusted p-Value
InfraRedWC (Control) vs. RedWT (Test)	−1.36	0.207
InfraRedPC (Control) vs. RedPT (Test)	4.23	0.001
InfraRedBC (Control) vs. RedBT (Test)	6.67	0.001

3.4. Statistical Comparison of Blue-Light Scans with and Without Markers

The paired *t*-test revealed no significant differences in the RMS values for the white mannequin scanned with or without fiducial markers ($p > 0.05$). Similarly, no differences were found in the RMS values for the pink mannequin scanned with and without markers ($p > 0.05$). Due to the lack of data, comparisons were not possible for the black mannequin (Table 4).

Table 4. T-value and P-value comparisons of RMS values in white (W), pink (P), and black (B) object color with (D) and without markers (C).

Samples	T-Value	Adjusted p-Value
BlueWC (Control) vs. BlueWT (Test)	−1.81	0.148
BluePC (Control) vs. BluePT (Test)	−1.88	0.092
BlueBC (Control) vs. BlueBT (Test)	-	-

4. Discussion

This study aimed to assess the effects of simulated skin color and fiducial marker use on the accuracy of 3D facial scans, comparing the performance of two light-based scanners (infrared light vs. blue light). Findings demonstrated that scans of simulated white skin achieved higher accuracy than those of simulated pink and black skin. Furthermore, the addition of fiducial markers significantly enhanced scan accuracy for simulated pink and black skin but had minimal impact on simulated white skin with the infrared scanner. Notably, the blue-light scanner showed lower accuracy than the infrared scanner and was unable to capture scans of the black mannequin head, even with fiducial markers. Thus, the null hypothesis was partially rejected.

These findings can be attributed to the optical properties of light interaction with different colors. Light is predominantly reflected from white surfaces, partially reflected from pink surfaces, and absorbed by black surfaces [20,22]. Allred et al. [20] demonstrated that lighter colors reflect more light, while darker colors reflect less, establishing a direct correlation between color and light reflection. Similarly, Bai et al. [23] found that darker materials, such as black carbon, absorb more light compared to lighter materials like elemental carbon. Their study, using thermal–optical carbon analyzers, showed that black carbon exhibited higher absorption values for both visible and infrared light. An additional explanation for the higher scanning accuracy observed in white mannequins compared to darker ones is the influence of ambient light on the scanning process [19]. White surfaces are more prone to creating shadows, which correlate with variations in depth and height, potentially contributing to the increased accuracy of scans from mannequins with simulated white skin.

The accuracies obtained in our study for the simulated white (RMS 0.177 ± 0.02), pink (RMS 0.361 ± 0.168), and black (RMS 0.590 ± 0.106) skin colors were significantly higher than those reported by Özsoy et al. [24] (RMS 0.78 to 3.42). These discrepancies are due to differences in experimental design. Özsoy et al.'s study involved real patients exhibiting various facial expressions and utilized different face scanning technologies, which may have contributed to the observed variations in accuracy.

Our study demonstrated higher accuracy compared to the findings of Knoops et al. [25], who evaluated three different scanning technologies: a full-body MRI scanner, a handheld scanner with two cameras, and a scanner equipped with a camera, infrared sensor, and two infrared lights, all compared against a dedicated facial scanner (structured light with three cameras). The reported RMS values for the MRI scanner, general surface scanner, and infrared-based camera were 1.11 ± 0.33, 0.71 ± 0.28, and 1.33 ± 0.46, respectively. The discrepancies between our study and theirs can be attributed to factors such as patient movement during scanning and the inherent differences in scanner technologies employed.

In this study, fiducial markers were utilized to assess their impact on scanning accuracy across different simulated skin tones. Our findings indicate that using markers significantly enhanced accuracy for the pink and black mannequins. This aligns with the results of Egri et al. (2022) [26], who investigated the effect of fiducial markers on surfaces of varying colors. They observed that markers improved scanning accuracy, particularly on dark backgrounds, due to the increased number of identifiable landmarks and the enhanced contrast between the markers and the surface. These findings support the notion that fiducial markers can serve as effective reference points, especially when scanning darker surfaces, where inherent contrast is limited.

In this study, nine facial landmarks were identified using fiducial markers, which were easily visible due to their distinct black and silver colors. From an optics standpoint, the high contrast between the markers and the skin surface enhanced their detectability by the scanner. This increased the number of reference points beyond the natural surface topography of the face, improving the scanner's triangulation accuracy. The stark contrast provided a clearer optical signal, reducing potential errors in capturing facial geometry and leading to more precise measurements [27]. As discussed previously, this study demonstrated that fiducial markers improved scanning accuracy for darker skin tones but had minimal impact on lighter skin tones. This is consistent with the findings of Varda et al. [21], who reported higher accuracy for scans of lighter-colored objects, although they did not provide an explanation for this observation. We hypothesize that lighter surfaces reflect more light, producing higher-quality data and sufficient accuracy without the need for additional markers. Conversely, darker skin tones absorb more light, resulting in reduced data capture. The use of fiducial markers on darker surfaces likely increased both the quantity and quality of reflected data, thereby enhancing scan accuracy.

In the present study, mannequin heads were used because they reduced confounding variables like facial movement [6,7]. Also, when scanning human subjects, if the time required to complete a scanner is longer, the likelihood of inaccuracies increases [4,5]. Furthermore, by using mannequin heads, the variable of the head position and movement were eliminated as a source of error [9].

An intriguing finding of this study is the inability of the blue-light-based scanner to effectively capture the black mannequin, while the infrared light scanner successfully accomplished this task. Infrared light, operating within the near-infrared spectrum (approximately 750 to 1400 nanometers), penetrates surfaces more deeply and experiences less absorption by dark materials, resulting in greater reflection. This property makes infrared light more effective for scanning objects with darker colors, including black [28]. In contrast, blue light, with wavelengths between 450 and 495 nanometers, is ideal for capturing fine surface details due to its shorter wavelength. However, blue light is more readily absorbed by dark surfaces, such as black, leading to diminished reflection and potential data loss during scanning [29], which could explain the lower accuracy of the blue-light scanner compared to the infrared-light scanner on darker surfaces.

This study has several limitations. First, the use of mannequins representing only three standardized skin tones, rather than real patients with diverse facial features and a broad spectrum of skin tones. While the mannequins provided realistic contours and anatomical landmarks, they lacked the complex characteristics of human skin, such as color variability, porosity, hair, and sweat. This controlled environment likely resulted in higher accuracy compared to what would be achieved with human subjects. However, the use of mannequins allowed for consistent and reproducible measurements, potentially mitigating some of the variability and inaccuracies associated with live facial scanning.

We recognize that the accuracy of 3D face scanners across different human skin tones requires additional investigation. Classifying human skin color presents unique challenges due to the variability in pigmentation, UV exposure, and superficial blood vessels, which can differ significantly even within the same individual. To minimize variability, we selected three basic colors for mannequins—white, pink, and black—representing very light, intermediate, and dark tones according to the individual typology angle (ITA) classification; thus, in this study, we covered half of six potential ranges of skin color [30].

This study has several strengths. First, the use of mannequins effectively eliminates head movements and micromovements of facial muscles, which can significantly impair the overall accuracy of the scanning process. Second, the incorporation of a rotating platform ensured a consistent speed of rotation for each sample, thereby standardizing the scanning procedure. Furthermore, this study is the first to evaluate the impact of simulated skin color on the accuracy of facial scanning, a critical factor in minimizing errors during facial scanning and the creation of virtual patient models.

Further studies to evaluate other scanners and scanning methods on different simulated skin tones and real patient faces are necessary. The clinical implications of this study are that non-white skin tones will benefit from the use of fiducial markers to achieve higher accuracy.

5. Conclusions

Within the limitations of this experimental in vitro study, it can be concluded that the color of the skin influences the accuracy of 3D-surface scans. White simulated skin produces higher accuracy scans and darker simulated skin colors produce lower accuracy scans. The incorporation of fiducial markers improves the accuracy of scans completed on darker skin colors, and infrared light scanners are more efficient for scanning black surfaces than blue-light scanners.

Author Contributions: Conceptualization, R.A.D.-R.; Formal analysis, R.A.D.-R.; Investigation, I.B. and A.A.; Methodology, R.A.D.-R.; Resources, R.A.D.-R. and G.E.R.; Software, I.B., A.A. and R.A.D.-R.; Supervision, R.A.D.-R.; Validation, I.B., A.A. and R.A.D.-R.; Visualization, R.A.D.-R.; Writing—original draft, I.B., A.A. and R.A.D.-R.; Writing—review and editing, I.B., A.A. and R.A.D.-R. All authors have read and agreed to the published version of the manuscript.

Funding: This research received no external funding.

Institutional Review Board Statement: Not applicable.

Informed Consent Statement: Not applicable.

Data Availability Statement: The raw data supporting the conclusions of this article will be made available by the authors on request.

Acknowledgments: The authors acknowledgment the support from the Laboratory for Periodontal Implant and Phototherapy (LA-PIP) and the Laboratory for Digital Implant and Prosthodontic Research (DIPRESLAB).

Conflicts of Interest: The authors declare no conflicts of interest.

References

1. Singh, P.; Bornstein, M.M.; Hsung, R.T.-C.; Ajmera, D.H.; Leung, Y.Y.; Gu, M. Frontiers in Three-Dimensional Surface Imaging Systems for 3D Face Acquisition in Craniofacial Research and Practice: An Updated Literature Review. *Diagnostics* **2024**, *14*, 423. [CrossRef] [PubMed]
2. Bootvong, K.; Liu, Z.; McGrath, C.; Hagg, U.; Wong, R.W.K.; Bendeus, M.; Yeung, S. Virtual Model Analysis as an Alternative Approach to Plaster Model Analysis: Reliability and Validity. *Eur. J. Orthod.* **2010**, *32*, 589–595. [CrossRef]
3. Santoro, M.; Galkin, S.; Teredesai, M.; Nicolay, O.F.; Cangialosi, T.J. Comparison of Measurements Made on Digital and Plaster Models. *Am. J. Orthod. Dentofac. Orthop.* **2003**, *124*, 101–105. [CrossRef] [PubMed]
4. Salloum, R. Revolutionizing Dentistry: Exploring the Potential of Facial Scanners for Precise Treatment Planning and Enhanced Patient Outcomes. *J. Prosthet. Dent.* **2024**, *132*, 1–5. [CrossRef] [PubMed]

5. Srinivasan, M.; Berisha, F.; Bronzino, I.; Kamnoedboon, P.; Leles, C.R. Reliability of a Face Scanner in Measuring the Vertical Dimension of Occlusion. *J. Dent.* **2024**, *146*, 105016. [CrossRef]
6. Quinzi, V.; Polizzi, A.; Ronsivalle, V.; Santonocito, S.; Conforte, C.; Manenti, R.J.; Isola, G.; Lo Giudice, A. Facial Scanning Accuracy with Stereophotogrammetry and Smartphone Technology in Children: A Systematic Review. *Children* **2022**, *9*, 1390. [CrossRef]
7. Raffone, C.; Gianfreda, F.; Pompeo, M.G.; Antonacci, D.; Bollero, P.; Canullo, L. Chairside Virtual Patient Protocol. Part 2: Management of Multiple Face Scans and Alignment Predictability. *J. Dent.* **2022**, *122*, 104123. [CrossRef] [PubMed]
8. Lippold, C.; Liu, X.; Wangdo, K.; Drerup, B.; Schreiber, K.; Kirschneck, C.; Moiseenko, T.; Danesh, G. Facial Landmark Localization by Curvature Maps and Profile Analysis. *Head Face Med.* **2014**, *10*, 54. [CrossRef]
9. Leung, M.Y.; Lo, J.; Leung, Y.Y. Accuracy of Different Modalities to Record Natural Head Position in 3 Dimensions: A Systematic Review. *J. Oral Maxillofac. Surg.* **2016**, *74*, 2261–2284. [CrossRef]
10. Probst, J.; Dritsas, K.; Halazonetis, D.; Ren, Y.; Katsaros, C.; Gkantidis, N. Precision of a Hand-Held 3D Surface Scanner in Dry and Wet Skeletal Surfaces: An Ex Vivo Study. *Diagnostics* **2022**, *12*, 2251. [CrossRef]
11. Xia, S.; Guo, S.; Li, J.; Istook, C. Comparison of Different Body Measurement Techniques: 3D Stationary Scanner, 3D Handheld Scanner, and Tape Measurement. *J. Text. Inst.* **2019**, *110*, 1103–1113. [CrossRef]
12. Renard, G.; Leid, J. Les dangers de la lumière bleue: La vérité! *J. Fr. D'Ophtalmol.* **2016**, *39*, 483–488. [CrossRef] [PubMed]
13. Schmit, J.; Olszak, A. High-Precision Shape Measurement by White-Light Interferometry with Real-Time Scanner Error Correction. *Appl. Opt.* **2002**, *41*, 5943. [CrossRef]
14. Rey, M.; Volpe, G.; Volpe, G. Light, Matter, Action: Shining Light on Active Matter. *ACS Photonics* **2023**, *10*, 1188–1201. [CrossRef]
15. Mizuno, K.; Ishii, J.; Kishida, H.; Hayamizu, Y.; Yasuda, S.; Futaba, D.N.; Yumura, M.; Hata, K. A Black Body Absorber from Vertically Aligned Single-Walled Carbon Nanotubes. *Proc. Natl. Acad. Sci. USA* **2009**, *106*, 6044–6047. [CrossRef]
16. Cheng, J.-C.; Pan, T.-S.; Hsiao, W.-C.; Lin, W.-H.; Liu, Y.-L.; Su, T.-J.; Wang, S.-M. Using Contactless Facial Image Recognition Technology to Detect Blood Oxygen Saturation. *Bioengineering* **2023**, *10*, 524. [CrossRef]
17. Petrides, G.; Clark, J.R.; Low, H.; Lovell, N.; Eviston, T.J. Three-Dimensional Scanners for Soft-Tissue Facial Assessment in Clinical Practice. *J. Plast. Reconstr. Aesthetic Surg.* **2021**, *74*, 605–614. [CrossRef]
18. Major, M.; Mészáros, B.; Würsching, T.; Polyák, M.; Kammerhofer, G.; Németh, Z.; Szabó, G.; Nagy, K. Evaluation of a Structured Light Scanner for 3D Facial Imaging: A Comparative Study with Direct Anthropometry. *Sensors* **2024**, *24*, 5286. [CrossRef]
19. Akl, M.A.; Mansour, D.E.; Zheng, F. The Role of Intraoral Scanners in the Shade Matching Process: A Systematic Review. *J. Prosthodont.* **2023**, *32*, 196–203. [CrossRef]
20. Allred, S.R.; Olkkonen, M. The Effect of Background and Illumination on Color Identification of Real, 3D Objects. *Front. Psychol.* **2013**, *4*, 821. [CrossRef]
21. Varda, K.; Bešlagić, E.; Zaimović-Uzunović, N.; Lemeš, S. Ambient Light and Object Color Influence on the 3D Scanning Process. In *New Technologies, Development and Application V*; Karabegović, I., Kovačević, A., Mandžuka, S., Eds.; Lecture Notes in Networks and Systems; Springer International Publishing: Cham, Switzerland, 2022; Volume 472, pp. 37–45. ISBN 978-3-031-05229-3.
22. Krithikadatta, J.; Gopikrishna, V.; Datta, M. CRIS Guidelines (Checklist for Reporting In-vitro Studies): A concept note on the need for standardized guidelines for improving quality and transparency in reporting in-vitro studies in experimental dental research. *J. Conserv. Dent.* **2014**, *17*, 301–304. [CrossRef] [PubMed]
23. Bai, Z.; Cui, X.; Wang, X.; Xie, H.; Chen, B. Light Absorption of Black Carbon Is Doubled at Mt. Tai and Typical Urban Area in North China. *Sci. Total Environ.* **2018**, *635*, 1144–1151. [CrossRef] [PubMed]
24. Özsoy, U.; Sekerci, R.; Hizay, A.; Yildirim, Y.; Uysal, H. Assessment of reproducibility and reliability of facial expressions using 3D handheld scanner. *J. Craniomaxillofac Surg.* **2019**, *47*, 895–901. [CrossRef]
25. Knoops, P.G.M.; Beaumont, C.A.A.; Borghi, A.; Rodriguez-Florez, N.; Breakey, R.W.F.; Rodgers, W.; Angullia, F.; Jeelani, N.U.O.; Schievano, S.; Dunaway, D.J. Comparison of Three-Dimensional Scanner Systems for Craniomaxillofacial Imaging. *J. Plast. Reconstr. Aesthetic Surg.* **2017**, *70*, 441–449. [CrossRef] [PubMed]
26. Egri, L.; Nabati, H.; Yu, J.Y. RainbowTag: A Fiducial Marker System with a New Color Segmentation Algorithm. In Proceedings of the 2022 IEEE International Conference on Connected Vehicle and Expo (ICCVE), Lakeland, FL, USA, 7–9 March 2022; pp. 1–6.
27. Dindaroğlu, F.; Kutlu, P.; Duran, G.; Görgülü, S.; Aslan, E. Accuracy and reliability of 3D stereophotogrammetry: A comparison to direct anthropometry and 2D photogrammetry. *Angle Orthod.* **2016**, *86*, 487–494. [CrossRef]
28. Paschotta, R. Infrared Light. RP Photonics Encyclopedia. Available online: https://www.rp-photonics.com/infrared_light.html (accessed on 30 September 2024).
29. McDonough, M. Blue Light. Encyclopedia Britannica. 2024. Available online: https://www.britannica.com/science/blue-light (accessed on 30 September 2024).
30. Del Bino, S.; Duval, C.; Bernerd, F. Clinical and Biological Characterization of Skin Pigmentation Diversity and Its Consequences on UV Impact. *Int. J. Mol. Sci.* **2018**, *19*, 2668. [CrossRef]

prosthesis

MDPI

Review

Additively Fabricated Permanent Crown Materials: An Overview of Literature and Update

Maram A. AlGhamdi

Department of Substitutive Dental Sciences, College of Dentistry, Imam Abdulrahman Bin Faisal University, Dammam 34212, Saudi Arabia; maalghamdi@iau.edu.sa

Abstract: Background/Objectives: With advancements in technology, three-dimensional (3D) printing has become widely used, offering many advantages. Recently, 3D printing has been utilized for the fabrication of permanent crowns. However, there is still a need for more information regarding the technology, materials, and factors that may affect the properties of 3D-printed permanent crowns. **Methods:** This review was conducted to collect and assess information regarding the performance of 3D printing technology for permanent crown fabrication. An electronic search was performed using various search engines (Scopus, PubMed, Google Scholar) up to December 2024, yielding 123 articles. After screening, 24 articles that specifically investigated 3D-printed crowns were included. **Results:** Based on the findings, two categories of materials for 3D-printed permanent crowns were identified: ceramic-based and resin-based. Among the technologies used, digital light processing (DLP) was the most common, reported in 11 studies, followed by stereolithography (SLA) in 7 studies, and lithography-based ceramic manufacturing (LCM) in 4 studies. **Conclusions:** Ceramic-based crowns demonstrated higher performance compared to resin-based crowns. However, resin-based crowns were found to be clinically acceptable. Ceramic-based crowns are recommended for permanent crown fabrication, while resin-based crowns require further investigation to address the limitations of the materials and technologies used.

Keywords: 3D printing; additive manufacturing; digital dentistry; zirconia crowns;resin crowns; prosthetic accuracy; mechanical properties; marginal fit; dental CAD/CAM; prosthodontics

1. Introduction

In the last few years, dentistry has seen considerable technological breakthroughs, with digital technologies transforming different parts of dental practice [1]. Among these advancements, three-dimensional (3D) printing has emerged as a transformative technology, opening up new options for manufacturing dental prostheses, especially those constructed of ceramic-based and resin-based materials [2]. Dental crowns have long been essential to restorative dentistry, protecting, strengthening, and improving the appearance of damaged or compromised teeth [3]. Traditionally, dental crown production has relied on labor-intensive techniques such as impression-taking, model-making, and manual craftsmanship [3]. Dental restorations are produced through three primary methods: conventional techniques, subtractive manufacturing (milling), and additive manufacturing (3D printing) [4]. Traditional techniques, including lost-wax casting, heat-pressed ceramics, and hand-layered porcelain, are labor-intensive and include several stages, such as wax pattern creation, investment, casting, and sintering [4]. These techniques yield superior restorations, but they are labor-intensive, reliant on the operator's skill, and susceptible to dimensional mistakes [5]. CNC milling, a subtractive manufacturing technique, established a digital workflow that markedly enhanced precision, consistency, and productivity [6]. Milling employs pre-sintered zirconia or resin blocks, which are shaped with high-speed burs. Due to the dense microstructure of the milled material [7], this technology has shown high accuracy (trueness ~10–50 μm) and better mechanical properties. Nevertheless, milling produces considerable material waste, tool degradation, and constraints in geometric intricacy [8]. Additive manufacturing (3D printing) constructs restorations incrementally, facilitating intricate designs, optimal material utilization, and mass customization [9].

While computer-aided design and computer-aided manufacturing (CAD/CAM) technologies have significantly improved the efficiency and precision of dental crown fabrication in recent decades [10], additive manufacturing is emerging as another tool that offers further opportunities for customization and material efficiency [11]. However, these technologies also present challenges, including the need for specialized software, proprietary systems limited to specific manufacturers, and material constraints such as the handling of zirconia-based ceramics. New advancements in dental crowns provide improved accuracy, personalization, and efficiency compared to older procedures [12].

Three-dimensional printing, also known as additive manufacturing, is the process of making 3D items layer by layer using digital models. Dentistry frequently uses this technology to create dental prostheses like crowns, bridges, and dentures [2], as well as surgical guides and orthodontic appliances [11]. The process starts with digital scanning of the patient's oral cavity, followed by the use of computer-aided design (CAD) software to design the crown. The digital model is subsequently transformed into a physical object utilizing various 3D printing technologies, such as stereolithography (SLA), digital light processing (DLP), and selective laser sintering [12,13].

Three-dimensional printing in dentistry has various advantages over traditional approaches. For starters, it enables the exact manufacturing of dental crowns with complex geometries that would be difficult or impossible to create using traditional methods [14]. This precision leads to better-fitting crowns, which can enhance patient comfort and lessen the need for dental modifications. Furthermore, 3D printing is typically faster and more efficient than traditional manufacturing procedures, allowing dentists to perform same-day crowns in some circumstances [15]. New technologies have led to the suggestion of innovative materials for 3D-printed crown fabrication, which fall into two categories: ceramic-based and resin-based. Dental materials used in crown fabrication are categorized according to worldwide criteria to guarantee quality and safety. Ceramic crowns are chiefly governed by ISO 6872, which delineates the standards for dental ceramics, encompassing their strength and translucency [16]. ISO 10477 describes the mechanical and physical properties of polymer-based materials used in both permanent and temporary restorations [7]. This is where resin-based crowns fit in. These categories establish defined criteria for assessing the appropriateness of 3D-printed items for clinical use.

Ceramic crowns are composed of advanced ceramic materials; these crowns have become increasingly popular in restorative dentistry [17]. Ceramic crowns offer exceptional esthetic benefits, closely mimicking the natural appearance of teeth through their remarkable translucency and color-matching capabilities [18]. Their biocompatibility is a significant advantage, with minimal risk of allergic reactions and excellent tissue tolerance [19]. Clinically, these crowns allow for more conservative tooth preparation, preserving more of the natural tooth structure compared to traditional alternatives [20]. Despite their numerous benefits, ceramic crowns present several challenges. The primary concern is their higher cost, which can be substantially pricier than traditional crown materials [21]. Mechanical limitations include a greater propensity for chipping or fracturing, particularly in areas of high occlusal stress [22,23].

Polymer-based crowns are dental restorative devices fabricated from advanced synthetic materials, offering an alternative approach to traditional crown fabrication [24,25]. These crowns utilize high-performance polymeric materials designed to address specific clinical challenges in dental crowns. Their lightweight nature and reduced weight compared to ceramic or metal crowns provide improved patient comfort and reduced stress on the underlying tooth structure [26]. However, their mechanical properties are generally inferior to ceramic or metal-based alternatives, with reduced hardness and wear resistance [23]. Long-term durability remains a concern, as polymeric materials may demonstrate higher susceptibility to degradation, discoloration, and dimensional changes under oral environmental conditions [27].

The use of 3D printing to produce definitive crowns has also been linked to higher patient satisfaction. Patients appreciate the ability to swiftly generate custom-fitted crowns that match the color and feel of their natural teeth [14]. Furthermore, the flexibility of 3D printing allows for simple alterations and modifications, ensuring that the final crown suits each patient's individual requirements [28].

As 3D printing technologies improve, several critical features must be thoroughly investigated to determine their suitability for clinical use [29]. These include the precision and fit of 3D-printed crowns, their mechanical qualities and longevity, surface characteristics and esthetic outcomes, and overall clinical performance [30]. Furthermore, factors such as manufacturing efficiency, cost implications, and the complexity of integrating these technologies into dental practices play a crucial role in their widespread adoption [31]. However, using 3D printing technology in dental crown production presents some obstacles [32].

This literature review provides a thorough review of the present state of 3D printing technologies for ceramic-based and polymer-based crown materials and fabrications. By evaluating recent studies and breakthroughs in this sector, we want to better understand the possible benefits, limitations, and future possibilities of these technologies. This review will look at various topics, including the accuracy and fit of 3D-printed crowns, their mechanical qualities, clinical performance, and manufacturing issues.

2. Materials and Methods

2.1. Search Strategy

A comprehensive literature search was conducted using the following electronic databases: PubMed/MEDLINE, Scopus, Web of Science, and Google Scholar. The search was limited to articles published in English from January 2014 to December 2024 to focus on the most recent developments in the 3D printing field. The search terms used included combinations of keywords such as "3D printing", "dentistry", "ceramic-based crowns", "resin-based crowns", "additive manufacturing", "dental crowns", "definitive crowns", and "Milling".

2.2. Inclusion and Exclusion Criteria

The inclusion criteria include original research (in vitro and clinical studies); reported 3D printing technologies for fabricating ceramic or resin-based crowns; studies evaluating the accuracy, fit, mechanical properties, or clinical performance of 3D-printed crowns; and full-text articles available in the English language only. Excluded studies were case reports, opinion articles, or conference abstracts; studies focusing solely on other dental applications of 3D printing (e.g., surgical guides, orthodontic appliances); and articles are not published in peer-reviewed journals or in English.

An initial search yielded 453 articles, from which 330 duplicates were removed. After screening 123 articles, 62 were excluded for being unrelated to permanent crowns. A total of 38 full-text articles were assessed for eligibility, leading to the exclusion of 14 studies due to irrelevance or unmet criteria. Ultimately, 24 full-text studies were included in this review (Figure 1).

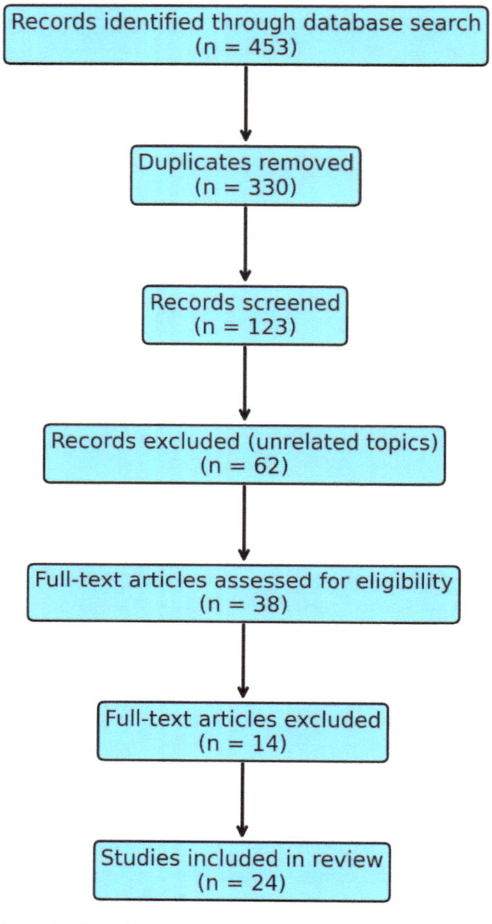

Figure 1. Flowchart illustrating the study selection process.

2.3. Data Extraction

A standardized data extraction form was created in Excel to collect relevant information from each included study. The following data were extracted in Table S1

Study characteristics (authors, year of publication, study design).

3D printing technology used.

Material(s) studied (zirconia, dental resin-based, or both).

Outcome measures (e.g., accuracy, fit, mechanical properties, clinical performance).

Key findings.

Strengths and limitations of the study.

2.4. Data Synthesis

The extracted data were synthesized to provide a comprehensive overview of the current state of research on 3D-printed ceramic-based and resin-based crowns. The synthesis involved summarizing the key findings, comparing the results across studies, and identifying common themes and discrepancies. The findings were categorized into key themes:

accuracy and fit, mechanical properties, clinical performance, surface characteristics and esthetics, manufacturing efficiency, and others. The results were synthesized within each theme to identify trends, consistencies, and discrepancies across studies.

3. Results

This review analyzed multiple studies on the application of 3D printing technologies for fabricating ceramic-based and resin-based dental crowns. The detailed characteristics of the included articles are presented in Table S1.

3.1. Printing Technologies

Approximately 11 out of 24 of the studies utilized DLP as the 3D printing technology, followed by SLA in 7 studies and lithography-based ceramic manufacturing (LCM) in 4 studies. The remaining studies reported using various other technologies: nanoparticle jetting (NPJ) in 1 study and inkjet printing in 1 study.

3.2. Crown Materials

A total of 15 studies used ceramic-based crown material, 7 studies used resin-based crown material, and 2 studies reported using both ceramic-based and resin-based crown materials.

3.3. Properties Investigated

The results from the selected studies can be categorized and discussed into several vital areas, including accuracy and fitness, mechanical properties, clinical performance, surface characteristics and esthetics, and manufacturing efficiency.

3.4. Accuracy and Fit

Several studies have evaluated the precision and fit of 3D-printed ceramic-based crowns. A study discovered that 3D-printed ceramic-based crowns had comparable trueness and a better fit than milled crowns [33]. Another study found that 3D-printed monolithic ceramic-based crowns had higher precision and better margin quality than conventional approaches [34]. This fits with what Revilla-León et al. found, which was that the fit of 3D-printed temporary dental crowns on the outside and inside was good enough for clinical use [35]. Different studies, like those by Refaie et al. and Lerner et al., which found bigger marginal gaps or lower trueness in 3D-printed crowns, show that there is still room for improvement [36,37]. According to a study, 3D-printed ceramic veneers were just as good at marginal adaptation (95 μm) and production accuracy (26 μm) as traditional methods [38].

For dental resin-based crowns, the authors discovered that ceramic-filled 3D-printed resin-based crowns fit and were just as accurate as traditionally made crowns, which suggests that they could be used in clinical settings [39]. Li et al. conducted a comparative analysis utilizing the 3D deviation and adaptation approach and discovered that SLAs with occlusal full-supporting structures demonstrated higher external 3D trueness and clinically acceptable performance than pillar supports [40].

The heterogeneity in results among research underscores the necessity for measuring methodology standardization, as underlined by Nawafleh et al. in their systematic review of crown margin measurements [41]. Future studies should concentrate on optimizing printing parameters and post-processing processes to ensure excellent accuracy and fitness.

3.5. Mechanical Properties

The mechanical integrity of 3D-printed crowns is crucial to their long-term clinical effectiveness. Several studies have focused on the mechanical qualities of 3D-printed crowns. A study discovered that both 3D-printed ceramic-based crowns and composite crowns had the right amount of fracture resistance. However, 3D-printed ceramic-based crowns were less reliable than milled ceramic-based crowns because they had flaws in the material [42]. This observation is reinforced by another study where two short-term clinical trials found no mechanical or biological complications with 3D-printed yttrium-stabilized tetragonal zirconia polycrystal (3Y-TZP) ceramic-based crowns [43]. A study discovered that milled ceramic-based crowns had stronger fracture resistance than 3D-printed crowns, but both groups produced clinically acceptable results [36]. This is consistent with a study that found the mechanical properties of 3D-printed dental ceramics were generally lower than those of conventionally produced ceramics [30]. When glued to ceramic-based abutments that are supported by implants, Zandinejad et al. found that there was no significant difference in how easily the crowns broke between those that were milled and those that were made with additive manufacturing [44,45].

The physical and mechanical properties of three-dimensionally printed crowns can be affected by the layer thickness, which can interfere with the choice of the 3D-printed resin-based solution for a desired clinical outcome [46]. Three-dimensionally printed materials may be suitable for long-term crowns, such as inlays, onlays, and laminate veneers, despite the observed decrease in mechanical properties after aging [47]. Another study also stated that 3D-printed composite resins have mechanical qualities comparable to commercially available composite resins [48]. It is possible that screw-retained, implant-supported crowns made from the tested definitive composite resin-based crowns could be good alternatives for premolar implant-supported crowns [49]. It was found that 3D-printed crowns were as true on the outside, inside, marginal area, and inside of the teeth's biting surface as CAD-CAM crowns, which means they met the standards for trueness [50]. Researchers compared fracture strength and hardness, reporting that milled materials' fracture strength increased with thickness, while 3D-printed materials' fracture strength varied [51].

3.6. Clinical Performance

In a short-term pilot study, Kao et al. looked at 3D-printed zirconia crowns made with selective laser melting (SLM) for restoring back teeth [52]. Over a 24-week follow-up period, the study reported that 100% of the crowns received satisfactory grades based on the quality evaluation system of the Modified California Dental Association. The crowns demonstrated excellent marginal adaptation and no adverse periodontal effects, despite minor increases in plaque and gingival indices during the early weeks. Three-dimensionally printed resin-based crowns performed similarly clinically to stainless steel crowns in primary molar crowns, with better esthetics and patient satisfaction [14]. However, as emphasized by Alharbi et al., long-term clinical trials are still required to properly prove the efficacy of 3D-printed dental restorations [30]. Another one-year recall study reported on the clinical performance of 3D-printed dental restorations, indicating that while short-term results are promising, randomized controlled studies with longer follow-up periods are crucially needed [53].

3.7. Surface Characteristics and Esthetics

Çakmak et al. evaluated different polishing techniques and found that coffee thermal cycling affected the surface roughness and stainability of the materials [54]. They recommended considering the polishing technique and material type to optimize the esthetic dental crown. In terms of esthetics, researchers compared the color stability and translucency of 3D-printed crowns with those of conventionally constructed ceramic crowns [55]. Another study compared multiple printing technologies and materials, providing a comprehensive understanding of their performance and finding variations in surface roughness and color stability [56]. This fluctuation is comparable with the findings of Chavali et al., who discovered that the rough surface of 3D-printed dental materials can vary greatly depending on printing conditions and post-processing processes [57].

Raszewski et al. (2023) and Shishehian et al. (2023) both show how important it is to prepare the surface of 3D-printed dental restorations before they are used to keep their color [58,59]. Both studies found that unpolished surfaces were highly susceptible to discoloration when exposed to common staining agents, such as coffee, tea, and orange juice. Polished surfaces, on the other hand, exhibited significantly better resistance to staining, highlighting the importance of proper polishing and curing to ensure long-lasting esthetic outcomes in clinical applications. The influence of material type and thickness on fracture resistance was observed. Recommendations for clinical practice include considering the material properties and thickness when selecting materials for dental crowns to enhance durability [60]. In another study, they additively manufactured resin-based crowns and found them more susceptible to simulated brushing and coffee thermal cycling than other materials [61]. Future studies should focus on refining these characteristics to achieve consistent, high-quality surface finishes.

3.8. Manufacturing Efficiency

Several studies have shown that 3D printing has the potential to reduce material waste compared to traditional milling technologies [12,62]. This conclusion is consistent with a comprehensive review by Dawood et al., who highlighted the potential of 3D printing to revolutionize dental manufacturing processes [2]. The increasing interest in 3D printing technologies highlights their significant potential in the future of implant dentistry [63]. These technologies offer benefits such as high material efficiency, the capability to create intricate geometric shapes, and the production of customized components from CAD files, making them a viable alternative for generating dental implants. Detailed cost–benefit evaluations are required to properly comprehend the economic ramifications of incorporating 3D printing technologies into dental practices [64].

3.9. Implications for Different Crown Materials

The review identifies unique patterns in ceramic-based and resin-based crowns. While 3D-printed ceramic-based crowns show potential, they still struggle to match the mechanical qualities of milled ceramic-based ones. This conclusion is congruent with the findings of Li et al., who highlighted the improved mechanical qualities of traditionally treated ceramic-based crowns [40]. In contrast, 3D-printed resin-based crowns, particularly for pediatric applications, perform similarly to or better than traditional options.

Kim et al. compared 3D-printed and conventional dental crowns clinically. They found that 3D-printed resin-based crowns performed similarly to traditional ones [65]. However, 3D-printed ceramic-based crowns showed lower fracture resistance in posterior areas compared to their conventional counterparts.

3.10. Technological Considerations

The research evaluated various 3D printing technologies, including stereolithography (SLA), digital light processing (DLP), and lithography-based ceramic manufacturing (LCM). According to Stansbury and Idacavage in their evaluation of 3D printing technologies for dental applications, the used technology appears to influence the end product's qualities [66].

3.11. Comparative Studies and Clinical Performance

Comparative studies have demonstrated that ceramic-based and resin-based crowns made with 3D printing methods perform well in clinical settings. For example, 3D-printed ceramic-based crowns have been shown to have good marginal adaptation and fracture resistance, comparable to or better than milling crowns [52]. Similarly, 3D-printed resin-based crowns have produced favorable clinical results regarding fit, durability, and patient satisfaction [52]. However, obstacles still exist, particularly in improving the surface roughness and long-term performance of 3D-printed crowns [15]. Additionally, future studies are needed to test other important characteristics such as fatigue [67] and resistance to acidic drinks [68].

3.12. Study Limitations

Despite several investigations into innovative technologies for permanent prostheses, a limited amount of research has examined crown configuration. The limited number of included studies constitutes a constraint of this research, compounded by the diverse technologies and materials employed (resin-based and ceramic-based) throughout the included articles, as well as the variations in assessed attributes among the studies. This study possesses multiple shortcomings requiring acknowledgment. The scope of materials and printing settings analyzed was limited, potentially failing to encompass the full range of alternatives present in the field. Secondly, the absence of extensive clinical data limits our capacity to ascertain the longevity and efficacy of these crowns in practical settings. Future studies should address those shortcomings to furnish a more thorough assessment of 3D-printed dental crowns. Despite these encouraging results, additional study is required to comprehensively understand the long-term clinical efficacy of 3D-printed crowns. Future investigations should prioritize systematic reviews and meta-analyses to deliver a more thorough evaluation of 3D-printed permanent crown materials. Integrating standardized procedures, including systematic risk of bias evaluations and extended clinical trials, would improve the reliability of results. Moreover, broadening the scope to include more types of fabrication techniques, material compositions, and clinical performance outcomes will enhance the evidence base.

4. Conclusions

Three-dimensional printing technology has substantially revolutionized the field of dental crowns, particularly in the production of ceramic-based and resin-based crowns. These advances have several benefits, including increased accuracy, adaptability, and efficiency. While current research shows promising outcomes, further studies are required to enhance these technologies for broader clinical application. This literature review focuses on the potential of 3D printing to alter dental crown practices, paving the door for more effective and personalized patient treatment.

Supplementary Materials: The following supporting information can be downloaded at: https://www.mdpi.com/article/10.3390/prosthesis7020035/s1, Table S1: Summary of the articles included in the review.

Funding: This research received no external funding.

Conflicts of Interest: The author declares no conflicts of interest.

References

1. Van Noort, R. The future of dental devices is digital. *Dent. Mater.* **2012**, *28*, 3–12. [CrossRef] [PubMed]
2. Dawood, A.; Marti, B.M.; Sauret-Jackson, V.; Darwood, A. 3D printing in dentistry. *Br. Dent. J.* **2015**, *219*, 521–529. [CrossRef] [PubMed]
3. Goodacre, C.J.; Bernal, G.; Rungcharassaeng, K.; Kan, J.Y. Clinical complications in fixed prosthodontics. *J. Prosthet. Dent.* **2003**, *90*, 31–41. [CrossRef]

4. Branco, A.C.; Colaço, R.; Figueiredo-Pina, C.G.; Serro, A.P. Recent advances on 3D-printed zirconia-based dental materials: A review. *Materials* **2023**, *16*, 1860. [CrossRef]
5. Alqutaibi, A.Y.; Alghauli, M.A.; Aljohani, M.H.; Zafar, M.S. Advanced additive manufacturing in implant dentistry: 3D printing technologies, printable materials, current applications and future requirements. *Bioprinting* **2024**, *42*, e00356. [CrossRef]
6. Alghauli, M.; Alqutaibi, A.Y.; Wille, S.; Kern, M. 3D-printed versus conventionally milled zirconia for dental clinical applications: Trueness, precision, accuracy, biological and esthetic aspects. *J. Dent.* **2024**, *144*, 104925. [CrossRef]
7. Gad, M.M.; Al Mahfoudh, H.A.; Al Mahfuth, F.A.; Hashim, K.A.; Khan, S.Q.; Al-Qarni, F.D.; Baba, N.Z.; Al-Harbi, F.A. A comparative study of strength and surface properties of permanent 3D-printed resins with CAD-CAM milled fixed dental prostheses. *J. Prosthodont.* 2024, *online ahead of print*.
8. Alghauli, M.A.; Almuzaini, S.; Aljohani, R.; Alqutaibi, A.Y. Influence of 3D printing orientations on physico-mechanical properties and accuracy of additively manufactured dental ceramics. *J. Prosthodont. Res.* 2025, *online ahead of print*. [CrossRef]
9. Alghauli, M.A.; Aljohani, R.; Almuzaini, S.; Aljohani, W.; Almutairi, S.; Alqutaibi, A.Y. Accuracy, marginal, and internal fit of additively manufactured provisional restorations and prostheses printed at different orientations. *J. Esthet. Restor. Dent.* 2024, *online ahead of print*.
10. Davidowitz, G.; Kotick, P.G. The use of CAD/CAM in dentistry. *Dent. Clin. N. Am.* **2011**, *55*, 559–570. [CrossRef] [PubMed]
11. Javaid, M.; Haleem, A. Current status and applications of additive manufacturing in dentistry: A literature-based review. *J. Oral. Biol Craniofacial Res.* **2019**, *9*, 179–185. [CrossRef]
12. Park, Y.; Kim, J.; Kang, Y.J.; Shim, E.Y.; Kim, J.H. Comparison of Fracture Strength of Milled and 3D-Printed Crown Materials According to Occlusal Thickness. *Materials* **2024**, *17*, 4645. [CrossRef]
13. Alshamrani, A.; Alhotan, A.; Kelly, E.; Ellakwa, A. Mechanical and biocompatibility properties of 3D-printed dental resin reinforced with glass silica and zirconia nanoparticles: In vitro study. *Polymers* **2023**, *15*, 2523. [CrossRef]
14. Lee, K.E.; Kang, H.S.; Shin, S.Y.; Lee, T.; Lee, H.S.; Song, J.S. Comparison of three-dimensional printed resin crowns and preformed stainless steel crowns for primary molar restorations: A randomized controlled trial. *J. Clin. Pediatr. Dent.* **2024**, *48*, 59–67.
15. Othman, A.; Sandmair, M.; Alevizakos, V.; von See, C. The fracture resistance of 3D-printed versus milled provisional crowns: An in vitro study. *PLoS ONE* **2023**, *18*, e0285760.
16. Hawsawi, R.A.; Miller, C.A.; Moorehead, R.D.; Stokes, C.W. Evaluation of reproducibility of the chemical solubility of dental ceramics using ISO 6872:2015. *J. Prosthet. Dent.* **2020**, *124*, 230–236. [PubMed]
17. McLaren, E.A.; Whitworth, J.M. Ceramics: Rationale for material selection. *Compend. Contin. Educ. Dent.* **2010**, *31*, 666–668. [PubMed]
18. Guess, P.C.; Kuli, M.; Wolkewitz, M.; Zhang, Y.; Strub, J.R. All-ceramic systems: Laboratory and clinical performance. *Dent. Clin. N. Am.* **2011**, *55*, 333–352.
19. Sailer, I.; Pjetursson, B.E.; Zwahlen, M.; Hämmerle, C.H. A systematic review of the survival and complication rates of all-ceramic and metal-ceramic reconstructions after an observation period of at least 3 years. Part I: Single crowns. *Clin. Oral Implant. Res.* **2007**, *18* (Suppl. S3), 73–85.
20. Edelhoff, D.; Sorensen, J.A. Tooth structure removal associated with various preparation designs for anterior teeth. *J. Prosthet. Dent.* **2002**, *87*, 503–509.
21. Kois, J.C.; Rye, A.H. Realistic expectations for ceramic restorations. *Compend. Contin. Educ. Dent.* **2000**, *21*, 664–670.
22. Lawn, B.R.; Deng, Y.; Lloyd, I.K.; Thompson, V.P. Mechanistic design of ceramic-based layer structures for mechanical property optimization. *J. Am. Ceram. Soc.* **2009**, *92*, 1792–1804.
23. Kelly, J.R. Ceramics in restorative and prosthetic dentistry. *Annu. Rev. Mater. Res.* **2014**, *44*, 443–470.
24. Ramakrishna, S.; Mayer, J.; Wintermantel, E.; Leong, K.W. Biomedical applications of polymer-composite materials: A review. *Compos. Sci. Technol.* **2001**, *61*, 1189–1224.
25. Miyazaki, T.; Nakamura, T.; Matsumura, H.; Ban, S.; Kobayashi, T. Current status of zirconia restoration in dental practice. *Dent. Mater. J.* **2013**, *32*, 1–12.
26. Wendler, M.; Lümkemann, N.; Rösch, P.; Stawarczyk, B. Mechanical properties of polymer-infiltrated ceramic-network materials. *Dent. Mater.* **2018**, *34*, 1007–1020.
27. Stawarczyk, B.; Thrun, H.; Eichberger, M.; Roos, M.; Gernet, W.; Keul, C. Mechanical properties of a newly developed core material for all-ceramic restorations. *Dent. Mater.* **2015**, *34*, 1–11.
28. Meng, M.; Wang, J.; Huang, H.; Liu, X.; Zhang, J.; Li, Z. 3D printing metal implants in orthopedic surgery: Methods, applications and future prospects. *J. Orthop. Translat.* **2023**, *42*, 94–112.
29. Mamo, H.B.; Adamiak, M.; Kunwar, A. 3D printed biomedical devices and their applications: A review on state-of-the-art technologies, existing challenges, and future perspectives. *J. Mech. Behav. Biomed. Mater.* **2023**, *143*, 105930.
30. Alharbi, N.; Osman, R.; Wismeijer, D. Effects of build direction on the mechanical properties of 3D-printed complete coverage interim dental restorations. *J. Prosthet. Dent.* **2016**, *115*, 760–767.

31. Alotaibi, K.F.; Kassim, A.M. Factors That Influence the Adoption of Digital Dental Technologies and Dental Informatics in Dental Practice. *Int. J. Online Biomed. Eng.* **2023**, *19*, 103–126.
32. Loges, K.; Tiberius, V. Implementation challenges of 3D printing in prosthodontics: A ranking-type delphi. *Materials* **2022**, *15*, 431. [CrossRef]
33. Abualsaud, R.; Alalawi, H. Fit, precision, and trueness of 3D-printed zirconia crowns compared to milled counterparts. *Dent. J.* **2022**, *10*, 215. [CrossRef]
34. Zhu, H.; Zhou, Y.; Jiang, J.; Wang, Y.; He, F. Accuracy and margin quality of advanced 3D-printed monolithic zirconia crowns. *J. Prosthet. Dent.* **2023**. [CrossRef]
35. Revilla-León, M.; Özcan, M. Additive manufacturing technologies used for processing polymers: Current status and potential application in prosthetic dentistry. *J. Prosthodont.* **2019**, *28*, 146–158. [PubMed]
36. Refaie, A.; Bourauel, C.; Fouda, A.M.; Keilig, L.; Singer, L. The effect of cyclic loading on the fracture resistance of 3D-printed and CAD/CAM milled zirconia crowns—An in vitro study. *Clin. Oral Investig.* **2023**, *27*, 6125–6133. [PubMed]
37. Lerner, H.; Nagy, K.; Pranno, N.; Zarone, F.; Admakin, O.; Mangano, F. Trueness and precision of 3D-printed versus milled monolithic zirconia crowns: An in vitro study. *J. Dent.* **2021**, *113*, 103792.
38. Ioannidis, A.; Park, J.M.; Hüsler, J.; Bomze, D.; Mühlemann, S.; Özcan, M. An in vitro comparison of the marginal and internal adaptation of ultrathin occlusal veneers made of 3D-printed zirconia, milled zirconia, and heat-pressed lithium disilicate. *J. Prosthet. Dent.* **2022**, *128*, 709–715. [PubMed]
39. Al-Ramadan, A.; Abualsaud, R.; Al-Dulaijan, Y.A.; Al-Thobity, A.M.; Alalawi, H. Accuracy and Fit of Ceramic Filled 3D-Printed Resin for Permanent Crown Fabrication: An In Vitro Comparative Study. *Prosthesis* **2024**, *6*, 1029–1041. [CrossRef]
40. Li, R.W.; Xu, T.; Wang, Y.; Sun, Y. Accuracy of zirconia crowns manufactured by stereolithography with an occlusal full-supporting structure: An in vitro study. *J. Prosthet. Dent.* **2023**, *130*, 902–907.
41. Nawafleh, N.A.; Mack, F.; Evans, J.; Mackay, J.; Hatamleh, M.M. Accuracy and reliability of methods to measure marginal adaptation of crowns and FDPs: A literature review. *J. Prosthodont.* **2013**, *22*, 419–428.
42. Handermann, R.; Zehender, N.; Rues, S.; Kobayashi, H.; Rammelsberg, P.; Schwindling, F.S. Load-bearing capacity of 3D-printed incisor partial-coverage crowns made from zirconia and composite. *J. Prosthodont. Res.* **2024**, *68*, 532–539.
43. Alghauli, M.A.; Alqutaibi, A.Y.; Wille, S.; Kern, M. The physical-mechanical properties of 3D-printed versus conventional milled zirconia for dental clinical applications: A systematic review with meta-analysis. *J. Mech. Behav. Biomed. Mater.* **2024**, *156*, 106601.
44. Zandinejad, A.; Methani, M.M.; Schneiderman, E.D.; Revilla-León, M.; Morton, D. Fracture resistance of additively manufactured zirconia crowns when cemented to implant supported zirconia abutments: An in vitro study. *J. Prosthodont.* **2019**, *28*, 893–897. [PubMed]
45. Zandinejad, A.; Revilla-León, M.; Methani, M.M.; Nasiry Khanlar, L.; Morton, D. The fracture resistance of additively manufactured monolithic zirconia vs. bi-layered alumina toughened zirconia crowns when cemented to zirconia abutments. Evaluating the potential of 3D printing of ceramic crowns: An in vitro study. *Dent. J.* **2021**, *9*, 115. [CrossRef]
46. Borella, P.S.; Alvares, L.A.; Ribeiro, M.T.; Moura, G.F.; Soares, C.J.; Zancopé, K.; Mendonça, G.; Rodrigues, F.P.; das Neves, F.D. Physical and mechanical properties of four 3D-printed resins at two different thick layers: An in vitro comparative study. *Dent. Mater.* **2023**, *39*, 686–692. [PubMed]
47. Di Fiore, A.; Stellini, E.; Alageel, O.; Alhotan, A. Comparison of mechanical and surface properties of two 3D printed composite resins for definitive restoration. *J. Prosthet. Dent.* **2024**, *132*, 839.e1–839.e7.
48. Tahayeri, A.; Morgan, M.; Fugolin, A.P.; Bompolaki, D.; Athirasala, A.; Pfeifer, C.S.; Bertassoni, L.E. 3D printed versus conventionally cured provisional crown and bridge dental materials. *Dent. Mater.* **2018**, *34*, 192–200. [PubMed]
49. Türksayar, A.A.D.; Demirel, M.; Donmez, M.B.; Olcay, E.O.; Eyüboğlu, T.F.; Özcan, M. Comparison of wear and fracture resistance of additively and subtractively manufactured screw-retained, implant-supported crowns. *J. Prosthet. Dent.* **2024**, *132*, 154–164.
50. Wang, W.; Yu, H.; Liu, Y.; Jiang, X.; Gao, B. Trueness analysis of zirconia crowns fabricated with 3-dimensional printing. *J. Prosthet. Dent.* **2019**, *121*, 285–291.
51. Park, S.; Cho, W.; Lee, H.; Bae, J.; Jeong, T.; Huh, J.; Shin, J. Strength and Surface Characteristics of 3D-Printed Resin Crowns for the Primary Molars. *Polymers* **2023**, *15*, 4241. [CrossRef]
52. Kao, C.T.; Liu, S.H.; Kao, C.Y.; Huang, T.H. Clinical evaluation of 3D-printed zirconia crowns fabricated by selective laser melting (SLM) for posterior teeth restorations: Short-term pilot study. *J. Dent. Sci.* **2023**, *18*, 715–721.
53. Verniani, G.; Casucci, A.; Val, M.; Ruggiero, G.; Manfredini, D.; Ferrari, M.; Ferrari Cagidiaco, E. A Randomized Controlled Clinical Trial on Press, Block Lithium Disilicate, and 3D Printed Partial Crowns in Posterior Teeth: One-Year Recall. *Prosthesis* **2024**, *6*, 887–895. [CrossRef]
54. Çakmak, G.; Oosterveen-Rüegsegger, A.L.; Akay, C.; Schimmel, M.; Yilmaz, B.; Donmez, M.B. Influence of polishing technique and coffee thermal cycling on the surface roughness and color stability of additively and subtractively manufactured resins used for definitive restorations. *J. Prosthodont.* **2024**, *33*, 467–474.

55. Park, J.M.; Ahn, J.S.; Cha, H.S.; Lee, J.H. Comparison of color stability and translucency between 3D-printed and conventionally fabricated ceramic crowns after artificial aging. *J. Prosthodont.* **2023**, *32*, 282–289.
56. Bozoğulları, H.N.; Temizci, T. Evaluation of the Color Stability, Stainability, and Surface Roughness of Permanent Composite-Based Milled and 3D Printed CAD/CAM Restorative Materials after Thermocycling. *Appl. Sci.* **2023**, *13*, 11895. [CrossRef]
57. Chavali, R.; Nejat, A.H.; Lawson, N.C. Machinability of CAD-CAM materials. *J. Prosthet. Dent.* **2017**, *118*, 194–199.
58. Raszewski, Z.; Chojnacka, K.; Mikulewicz, M. Effects of surface preparation methods on the color stability of 3D-printed dental restorations. *J. Funct. Biomater.* **2023**, *14*, 257. [CrossRef] [PubMed]
59. Shishehian, A.; Firouz, F.; Khazaee, S.; Rajabi, H.; Farhadian, M.; Niaghiha, F. Evaluating the color stability of 3D-printed resins against various solutions. *Eur. J. Transl. Myol.* **2023**, *33*, 11493. [PubMed]
60. Zimmermann, M.; Ender, A.; Egli, G.; Özcan, M.; Mehl, A. Fracture load of CAD/CAM-fabricated and 3D-printed composite crowns as a function of material thickness. *Clin. Oral Investig.* **2019**, *23*, 2777–2784.
61. Çakmak, G.; Donmez, M.B.; de Paula, M.S.; Akay, C.; Fonseca, M.; Kahveci, Ç.; Abou-Ayash, S.; Yilmaz, B. Surface roughness, optical properties, and microhardness of additively and subtractively manufactured CAD-CAM materials after brushing and coffee thermal cycling. *J. Prosthodont.* **2023**, *34*, 68–77.
62. Camargo, B.; Willems, E.; Jacobs, W.; Van Landuyt, K.; Peumans, M.; Zhang, F.; Vleugels, J.; Van Meerbeek, B. 3D printing and milling accuracy influence full-contour zirconia crown adaptation. *Dent. Mater.* **2022**, *38*, 1963–1976.
63. Nikoyan, L.; Patel, R. Intraoral Scanner, Three-Dimensional Imaging, and Three-Dimensional Printing in the Dental Office. *Dent. Clin. N. Am.* **2020**, *64*, 365–378.
64. Ishida, Y.; Miyasaka, T. Dimensional accuracy of dental casting patterns created by 3D printers. *Dent. Mater. J.* **2016**, *35*, 250–256. [CrossRef] [PubMed]
65. Kim, N.; Kim, H.; Kim, I.H.; Lee, J.; Lee, K.E.; Lee, H.S.; Kim, J.H.; Song, J.S.; Shin, Y. Novel 3D printed resin crowns for primary molars: In vitro study of fracture resistance, biaxial flexural strength, and dynamic mechanical analysis. *Children* **2022**, *9*, 1445. [CrossRef] [PubMed]
66. Stansbury, J.W.; Idacavage, M.J. 3D printing with polymers: Challenges among expanding options and opportunities. *Dent. Mater.* **2016**, *32*, 54–64. [CrossRef] [PubMed]
67. Rodríguez-Ivich, J.; Razaghy, M.; Henriques, B.; Magne, P. Accelerated Fatigue Resistance of Bonded Composite Resin and Lithium Disilicate Screw-Retained Incisor Crowns with Long and Short Titanium Bases. *Int. J. Periodontics Restor. Dent.* **2022**, *42*, 459–469. [CrossRef]
68. Poggio, C.; Dagna, A.; Chiesa, M.; Colombo, M.; Scribante, A. Surface roughness of flowable resin composites eroded by acidic and alcoholic drinks. *J. Conserv. Dent. Endod.* **2012**, *15*, 137–140. [CrossRef]

MDPI AG
Grosspeteranlage 5
4052 Basel
Switzerland
Tel.: +41 61 683 77 34

Prosthesis Editorial Office
E-mail: prosthesis@mdpi.com
www.mdpi.com/journal/prosthesis

www.ingramcontent.com/pod-product-compliance
Lightning Source LLC
Chambersburg PA
CBHW041412210926
51850CB00013B/3255
* 9 7 8 3 7 2 5 8 4 1 1 7 2 *